T0141834

SPACE TECHNOLOGY LIBRARY
Published jointly by Microcosm Press and Springer

More information about this series at http://www.springer.com/series/6575

James Miller

Planetary Spacecraft Navigation

James Miller
Porter Ranch
CA, USA

Space Technology Library
ISBN 978-3-030-07678-8 ISBN 978-3-319-78916-3 (eBook)
https://doi.org/10.1007/978-3-319-78916-3

Softcover re-print of the Hardcover 1st edition 2019

This Springer imprint is published by the registered company Springer Nature Switzerland AG
The registered company address is: Gewerbestrasse 11, 6330 Cham, Switzerland

Preface

This book is based on my 50 years of experience in navigation of Earth entry vehicles, Minuteman ballistic missiles, and planetary spacecraft. At the Jet Propulsion Laboratory, I worked on the Mariner 6, Viking, Pioneer, Galileo, and Near Earth Asteroid Rendezvous missions. At KinetX Inc. I worked on the MESSENGER and New Horizons missions. In writing this book I have drawn on engineering memoranda, conference papers, and publications I have written as well as many notes that I have accumulated over many years. My purpose is to present a book that will describe how navigation is done. The emphasis is on mathematics that have been coded in computer programs used for mission operations. Therefore, the derivations are given in detail since relatively little mathematics actually makes it into operational software. The mathematics that do are generally straightforward, but the programs are large and complex and must be virtually error free. In writing this book, I have frequently checked the mathematics by looking at computer code that I am confident is correct.

Chapter 1 contains the equations of motion exclusive of the force models. Since navigation is concerned with everything that moves, the equations of motion include the kinetic theory of gasses and propagation of electromagnetic waves. The force models are given in Chap. 2. Since motion is in a straight line without force being applied, the force models enable spacecraft to go somewhere. Chapter 3 describes the procedure for designing the trajectory a spacecraft will follow. A detailed derivation of Kepler's equation and Lambert's theorem is given. Just about all trajectory design is based on these two individuals' work aside from Newton of course. Trajectory optimization is described in Chap. 4. Most trajectory optimization is performed by developing an intuitive feel for the problem being solved. Sometimes it is necessary to perform a detailed constrained parameter optimization when there are many more control parameters than constraints. The first constrained parameter optimization using a computer program was the Viking orbit insertion maneuver. Previously, the trajectory designs were mostly Hohmann transfers. Chapter 5 describes the probability and statistics needed for navigation analysis. This is probably the most important navigation design function since it relates directly to the probability of achieving mission success. Orbit determination is described in Chap. 6. At the beginning of planetary spacecraft navigation, orbit determination was a major problem. The Mariner class spacecraft had to contend with orbit determination errors of several thousand kilometers. Missions were flown so we could do orbit determination, not science. Today, the orbit determination error at Mars is on the order of tens of kilometers. This progress may be attributed to the introduction of VLBI. There is still the problem of orbit convergence and this problem is addressed in some detail. The measurements and calibrations are described in Chap. 7. Navigation is primarily concerned with the physical quantity being measured and not the hardware required to perform the measurement. However, some detailed knowledge of the measurement implementation is necessary to write navigation software. The navigation system is described in Chap. 8. The navigation system and navigation operations procedures are constantly evolving. An overview is given that applies to all navigation systems. It is conservatively estimated that there is at least one navigation system for each person doing navigation operations. The final chapter is anecdotal and describes some navigation analyses I performed over the years. My purpose is to describe the type of analyses the reader would be expected to perform if he or she pursued a career in navigation. I hope I have succeeded in convincing the reader that navigation is a lot of fun.

Planetary spacecraft navigation is the result of the work of many individuals including the author of this book. The person who had the original idea is often not known. I have mentioned some individuals in the text who I am aware of and made some significant contribution. Some are contemporary and known personally, but often these individuals are mathematicians and have lived over a 100 years ago. I have made little effort to search the literature and track down the original source. Most of my acknowledgments are anecdotal and the source is discussions in the coffee room that have not been verified. There are a few who have contributed directly to the writing of this book. My wife, Dr. Connie Weeks, Professor Emeritus of Mathematics, Loyola Marymount University, is the source of a large amount of my limited knowledge of mathematics. I cannot recall a question about mathematics where she did not have an immediate answer. The person most responsible for this book being written is Dr. Gerald Hintz who teaches at the University of Southern California. I have known him for about 50 years and since I started writing this book 20 years ago, he has kept encouraging me to finish. Writing this book has not been as difficult as I thought it would be, but rather an enjoyable experience. I must also acknowledge my mother, Eunice Miller, who thought it would be a good idea for me to be the first one on either side of my family to go to college. My sisters, Peggy Joyce, Nan Elizabeth, and Linda Lee, also got behind this major effort and I will be forever indebted to them.

I am particularly indebted to the editors at Springer International Publishing. Hannah Kaufman and Maury Solomon showed great patience in leading me through the process. The production people headed by Batmanadan Karthikeyan transformed my manuscript into what I regard as a work of art.

Porter Ranch, CA, USA James Miller

Contents

Chapter 1
Equations of Motion

1.1 Introduction

The equations of motion describe the path that a spacecraft, planet, satellite, molecule, electromagnetic wave, or any body will follow. In space, the path that a spacecraft follows is called a trajectory and for a planet it is called an ephemeris. For the purpose of navigation, a planet is defined as any object that orbits the sun and, thus, includes comets and asteroids. A satellite is any body that orbits a planet. Flight operations are generally conducted using solutions of Newton's equation of motion obtained by numerical integration. Analytic solutions of Newton's equation of motion provide some insight into trajectory design and navigation analysis, but these solutions are seldom used in the conduct of flight operations. For spacecraft near the Sun and Jupiter, and for the planet ephemerides, Newton's equations of motion are augmented with terms from the n-body solution of General Relativity.

In this chapter, Newton's equations of motion are applied to molecules in a container to obtain the kinetic theory of gasses and a rigid body to obtain the rotational equations of motion. Equations for the motion of a spacecraft and photon are developed from the equation of geodesics which describes motion in the vicinity of a massive body obtained from the general theory of relativity. Finally, numerical integration of the equations of motion is described.

1.2 Particle Dynamics

A body in space will continue to move in a straight line at a constant velocity unless acted on by some external force. The same body in the atmosphere or on the surface of the Earth will move in a given direction at constant speed provided that there are no forces acting on the body. Since the Earth's gravity combined with atmospheric drag will result in force components along the direction of motion and normal to

J. Miller, *Planetary Spacecraft Navigation*, Space Technology Library 37,
https://doi.org/10.1007/978-3-319-78916-3_1

the direction of motion, the case of straight line or rectilinear motion can only be approximated. Newton's equation of motion describes the departure from rectilinear motion of a body when acted on by external forces and is given by,

$$\mathbf{F} = m\,\mathbf{a} \tag{1.1}$$

where m is the mass of the body, **F** is the applied force vector, and **a** is the resultant acceleration vector. Once the applied forces have been characterized, Newton's equation is the only equation that one needs to solve to determine the motion of a body provided that the effects of General Relativity and quantum mechanics are small enough to be ignored.

When one body exerts a force on another body, there is an equal and opposite reaction of the second body exerting the same force on the first body. This property of force results in the conservation of certain mathematical properties called energy and momentum when Newton's equation of motion is applied. Consider the collision of two elastic spheres. As the spheres collide, they are compressed by a force that acts along the line joining their centers. As the spheres separate, this same force acts like a spring and the spheres are returned to their previous spherical shape. The geometry of the collision is shown in Fig. 1.1.

The spheres have velocities $\mathbf{V}_1$ and $\mathbf{U}_1$ before the collision and velocities $\mathbf{V}_2$ and $\mathbf{U}_2$ after the collision and receive incremental changes in velocity of $\Delta\mathbf{V}_1$ and $\Delta\mathbf{U}_1$, respectively. The masses of the two bodies are m_v and m_u. The x coordinate axis is along the line joining the centers at impact and $\mathbf{V}_1$ is in the $x - y$ plane. During the time interval of contact, the motion is given by Eq. (1.1).

$$\mathbf{F_u}dt = m_u d\mathbf{U} \tag{1.2}$$

$$\mathbf{F_v}dt = m_v d\mathbf{V} \tag{1.3}$$

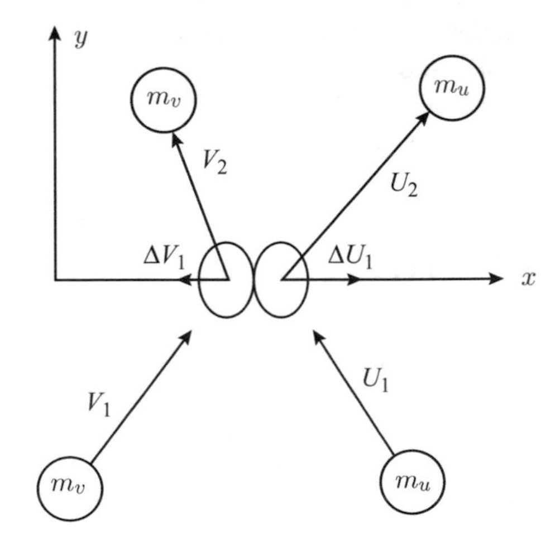

Fig. 1.1 Particle dynamics

Since the forces are equal and opposite and $\mathbf{F_u}$ is the negative of $\mathbf{F_v}$, the following result is obtained by integrating Eq. (1.2) and Eq. (1.3) and adding the results,

$$m_u(\mathbf{U_2} - \mathbf{U_1}) + m_v(\mathbf{V_2} - \mathbf{V_1}) = 0 \tag{1.4}$$

If momentum is defined as the product of mass and velocity, Eq. (1.4) reveals that momentum is conserved during impact. The momentum increase of one body is equal to the momentum decrease of the other body. An equally important result may be obtained if the components of the vectors are examined. Before the impact, the velocities are given by

$$\mathbf{V_1} = [V_{x1}, V_{y1}, 0]$$
$$\mathbf{U_1} = [U_{x1}, U_{y1}, U_{z1}]$$

and after the impact the velocities are

$$\mathbf{V_2} = [V_{x1} + \Delta V_{x1}, V_{y1}, 0]$$
$$\mathbf{U_2} = [U_{x1} + \Delta U_{x1}, U_{y1}, U_{z1}]$$

Taking the square of the velocity magnitudes before and after impact and differencing,

$$V_2^2 - V_1^2 = 2V_{x1}\Delta V_{x1} + \Delta V_{x1}^2$$
$$U_2^2 - U_1^2 = 2U_{x1}\Delta U_{x1} + \Delta U_{x1}^2$$

Multiplying the x component of Eqs. (1.2) and (1.3) by U_x and V_x, respectively, the following equations are obtained.

$$F_u\, dx = m_u\, U_x\, dU_x \tag{1.5}$$
$$F_v\, dx = m_v\, V_x\, dV_x \tag{1.6}$$

Integrating the force over the distance traveled during the interval of contact and adding the results from Eqs. (1.5) and (1.6)

$$m_u\left[\frac{(U_{x1} + \Delta U_{x1})^2}{2} - \frac{U_{x1}^2}{2}\right] + m_v\left[\frac{(V_{x1} + \Delta V_{x1})^2}{2} - \frac{V_{x1}^2}{2}\right] = 0$$

$$m_u\left[\frac{2U_{x1}\Delta U_{x1} + \Delta U_{x1}^2}{2}\right] + m_v\left[\frac{2V_{x1}\Delta V_{x1} + \Delta V_{x1}^2}{2}\right] = 0$$

and

$$m_u\left[\frac{U_2^2 - U_1^2}{2}\right] + m_v\left[\frac{V_2^2 - V_1^2}{2}\right] = 0 \tag{1.7}$$

which may also be written

$$\frac{1}{2}\, m_u\, U_2^2 + \frac{1}{2}\, m_v\, V_2^2 = \frac{1}{2}\, m_u\, U_1^2 + \frac{1}{2}\, m_v\, V_1^2 \tag{1.8}$$

Energy is defined as force acting over distance and Eq. (1.7) and Eq. (1.8) show that the energy of a particle is one half of the product of mass and the magnitude of velocity squared and reveals that energy is also conserved during impact.

Particle impacts have been assumed to be impulsive. The force is assumed to be constant over a small interval of time or distance. The integral is defined as the sum of an infinite number of impacts as the time interval approaches zero and is referred to as a Riemann sum. Consider the sum

$$U_{n+1}, V_{n+1} = U_1, V_1 + \Delta U_1, \Delta V_1 + \Delta U_2, \Delta V_2 + \Delta U_3, \Delta V_3 + \cdots \Delta U_n, \Delta V_n$$

If U_{i+1}, V_{i+1} is the result of the velocity change from U_i, V_i, then U_{i+1}, V_{i+1} will have the same energy and momentum as U_i, V_i and by extension U_{n+1}, V_{n+1} will have the same energy and momentum as U_1, V_1 even if the force vectors are different for each interval. Thus, the conservation of momentum and energy may be extended to all collisions, electrostatic interactions, gravitational interactions, and electromagnetic emanations. They all obey Newton's action equals reaction. When we fire a rocket engine or hit a baseball with a bat or fall out of a tree, the energy is conserved by the molecules colliding with one another, the gravitational force on the molecules or the electrostatic repulsion and attraction of the protons and electrons comprising the molecule. This may be an oversimplified view of the universe but will enable us to navigate anywhere in the solar system.

1.3 n-Body Equations of Motion

For a system of n bodies, the resultant gravitational force on the i'th body is the sum of the individual contributions from the other j bodies and

$$\mathbf{F_i} = G\sum_{j=1}^{n} m_i m_j \frac{\mathbf{r_j} - \mathbf{r_i}}{r_{ij}^3} + m_i m_j \nabla O_i(\mathbf{r_j} - \mathbf{r_i}) + m_i m_j \nabla O_j(\mathbf{r_i} - \mathbf{r_j}) + \mathrm{O}(\Delta r^3)$$

where G is the gravitational constant, $\mathbf{r}$ and m are the position and mass of a body, and ∇O refers to the oblateness. When $i = j$, the terms vanish and may be omitted. The acceleration of body i is obtained by simply dividing through by m_i.

$$\mathbf{a_i} = Gm_j \sum_{j=1}^{n} \frac{\mathbf{r_j} - \mathbf{r_i}}{r_{ij}^3} + \nabla O_i(\mathbf{r_j} - \mathbf{r_i}) + \nabla O_j(\mathbf{r_i} - \mathbf{r_j}) \tag{1.9}$$

The center of the coordinate system is at rest and may be determined from the following equation.

$$\mathbf{r_0} = \frac{1}{m_t} \sum_{i=1}^{n} m_i \mathbf{r_i}$$

Taking the derivative with respect to time,

$$\dot{\mathbf{r}}_\mathbf{0} = \frac{1}{m_t} \sum_{i=1}^{n} m_i \dot{\mathbf{r}}_\mathbf{i}$$

Since the exchange of momentum of body i with body j results in no change in the sum, the total momentum exchange of all the bodies must be zero and the summation terms remain constant as a function of time. Since m_t is the total system mass and $\mathbf{r}_0$ is the center of mass, the center of mass may be stationary or move at a constant velocity with respect to inertial space. The n-body equations of motion given by Eq. (1.9) are referred to as the barycentric formulation. The barycenter is the center of mass of the system containing n bodies. These equations are generally integrated numerically to obtain planetary and satellite ephemerides. Since the mass of one or more of the bodies may be assumed to be zero, a spacecraft or other point mass object may be included in the system of equations to be integrated. For high-precision ephemerides, other force models and additional terms from the general theory of relativity may be included in the integration.

For a spacecraft that is orbiting or flying close to a gravitating body, an alternative form of the n-body equations is often used. An increase in the accuracy of integration may be obtained if the coordinate system is located at the center of mass of the nearby dominant body. For this configuration, bodies that are far away from the spacecraft enter as a tidal acceleration that is differenced before the integration thus potentially improving the numerical accuracy. This formulation is referred to as planetocentric and may be obtained by simply subtracting the barycentric acceleration of the body that is close to the spacecraft from all the other bodies including the spacecraft. For convenience, assume that the spacecraft is body number one and the central body is body number 2.

$$\mathbf{a_2} = 0 \quad \mathbf{r_2} = 0$$

$$\mathbf{a_1} = Gm_j \sum_{j=2}^{n} \left\{ \frac{\mathbf{r_j} - \mathbf{r_1}}{r_{1j}^3} + \nabla O_1(\mathbf{r_j} - \mathbf{r_1}) + \nabla O_j(\mathbf{r_1} - \mathbf{r_j}) \right\}$$

$$-Gm_j \sum_{j=2}^{n} \left\{ \frac{\mathbf{r_j}}{r_j^3} + \nabla O_1(\mathbf{r_j}) + \nabla O_j(-\mathbf{r_j}) \right\}$$

1.4 Translational Variational Equations

Orbit determination, trajectory optimization, trajectory design, and propulsive maneuver design require partial derivatives of spacecraft state, planet ephemerides, and planet attitude with respect to constant dynamic parameters. The constant parameters (q) include initial conditions, gravity harmonics, solar pressure model parameters, propulsive maneuver model parameters, and other force model parameters. For orbit determination, the partial derivatives of measurements (Z) with respect to q may be obtained by application of the chain rule.

$$\frac{\partial Z}{\partial q} = \frac{\partial Z}{\partial(\mathbf{r}, \mathbf{v})} \frac{\partial(\mathbf{r}, \mathbf{v})}{\partial q}$$

where **r** and **v** are the spacecraft position and velocity at some time t referred to as the spacecraft state. The partial derivative of Z with respect to spacecraft state is called the data partial. The measurement and data partials are a function of the spacecraft state and constant parameters (q). The partial derivatives of spacecraft state with respect to q are called the variational partials. If Z is replaced by target parameters, the partial derivatives needed for trajectory design or optimization are obtained. If spacecraft state is replaced by planet state, the planetary variational partials are obtained. If spacecraft state is replaced by planet attitude, the rotational variational partials are obtained.

The translational variational partial derivatives are obtained by integrating the partial derivatives of acceleration with respect to the dynamic parameters. The spacecraft acceleration is a function of spacecraft state and q.

$$\mathbf{A} = f(\mathbf{r}, \mathbf{v}, q)$$

Differentiating with respect to q, we obtain

$$\frac{\partial \mathbf{A}}{\partial q} = \frac{\partial \mathbf{A}}{\partial \mathbf{r}} \frac{\partial \mathbf{r}}{\partial q} + \frac{\partial \mathbf{A}}{\partial \mathbf{v}} \frac{\partial \mathbf{v}}{\partial q} + \frac{\partial \mathbf{A}}{\partial q}\Big|_{\mathbf{r},\mathbf{v} \text{ constant}} \tag{1.10}$$

The acceleration (**A**) and the partial derivatives of **A** with respect to r, v, and q are described below in the chapter on Force Models. The equations are given for **A** but are actually the force on the body when the acceleration is multiplied by the mass of the body. Since, in the limit as the mass of the body goes to zero, the force also goes to zero, and the ratio is the acceleration, it is convenient to compute the acceleration directly for a spacecraft. The mass of the spacecraft is small relative to a planet and can be ignored.

The partial derivatives of spacecraft state with respect to the dynamic parameters are obtained by numerical integration.

$$\frac{\partial \mathbf{r}}{\partial q} = \int_0^t \frac{\partial \mathbf{v}}{\partial q}\, dt$$

$$\frac{\partial \mathbf{v}}{\partial q} = \int_0^t \frac{\partial \mathbf{A}}{\partial q}\, dt$$

Recall that $\mathbf{r}_0$ and $\mathbf{v}_0$, the initial state, are the first six elements of q. Determination of the variational partial derivatives by numerical integration requires derivation of partial derivatives for many force models. The gravitational partial derivatives are particularly difficult to derive and program. However, orbit determination requires precise partial derivatives to assure convergence. Trajectory optimization can usually be performed with less accurate partial derivatives. Finite-difference partial derivatives may be obtained with a simple algorithm that only requires propagation of the initial state. For example, consider the case of a spacecraft launched from Earth with an initial condition $\mathbf{r}_0$, $\mathbf{v}_0$. The spacecraft state at some later time may be obtained by numerical integration or conic orbit element propagation of the initial state.

$$\mathbf{r}, \mathbf{v} = f(t, \mathbf{r}_0, \mathbf{v}_0, q)$$

The variational partial derivatives of spacecraft state at some later time with respect to state at the initial time or epoch may be computed by finite difference.

$$\frac{\partial \mathbf{r}, \mathbf{v}}{\partial \mathbf{r}_0, \mathbf{v}_0} = \begin{bmatrix} \frac{f(t,\, \mathbf{r}_0+\Delta r_x,\, \mathbf{v}_0,\, q)-f(t,\, \mathbf{r}_0, \mathbf{v}_0, q)}{\Delta r_x} \\ \frac{f(t,\, \mathbf{r}_0+\Delta r_y,\, \mathbf{v}_0,\, q)-f(t,\, \mathbf{r}_0, \mathbf{v}_0, q)}{\Delta r_y} \\ \frac{f(t,\, \mathbf{r}_0+\Delta r_z,\, \mathbf{v}_0,\, q)-f(t,\, \mathbf{r}_0, \mathbf{v}_0, q)}{\Delta r_z} \\ \frac{f(t,\, \mathbf{r}_0,\, \mathbf{v}_0+\Delta v_x,\, q)-f(t,\, \mathbf{r}_0, \mathbf{v}_0, q)}{\Delta v_x} \\ \frac{f(t,\, \mathbf{r}_0,\, \mathbf{v}_0+\Delta v_y,\, q)-f(t,\, \mathbf{r}_0, \mathbf{v}_0, q)}{\Delta v_y} \\ \frac{f(t,\, \mathbf{r}_0,\, \mathbf{v}_0+\Delta v_z,\, q)-f(t,\, \mathbf{r}_0, \mathbf{v}_0, q)}{\Delta v_z} \end{bmatrix}^T$$

The 6 by 6 matrix in the brackets is called the state transition matrix.

1.5 Rotational Equations of Motion

A rigid body may be regarded as a collection of point masses that are constrained to not move with respect to one another. Newton's equation may be applied to each point mass with the appropriate constraints and summed or integrated over the body to obtain the rotational equations of motion. Consider the rigid body shown in Fig. 1.2. The body is constrained to rotate about the z axis or axle and a force (F_0) is applied in a direction tangential or perpendicular to the radius vector (R_0) drawn to the point of application. The force may be the result of a small thruster

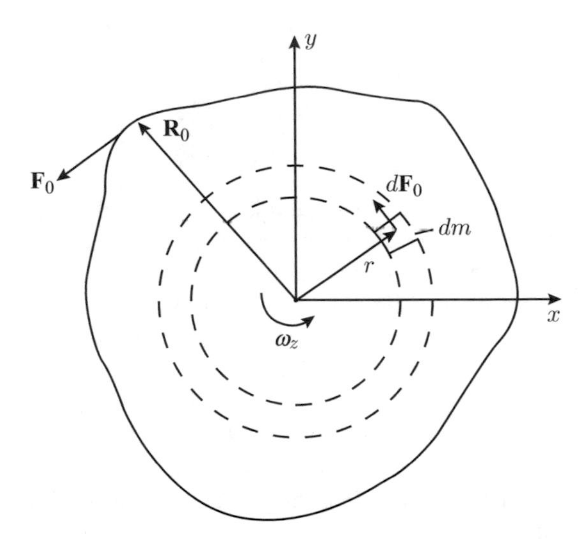

Fig. 1.2 Rotating body

or a hand crank as long as it continues to act tangentially. The applied force is distributed among all the mass elements that make up the rigid body. Since the axle constrains the particles in the body to move only in the tangential or (F_θ) direction, we may disregard forces in the z and r directions since they result in no motion or energy transfer. A typical mass element is shown in Fig. 1.2 with mass dm and it is accelerated by a force dF_θ. The motion of this mass element is governed by Newton's equation of motion.

$$dF_\theta = dm\ a_\theta$$

Multiplying by the local tangential velocity ($\omega_z r$) and substituting ($\dot{\omega}_z r$) for the tangential acceleration,

$$\omega_z r\ dF_\theta = \omega_z r\ \dot{\omega}_z r\ dm$$

The term on the left is the rate of accumulation of energy or power being applied to the mass element. Integrating over the entire body gives the total power being applied to the rigid body which must equal the externally applied power.

$$\int \omega_z r\ dF_\theta = \omega_z\ F_0 R_0$$

Dividing out the ω_z term and integrating,

$$\int r\ dF_\theta = F_0\ R_0 = \int r^2\ dm\ \dot{\omega}_z$$

The applied moment is defined by

$$M_z = \int r\, dF_\theta = F_0\, R_0$$

and the moment of inertia about the z axis is defined by,

$$I_{zz} = \int r^2\, dm = \int \rho\, (x^2 + y^2)\, dV$$

where ρ is the density. The rotational equation of motion, where the motion is constrained to a single axis, is thus

$$M_z = I_{zz}\, \dot{\omega}_z$$

When the constraint on the axis of rotation is removed, the torque about the z axis will cross couple into angular accelerations about axes normal to the z axis. These angular accelerations about the x and y axes are obtained by integrating along the z axis and the components of the force are resolved in the x and y directions.

$$\omega_z x\, dF_\theta = \omega_z x\, (-\dot{\omega}_x z)\, dm$$
$$\omega_z(-y\, dF_\theta) = \omega_z y\, \dot{\omega}_y z\, dm$$

Proceeding as above,

$$M_z = I_{zx}\, \dot{\omega}_x + I_{zy}\, \dot{\omega}_y + I_{zz}\, \dot{\omega}_z$$

where,

$$I_{zx} = -\int xz\, dm = -\int \rho\, xz\, dV$$
$$I_{zy} = -\int yz\, dm = -\int \rho\, yx\, dV$$

If an infinitesimal rotation is allowed over an interval of time dt, angular momentum is accumulated which is analogous to linear momentum and is also conserved.

$$dH_z = \int M_z\, dt = I_{zx}\, d\omega_z + I_{zy}\, d\omega_z + I_{zz}\, d\omega_z$$

After an interval of time, angular momentum is accumulated and the one-dimensional result may be extended to three dimensions.

$$\mathbf{H} = I\,\mathbf{\Omega}$$

$$\mathbf{\Omega} = [\omega_x,\ \omega_y,\ \omega_z]$$

$$\mathbf{H} = [H_x,\ H_x,\ H_x]$$

$$I = \begin{bmatrix} I_{xx} & I_{xy} & I_{xz} \\ I_{yx} & I_{yy} & I_{yz} \\ I_{zx} & I_{zy} & I_{zz} \end{bmatrix}$$

and the complete inertia tensor is defined by,

$$I_{xx} = \int \rho\,(y^2 + z^2)\,dV$$

$$I_{yy} = \int \rho\,(x^2 + z^2)\,dV$$

$$I_{zz} = \int \rho\,(x^2 + y^2)\,dV$$

$$I_{xy} = I_{yx} = -\int \rho\,xy\,dV$$

$$I_{xz} = I_{zx} = -\int \rho\,xz\,dV$$

$$I_{yz} = I_{zy} = -\int \rho\,yz\,dV$$

and

$$\dot{\mathbf{H}} = \mathbf{M}$$

The coordinate axes are fixed on the body and the accelerations are given in inertial space. When the body fixed axes rotate, the inertia tensor or the accelerations computed with respect to body fixed axes must be allowed to vary. The standard convention is to integrate in body fixed coordinates. Keeping the axes fixed on the body makes the inertia tensor constant but requires the introduction of angles to describe the orientation of the body fixed axes in inertial space. The time derivative of angular momentum is the sum of two parts. The body fixed angular momentum and the time derivative of the coordinate axes given by,

$$\dot{\mathbf{H}} = \dot{H}_x\,\hat{\mathbf{x}} + \dot{H}_y\,\hat{\mathbf{y}} + \dot{H}_z\,\hat{\mathbf{z}} + H_x\,\dot{\hat{\mathbf{x}}} + H_y\,\dot{\hat{\mathbf{y}}} + H_z\,\dot{\hat{\mathbf{z}}}$$

An elementary property of vectors may be used for a more compact form of the equations.

$$\dot{\hat{\mathbf{x}}} = \mathbf{\Omega} \times \hat{\mathbf{x}}$$

$$\dot{\hat{\mathbf{y}}} = \mathbf{\Omega} \times \hat{\mathbf{y}}$$

$$\dot{\hat{\mathbf{z}}} = \mathbf{\Omega} \times \hat{\mathbf{z}}$$

Since,

$$H_x \; \mathbf{\Omega} \times \hat{\mathbf{x}} + H_y \; \mathbf{\Omega} \times \hat{\mathbf{y}} + H_z \; \mathbf{\Omega} \times \hat{\mathbf{z}} = \mathbf{\Omega} \times (Hx\,\hat{\mathbf{x}} + Hy\,\hat{\mathbf{y}} + Hz\,\hat{\mathbf{z}}) = \mathbf{\Omega} \times \mathbf{H}$$

the final form of Euler's equation is given by

$$\mathbf{M} = I\,\dot{\mathbf{\Omega}} + \mathbf{\Omega} \times \mathbf{H} \tag{1.11}$$

The selection of body fixed axes can result in a considerable simplification of Euler's equation. The initial selection of body fixed axes is arbitrary and often governed by other considerations such as the location of landmarks. Once the body axes have been defined and the inertia tensor obtained by integration or solution of the gravity potential, it may be convenient to redefine the direction of the body fixed axes. Consider a new body fixed coordinate system defined by a simple orthogonal rotation (R). The new moment and angular momentum vectors are primed.

$$\mathbf{M}' = R\,\mathbf{M}$$

$$\dot{\mathbf{\Omega}}' = R\,\dot{\mathbf{\Omega}}$$

$$\mathbf{\Omega}' \times \mathbf{H}' = R\,(\mathbf{\Omega} \times \mathbf{H})$$

Making the above substitutions into Eq. (1.11) gives

$$R^T\,\mathbf{M}' = I\,R^T\,\dot{\mathbf{\Omega}}' + R^T\,(\mathbf{\Omega}' \times \mathbf{H}')$$

and

$$\mathbf{M}' = [R\,I\,R^T]\,\dot{\mathbf{\Omega}}' + \mathbf{\Omega}' \times \mathbf{H}'$$

The matrix I is positive definite and so is the matrix in brackets. The inertia tensor may be diagonalized by solving for its eigenvalues and the matrix of eigenvectors,

$$I = [T\,\lambda\,T^T]$$

and λ is a diagonal matrix of eigenvalues. If the rotation matrix R is selected to be the transpose or inverse of T,

$$\mathbf{M}' = I'\,\dot{\mathbf{\Omega}}' + \mathbf{\Omega}' \times \mathbf{H}' \tag{1.12}$$

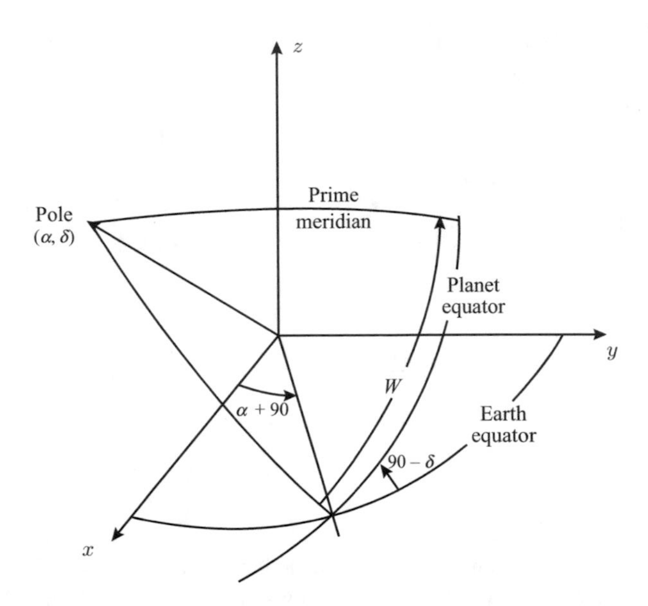

Fig. 1.3 Transformation to planetary body fixed coordinates

The new primed coordinates define principal axes and the inertia tensor is diagonal. The principal axes are often used for body fixed axes and every rigid body has at least one set of principal axes. For a sphere or a cube, every axis set is principal.

The choice to integrate the rotational equations of motion in body fixed axes requires the introduction of angles to describe the orientation of the body fixed axes in inertial space. The conventional method for describing the attitude of a rotating body is a set of Euler angles. The Euler angles define consecutive rotations about the body fixed axes that transform a vector from a reference frame to body fixed axes. For a body in space, it is convenient to select the same Euler angle set as is used to define the pole and prime meridian of a planet. The Euler angles are right ascension (α) and declination (δ) of the pole and the angle W from the intersection of the body equator with the Earth's equator at epoch January 1, 2000 (J2000) to the prime meridian as shown in Fig. 1.3. The first rotation is a right-hand rotation about the z axis through the angle right ascension (α). The second rotation is a right-hand rotation about the y axis through the angle $90° - \delta$ that places the z axis in the direction of the pole. The third rotation is another right-hand rotation about the z axis through the angle $90° + W$ to place the x axis on the prime meridian of the body.

The rotations from inertial space to body fixed axes are

$$\mathrm{T} = \begin{bmatrix} -\sin W & \cos W & 0 \\ -\cos W & -\sin W & 0 \\ 0 & 0 & 1 \end{bmatrix} \begin{bmatrix} \sin\delta & 0 & -\cos\delta \\ 0 & 1 & 0 \\ \cos\delta & 0 & \sin\delta \end{bmatrix} \begin{bmatrix} \cos\alpha & \sin\alpha & 0 \\ -\sin\alpha & \cos\alpha & 0 \\ 0 & 0 & 1 \end{bmatrix}$$

and may be combined to obtain

$$
T = \begin{bmatrix} -\sin W \sin\delta \cos\alpha & -\sin W \sin\delta \sin\alpha & \sin W \cos\delta \\ -\cos W \sin\alpha & +\cos W \cos\alpha & \\ & & \\ -\cos W \sin\delta \cos\alpha & -\cos W \sin\delta \sin\alpha & \cos W \cos\delta \\ +\sin W \sin\alpha & -\sin W \cos\alpha & \\ & & \\ \cos\delta \cos\alpha & \cos\delta \sin\alpha & \sin\delta \end{bmatrix}
$$

The body fixed spin rates are related to the Euler angle rates by

$$
\begin{bmatrix} \omega_x \\ \omega_y \\ \omega_z \end{bmatrix} = \begin{bmatrix} \sin \mathrm{W} \cos\delta & -\cos \mathrm{W} & 0 \\ \cos \mathrm{W} \cos\delta & \sin \mathrm{W} & 0 \\ \sin\delta & 0 & 1 \end{bmatrix} \begin{bmatrix} \dot{\alpha} \\ \dot{\delta} \\ \dot{W} \end{bmatrix}
$$

A unit vector in the direction of each Euler angle rate is transformed to body fixed axes and the resultant vectors are assembled into the above transformation matrix. This transformation is not orthogonal. The inverse transformation that relates the Euler angle rates to the body fixed spin rates is obtained by matrix inversion and is given by

$$
\begin{bmatrix} \dot{\alpha} \\ \dot{\delta} \\ \dot{W} \end{bmatrix} = \begin{bmatrix} \sin \mathrm{W} \sec\delta & \cos \mathrm{W} \sec\delta & 0 \\ -\cos \mathrm{W} & \sin \mathrm{W} & 0 \\ -\sin \mathrm{W} \tan\delta & -\cos \mathrm{W} \tan\delta & 1 \end{bmatrix} \begin{bmatrix} \omega_x \\ \omega_y \\ \omega_z \end{bmatrix}
$$

The body fixed spin vector may be obtained by integration of Euler's equation. First, Euler's equation must be solved for the angular acceleration. Thus, we have for Euler's equation,

$$
\dot{\boldsymbol{\Omega}} = I^{-1}\{\mathbf{M} - \boldsymbol{\Omega} \times \mathbf{H}\} \tag{1.13}
$$

$$
\mathbf{H} = I\,\boldsymbol{\Omega}
$$

and the angular rates are obtained by integrating

$$
\boldsymbol{\Omega}(t) = \int \dot{\boldsymbol{\Omega}}\, dt
$$

The attitude may be obtained as a function of time by integrating the Euler angle rates computed from the spin vector.

$$\mathbf{\Phi}(t) = \int \dot{\mathbf{\Phi}}\, dt$$

The attitude, shown here as the data vector $\mathbf{\Phi}$, is defined by the Euler angles,

$$\mathbf{\Phi} = \begin{bmatrix} \alpha \\ \delta \\ W \end{bmatrix}$$

and the rates are

$$\dot{\mathbf{\Phi}} = \begin{bmatrix} \sin W \sec\delta & \cos W \sec\delta & 0 \\ -\cos W & \sin W & 0 \\ -\sin W \tan\delta & -\cos W \tan\delta & 1 \end{bmatrix} \mathbf{\Omega}$$

1.6 Rotational Variational Equations

From the equations of motion, the angular acceleration may be obtained as an explicit function of the attitude, spin rate, and dynamic parameters (q).

$$\dot{\mathbf{\Omega}} = f_{\dot{\Omega}}(\boldsymbol{\phi}, \mathbf{\Omega}, q) \tag{1.14}$$

where the dynamic parameters (q) consist of the initial attitude and spin rate, moments of inertia, and the applied moment.

$$q = (\boldsymbol{\phi}_o, \mathbf{\Omega_o}, I_e, \mathbf{M}) \tag{1.15}$$

where I_e denotes a column matrix containing the six unique inertia tensor elements. The Euler angle attitude rates may be explicitly related to the attitude and body fixed spin rate and are given by

$$\dot{\boldsymbol{\phi}} = f_{\dot{\phi}}(\boldsymbol{\phi}, \mathbf{\Omega}) \tag{1.16}$$

The variational equations are obtained by differentiating Equation 1.14 with respect to the dynamic parameters.

$$\frac{\partial \dot{\boldsymbol{\Omega}}}{\partial q} = \frac{\partial f_{\dot{\Omega}}}{\partial \boldsymbol{\phi}} \frac{\partial \boldsymbol{\phi}}{\partial q} + \frac{\partial f_{\dot{\Omega}}}{\partial \boldsymbol{\Omega}} \frac{\partial \boldsymbol{\Omega}}{\partial q} + \frac{\partial f_{\dot{\Omega}}}{\partial q} \tag{1.17}$$

$$\frac{\partial \dot{\boldsymbol{\phi}}}{\partial q} = \frac{\partial f_{\dot{\phi}}}{\partial \boldsymbol{\phi}} \frac{\partial \boldsymbol{\phi}}{\partial q} + \frac{\partial f_{\dot{\phi}}}{\partial \boldsymbol{\Omega}} \frac{\partial \boldsymbol{\Omega}}{\partial q} \tag{1.18}$$

Thus, we have, for the case of free-body rotation or a constant applied moment in body fixed coordinates,

$$\frac{\partial f_{\dot{\Omega}}}{\partial \boldsymbol{\phi}} = 0 \tag{1.19}$$

$$\frac{\partial f_{\dot{\Omega}}}{\partial \boldsymbol{\Omega}} = -I^{-1} \left\{ \begin{bmatrix} 0 & h_z & -h_y \\ -h_z & 0 & h_x \\ h_y & -h_x & 0 \end{bmatrix} - \begin{bmatrix} 0 & \omega_z & -\omega_y \\ -\omega_z & 0 & \omega_x \\ \omega_y & -\omega_x & 0 \end{bmatrix} I \right\} \tag{1.20}$$

$$\frac{\partial f_{\dot{\phi}}}{\partial \boldsymbol{\Omega}} = \begin{bmatrix} sin\mathrm{W} \sec\delta & cos\mathrm{W} \sec\delta & 0 \\ -cos\mathrm{W} & sin\mathrm{W} & 0 \\ -sin\mathrm{W} \tan\delta & -cos\mathrm{W} \tan\delta & 1 \end{bmatrix} \tag{1.21}$$

$$\frac{\partial f_{\dot{\phi}}}{\partial \boldsymbol{\phi}} = \begin{bmatrix} 0 & \dfrac{(\omega_x \sin W + \omega_y \cos W) \tan\delta}{\cos\delta} & \dfrac{\omega_x \cos W - \omega_y \sin W}{\cos\delta} \\ 0 & 0 & \omega_x \sin W + \omega_y \cos W \\ 0 & \dfrac{-\omega_x \sin W - \omega_y \cos W}{\cos^2\delta} & (-\omega_x \cos W + \omega_y \sin W) \tan\delta \end{bmatrix} \tag{1.22}$$

With respect to attitude at epoch ($q = \boldsymbol{\phi}_o, \boldsymbol{\Omega}_o$)
The variational equations that relate the current attitude to the attitude at epoch are given by

$$\frac{\partial \dot{\boldsymbol{\phi}}}{\partial \boldsymbol{\phi}_o} = \frac{\partial f_{\dot{\phi}}}{\partial \boldsymbol{\phi}} \frac{\partial \boldsymbol{\phi}}{\partial \boldsymbol{\phi}_o} \tag{1.23}$$

$$\frac{\partial \dot{\boldsymbol{\phi}}}{\partial \boldsymbol{\Omega}_\mathbf{o}} = \frac{\partial f_{\dot{\phi}}}{\partial \boldsymbol{\phi}} \frac{\partial \boldsymbol{\phi}}{\partial \boldsymbol{\Omega}_\mathbf{o}} + \frac{\partial f_{\dot{\phi}}}{\partial \boldsymbol{\Omega}} \frac{\partial \boldsymbol{\Omega}}{\partial \boldsymbol{\Omega}_\mathbf{o}} \tag{1.24}$$

$$\frac{\partial \dot{\boldsymbol{\Omega}}}{\partial \boldsymbol{\phi}_0} = 0$$

$$\frac{\partial \dot{\mathbf{\Omega}}}{\partial \mathbf{\Omega_o}} = \frac{\partial f_{\dot{\Omega}}}{\partial \mathbf{\Omega}} \frac{\partial \mathbf{\Omega}}{\partial \mathbf{\Omega_o}} \tag{1.25}$$

We obtain these partial derivatives as a function of time by integrating as in

$$\frac{\partial \boldsymbol{\phi}}{\partial \boldsymbol{\phi}_o} = \int \frac{\partial \dot{\boldsymbol{\phi}}}{\partial \boldsymbol{\phi}_o} dt \tag{1.26}$$

$$\frac{\partial \boldsymbol{\phi}}{\partial \mathbf{\Omega_o}} = \int \frac{\partial \dot{\boldsymbol{\phi}}}{\partial \mathbf{\Omega_o}} dt \tag{1.27}$$

$$\frac{\partial \mathbf{\Omega}}{\partial \boldsymbol{\phi}_o} = 0 \tag{1.28}$$

$$\frac{\partial \mathbf{\Omega}}{\partial \mathbf{\Omega_o}} = \int \frac{\partial \dot{\mathbf{\Omega}}}{\partial \mathbf{\Omega_o}} dt \tag{1.29}$$

With respect to elements of inertia tensor ($q = I_e$)
The variational equations for the elements of the inertia tensor are given by

$$\frac{\partial \dot{\mathbf{\Omega}}}{\partial I_e} = \frac{\partial f_{\dot{\Omega}}}{\partial \mathbf{\Omega}} \frac{\partial \mathbf{\Omega}}{\partial I_e} + \frac{\partial f_{\dot{\Omega}}}{\partial I_e} \tag{1.30}$$

$$\frac{\partial \dot{\boldsymbol{\phi}}}{\partial I_e} = \frac{\partial f_{\phi}}{\partial \boldsymbol{\phi}} \frac{\partial \boldsymbol{\phi}}{\partial I_e} + \frac{\partial f_{\dot{\phi}}}{\partial \mathbf{\Omega}} \frac{\partial \mathbf{\Omega}}{\partial I_e} \tag{1.31}$$

where I_e is a column matrix, restricted to the independent elements of I,

$$I_e = \left[I_{xx}\ I_{yy}\ I_{zz}\ I_{xy}\ I_{xz}\ I_{yz}\right]^{\mathrm{T}}$$

The matrix that defines the partial derivative of angular acceleration with respect to the elements of the moment of inertia tensor may be obtained by differentiating the equations of motion (Eq. 1.13).

$$I \frac{\partial \dot{\mathbf{\Omega}}}{\partial I_e} + \frac{\partial I}{\partial I_e} \dot{\mathbf{\Omega}} + \frac{\partial (\mathbf{\Omega} \times \mathbf{H})}{\partial \mathbf{H}} \frac{\partial \mathbf{H}}{\partial I_e} = 0 \tag{1.32}$$

The terms in the above matrix equation are given by

$$\frac{\partial I}{\partial I_e} \dot{\mathbf{\Omega}} = \begin{bmatrix} \dot{\omega}_x & 0 & 0 & \dot{\omega}_y & \dot{\omega}_z & 0 \\ 0 & \dot{\omega}_y & 0 & \dot{\omega}_x & 0 & \dot{\omega}_z \\ 0 & 0 & \dot{\omega}_z & 0 & \dot{\omega}_x & \dot{\omega}_y \end{bmatrix}$$

$$\frac{\partial (\mathbf{\Omega} \times \mathbf{H})}{\partial \mathbf{H}} = \begin{bmatrix} 0 & -\omega_z & \omega_y \\ \omega_z & 0 & -\omega_x \\ -\omega_y & \omega_x & 0 \end{bmatrix}$$

$$\frac{\partial \mathbf{H}}{\partial I_e} = \begin{bmatrix} \omega_x & 0 & 0 & \omega_y & \omega_z & 0 \\ 0 & \omega_y & 0 & \omega_x & 0 & \omega_z \\ 0 & 0 & \omega_z & 0 & \omega_x & \omega_y \end{bmatrix}$$

The above partial derivatives are obtained by performing the indicated matrix multiplication to obtain the individual equations, differentiating, and then reassembling the result in matrix form. Solving for the partial derivative of angular acceleration with respect to the elements of the inertia tensor, we obtain

$$\frac{\partial f_{\dot{\Omega}}}{\partial I_e} = -I^{-1} \begin{bmatrix} 0 & -\omega_z & \omega_y \\ \omega_z & 0 & -\omega_x \\ -\omega_y & \omega_x & 0 \end{bmatrix} \begin{bmatrix} \omega_x & 0 & 0 & \omega_y & \omega_z & 0 \\ 0 & \omega_y & 0 & \omega_x & 0 & \omega_z \\ 0 & 0 & \omega_z & 0 & \omega_x & \omega_y \end{bmatrix}$$
$$-I^{-1} \begin{bmatrix} \dot{\omega}_x & 0 & 0 & \dot{\omega}_y & \dot{\omega}_z & 0 \\ 0 & \dot{\omega}_y & 0 & \dot{\omega}_x & 0 & \dot{\omega}_z \\ 0 & 0 & \dot{\omega}_z & 0 & \dot{\omega}_x & \dot{\omega}_y \end{bmatrix} \tag{1.33}$$

The inertia tensor partial derivatives as a function of time are obtained by integrating the rates and

$$\frac{\partial \mathbf{\Omega}}{\partial I_e} = \int \frac{\partial \dot{\mathbf{\Omega}}}{\partial I_e} dt \tag{1.34}$$

$$\frac{\partial \boldsymbol{\phi}}{\partial I_e} = \int \frac{\partial \dot{\boldsymbol{\phi}}}{\partial I_e} dt \tag{1.35}$$

With respect to applied moment ($q = \mathbf{M}$)
The variational equations for a constant body fixed applied moment are given by

$$\frac{\partial \dot{\mathbf{\Omega}}}{\partial \mathbf{M}} = \frac{\partial f_{\dot{\Omega}}}{\partial \Omega} \frac{\partial \mathbf{\Omega}}{\partial \mathbf{M}} + \frac{\partial f_{\dot{\Omega}}}{\partial \mathbf{M}} \tag{1.36}$$

$$\frac{\partial \dot{\boldsymbol{\phi}}}{\partial \mathbf{M}} = \frac{\partial f_{\dot{\phi}}}{\partial \phi} \frac{\partial \boldsymbol{\phi}}{\partial \mathbf{M}} + \frac{\partial f_{\dot{\phi}}}{\partial \Omega} \frac{\partial \Omega}{\partial \mathbf{M}} \tag{1.37}$$

The final matrix that is needed in the above variational equations defines the direct effect of the applied moment on the angular acceleration and is given by

$$\frac{\partial f_{\dot{\Omega}}}{\partial \mathbf{M}} = I^{-1} \tag{1.38}$$

We obtain the applied moment partial derivatives as a function of time by integrating the rates and

$$\frac{\partial \mathbf{\Omega}}{\partial \mathbf{M}} = \int \frac{\partial \dot{\mathbf{\Omega}}}{\partial \mathbf{M}} dt \tag{1.39}$$

$$\frac{\partial \boldsymbol{\phi}}{\partial \mathbf{M}} = \int \frac{\partial \dot{\boldsymbol{\phi}}}{\partial \mathbf{M}} dt \tag{1.40}$$

The above rotational variational equations may be integrated with the equations of motion to describe the body attitude state (Euler angle and spin vector) and partial derivatives of state with respect to initial attitude, moments of inertia, and applied moments as a function of time.

1.7 Kinetic Theory of Gases

An interesting application of Newton's equation of motion to a system of particles is the kinetic theory of gasses that results in the ideal gas law and several other laws that govern the behavior of gasses. The motion of a gas molecule, neglecting rotations, may be described by assuming the molecules of gas to be rigid spheres. It was postulated by Maxwell, Boltzmann, and others that the simplicity of the experimental behavior of gasses implied simplicity on the molecular scale. An extension of the principals governing particle dynamics may be applied to molecules bouncing around in a container to obtain the gas laws. The motion of gas molecules is of interest to spacecraft navigation for two reasons. The rapid expulsion of gas molecules from a rocket engine provides the thrust force that changes the velocity of a spacecraft and the incidental expulsion of gas molecules from a spacecraft causes small accelerations that must be accounted for in solving for the trajectory of the spacecraft. The former is essential for the success of the mission and the latter is a nuisance to navigation. An example of the latter is the venting from a parachute that may be carried on a spacecraft or the expulsion of gas from propulsion system leaks.

Consider a gas molecule of mass μ within a container that comes into contact with a wall. The mass of the molecule is approximately the product of the molecular weight (M) times the mass of a proton or neutron (μ_0). The exact relationship involves the mass associated with the binding energy and as a standard μ_0 is taken as one sixteenth of the mass of the O^{16} molecule. The molecular weight (M) is dimensionless and generally nearly an integer representing the total number of protons and neutrons in a molecule. For a molecule moving in the v_x direction, Newton's equation may be used to describe the motion during the time that the molecule interacts with the wall.

$$F_i = \mu_i \frac{dv_x}{dt}$$

$$\int_t^{t+\Delta t} F_i \, dt = \mu_i \int_{v_x}^{-v_x} dv_x$$

$$F_i \Delta t_i = -2\mu_i v_x$$

A thin layer of gas next to a wall of thickness δx and area A will contain δN molecules moving at an average speed v_x and half of these molecules will be moving in the plus v_x direction and half will be moving in the $-v_x$ direction. Thus, $\frac{1}{2}\delta N$ molecules will strike the wall in the time interval δt. Summing all the impacts over the time interval δt and volume $A\delta x$ gives

$$\frac{1}{2}\delta N \; \delta x \; F_i \Delta t_i = \frac{1}{2}\delta N \; (-2\mu_i \; v_x^2) \; \delta t$$

$$v_x = \frac{\delta x}{\delta t}$$

Over the time interval δt, the total force exerted by the gas on the wall must be opposite and equal to the total force exerted by the wall on the gas. This is a direct consequence of the requirement that the momentum exchange at the surface of the wall must be zero. Consider a thin massless rigid coating that is applied to the wall at the interface. The momentum of this coating, which is sum of the gas molecules pushing outward and the container wall pushing inward, must be zero. This momentum balance is given by

$$\frac{1}{2}\delta N \; F_i \Delta t_i = -F \delta t$$

which simplifies to

$$F \delta x = \delta N \; \mu_i \; v_x^2$$

The number density (N) is the number of molecules per unit volume and is given by,

$$N = \frac{\delta N}{A \delta x}$$

and

$$P = N \; \mu_i \; v_x^2$$

$$F = PA$$

The square of the magnitude of the velocity is simply the sum of the squares of the components.

$$v^2 = v_x^2 + v_y^2 + v_z^2$$

Due to symmetry, the average magnitudes of the velocity components must be equal and $v_x = v_y = v_z$. The gas law derived from kinetic theory is now,

$$P = \frac{1}{3} N \; \mu_i \; v^2 \tag{1.41}$$

The agreement with the ideal gas law derived from experimentation would be complete if the temperature of the gas is related to the kinetic energy of the gas molecules. It may be argued from first principals that temperature is a measure of the kinetic energy of the gas molecule. The liquid in a thermometer rises in response to the expansion of the liquid and this is directly proportional to the kinetic energy of the liquids molecules. It may be concluded that temperature is proportional to energy of a gas molecule and the constant of proportionality is assumed to be $\frac{3}{2}k$ where k is Boltzmann's constant. The scaling of k by three halves is arbitrary and designed to yield a familiar form for the end result.

$$\frac{E_\mu}{T} = \frac{3}{2}k$$

$$E_\mu = \frac{1}{2}\mu_i\, v^2$$

Making these substitutions, the kinetic theory description of an ideal gas becomes

$$P = NkT$$

The number density N is simply the total number of gas molecules N_t enclosed within the volume (V) divided by the volume.

$$N = \frac{N_t}{V}$$

The total number of molecules is equal to the total mass (m) dived by the mass of one molecule.

$$N_t = \frac{m}{M\mu_0}$$

Making these substitutions, the gas law becomes

$$PV = \frac{m}{M}\frac{k}{\mu_0}T \tag{1.42}$$

The conventional experimental form of the ideal gas law is

$$PV = \frac{m}{M}RT \tag{1.43}$$

or

$$P = \frac{\rho}{M}RT \tag{1.44}$$

where the density is given by,

$$\rho = \frac{m}{V}$$

The kinetic theory gas constant and experimental gas constant are related by,

$$R = \frac{k}{\mu_0}$$

The number of molecules contained in a sample of gas of mass m in grams equal to the molecular weight (M) is a constant (N_0) called Avogadro's number. The reciprocal of Avogadro's number is numerically equal to μ_0, one sixteenth of the mass of O^{16}, or approximately the mass of atomic hydrogen. The ratio of the mass to the molecular weight is called the mole fraction (n) and Avogadro's number is simply the number of molecules in one mole. The relationship of Boltzmann's constant to the universal gas constant is also given by,

$$R = N_0\, k \tag{1.45}$$

A mixture of gases with different molecular weights will reach an equilibrium and all molecules will have the same energy and consequently the same temperature. Molecules of higher than average energy transfer energy to the walls of the container when they collide with the walls and conversely molecules of lower than average energy will receive energy from the wall. As the gasses mix, they will converge to the same average energy. Since the two groups of molecules with different masses have the same temperature and move independently, the total pressure may be obtained by summing the pressures that each group of molecules would exert if it occupied the container alone. This observation of the behavior of gas mixtures is called Dalton's law of partial pressures.

If a container of gas is vented to space, the momentum of the gas molecules will exert a force on the container which is transmitted to the spacecraft resulting in an acceleration of the spacecraft. From Newton's equation, we have for a small quantity of gas that is vented,

$$\delta F = \delta m \frac{dv}{dt}$$

The average velocity of the gas molecules is constant and

$$v^2 = \frac{3kT}{\mu_i} = \frac{3R\mu_0 T}{\mu_i}$$

and since $\mu_i = M\mu_0$,

$$v = \sqrt{\frac{3RT}{M}} \tag{1.46}$$

Since v may be assumed to be constant, Newton's equation of motion may be put in the form,

$$F = \sqrt{\frac{3RT}{M}} \frac{dm}{dt} \tag{1.47}$$

Analysis of the propulsion system gas leaks for the 1975 Viking mission to Mars provides a typical example of the application of the kinetic theory of gasses to navigation of a spacecraft. The gas leak rate is specified to be less than 100 standard cubic centimeters of Helium per hour. The problem for navigation is to place an upper bound on the acceleration of the Viking spacecraft that may be expected. The density of helium at standard temperature and pressure is $\rho = 0.166\,\mathrm{kg/m^3}$, where $\mathrm{P} = 1.013 \times 10^5\,\mathrm{nt/m^2}$ (1 atmosphere), M = 4, $\mathrm{R} = 8317\,\mathrm{m^2/s^2K}$, and T = 293 K. The velocity of the vented helium molecules, assuming that all molecules are vented in the same direction, is v = 1351 m/s. The mass flow rate is simply the density times the volumetric flow rate.

$$\frac{dV}{dt} = 100\,\mathrm{cm^3/h} = 2.78 \times 10^{-8}\,\mathrm{m^3/s}$$

$$\frac{dm}{dt} = \rho \frac{dV}{dt} = 4.62 \times 10^{-9}\,\mathrm{kg/s}$$

The net force on the spacecraft is 6.25×10^{-6} nt. The acceleration of the spacecraft of mass $m_{sc} = 3468$ kg is given by

$$a_{sc} = \frac{F}{m_{sc}} = 1.8 \times 10^{-9}\,\mathrm{m/s^2}$$

The acceleration computed in this manner from the kinetic theory of gasses will yield a result that is a little high. The vented molecules do not all vent in the same direction and there is some loss of energy as the molecules leave the container. These inefficiencies reduce the total acceleration.

1.8 General Relativity Equations of Motion

The General Theory of Relativity, which includes Special Relativity, replaces classical Newtonian theory. For spacecraft navigation, General Relativity enters into the equations of motion and computation of measurements. The effect is small and generally could be ignored for trajectory design but can be observed when performing orbit determination. For this reason, Generally Relativity is formulated as a perturbation to Newtonian theory. The effect of General Relativity on orbit determination is ubiquitous. The orbit of Mercury is perturbed, radio signals from the spacecraft are bent and delayed if they pass near the sun, and clocks slow down if they are near a massive body or have significant velocity with respect to the solar system barycenter. These perturbations may be determined by integrating the equation of geodesics or computed from simple formulas.

The details of the solution of the Einstein field equations are omitted here and the solutions for the equations of motion are initiated from the metric tensor. The metric tensor is the solution to the field equations. The equation of geodesics operates on the metric tensor to generate the equations of motion. The application of Einstein's summation notation is fairly straightforward. The resultant equations of motion require no further understanding of relativity theory except for some simple applications of Special Relativity. A detailed understanding of General Relativity is not needed to navigate spacecraft.

1.8.1 Einstein Field Equation

The Einstein field equation relates the curvature of space to the distribution of mass, energy, and stress. The mathematical statement of general relativity is simply

$$\mathbf{G} = 8\pi\ \mathbf{T}$$

This equation replaces Newton's law of gravitation. **G** is the Einstein tensor that describes the geometry and **T** is the stress energy tensor that describes the distribution of mass, energy, and stress in the same coordinate system as the geometry. The Einstein tensor is a specific description of curved space that is extracted from the general Riemann tensor in a way to satisfy the basic postulates of general relativity. The Einstein field equation has been solved exactly for the case of a particle moving in the spherically symmetric gravitational field of a body. For this distribution of matter, the stress energy tensor inside the body is given by

$$\mathbf{T} = \begin{bmatrix} p & 0 & 0 & 0 \\ 0 & p & 0 & 0 \\ 0 & 0 & p & 0 \\ 0 & 0 & 0 & \rho \end{bmatrix}$$

Outside the body, **T** is equal to zero. The variable ρ is the scalar invariant density of matter and the variable p is the pressure that is obtained in hydrostatic equilibrium. The solution was obtained by Schwarzschild about a month after Einstein published the theory of general relativity and is given by the following metric tensor.

$$g_{ij} = \begin{bmatrix} g_{11} & g_{12} & g_{13} & g_{14} \\ g_{21} & g_{22} & g_{23} & g_{24} \\ g_{31} & g_{32} & g_{33} & g_{34} \\ g_{41} & g_{42} & g_{43} & g_{44} \end{bmatrix}$$

where

$$g_{11} = g_{rr} = -\frac{1}{c^2}\left(1 - \frac{2m}{r}\right)^{-1}$$

$$g_{22} = g_{\theta\theta} = -\frac{1}{c^2}r^2$$

$$g_{33} = g_{\phi\phi} = -\frac{1}{c^2}r^2\sin^2\theta$$

$$g_{44} = g_{tt} = \left(1 - \frac{2m}{r}\right)$$

$$m = \frac{\mu}{c^2}$$

and all the offdiagonal elements are zero. The metric tensor takes the place of the potential in classical theory.

1.8.2 Geodesic Equation

The geodesic equation describes the acceleration of a particle in space-time coordinates and takes the place of the gradient in classical theory.

$$\frac{d^2x^\alpha}{ds^2} + \Gamma^\alpha_{uv}\frac{dx^u}{ds}\frac{dx^v}{ds} = 0 \tag{1.48}$$

The Christoffel symbols ($\Gamma's$) establish the connection between curved space and the observed world. For this reason, they are sometimes called connection coefficients. The Christoffel symbols are obtained by integrating the line element between two points and then solving for the path that gives the minimum time which is the speed of light and a straight line in curved space. The Christoffel symbols are given by

$$\Gamma^u_{\alpha\beta} = g^{uv}\Gamma_{v\alpha\beta}$$

$$\Gamma_{u\alpha\beta} = \frac{1}{2}\left(\frac{\partial g_{u\alpha}}{\partial x^\beta} + \frac{\partial g_{u\beta}}{\partial x^\alpha} - \frac{\partial g_{\alpha\beta}}{\partial x^u}\right)$$

In Einstein summation notation, raising both indices of the metric tensor is matrix inversion. The Christoffel symbols for the Schwarzschild solution are given by

$$\Gamma^r_{rr} = \Gamma^1_{11} = \frac{-m}{r^2}\left(1 - \frac{2m}{r}\right)^{-1}$$

$$\Gamma^r_{\phi\phi} = \Gamma^1_{33} = -r\left(1 - \frac{2m}{r}\right)\sin^2\theta$$

$$\Gamma^r_{tt} = \Gamma^1_{44} = \frac{mc^2}{r^2}\left(1 - \frac{2m}{r}\right)$$

$$\Gamma^r_{\theta\theta} = \Gamma^1_{22} = -r\left(1 - \frac{2m}{r}\right)$$

$$\Gamma^t_{rt} = \Gamma^4_{14} = \frac{m}{r^2}\left(1 - \frac{2m}{r}\right)^{-1}$$

$$\Gamma^\phi_{r\phi} = \Gamma^3_{13} = \frac{1}{r}$$

$$\Gamma^\phi_{\theta\phi} = \Gamma^3_{23} = \cot\theta$$

$$\Gamma^\theta_{r\theta} = \Gamma^2_{12} = \frac{1}{r}$$

$$\Gamma^\theta_{\phi\phi} = \Gamma^2_{33} = -\sin\theta\cos\theta$$

The equations of motion are obtained by substituting the Christoffel symbols into the geodesic equation. Since the motion is planar, we may rotate to a coordinate system such that the motion is in the x-y plane. The θ dependency is thus removed and for $\theta = 2\pi$, we obtain from the geodesic equation

$$\frac{d^2r}{ds^2} = \frac{m}{r^2}\left(1 - \frac{2m}{r}\right)^{-1}\left(\frac{dr}{ds}\right)^2 + r\left(1 - \frac{2m}{r}\right)\left(\frac{d\phi}{ds}\right)^2 - \frac{mc^2}{r^2}\left(1 - \frac{2m}{r}\right)\left(\frac{dt}{ds}\right)^2$$

$$\frac{d^2\phi}{ds^2} = \frac{2}{r}\frac{dr}{ds}\frac{d\phi}{ds}$$

A clock carried on the particle will provide a measure of proper time defined by the line element

$$ds^2 = -\left\{(1 - \frac{2m}{r})^{-1}\,dr^2 + r^2\,d\theta^2 + r^2\sin^2\theta\,d\phi^2\right\} + (1 - \frac{2m}{r})\,c^2dt^2$$

For $ds^2 = c^2\,d\tau^2$, this equation yields

$$\left(\frac{dt}{d\tau}\right)^2 = \left(1 - \frac{2m}{r}\right)^{-1} + \frac{1}{c^2}\left(1 - \frac{2m}{r}\right)^{-2}\left(\frac{dr}{d\tau}\right)^2 + \frac{r^2}{c^2}\left(1 - \frac{2m}{r}\right)^{-1}\left(\frac{d\phi}{d\tau}\right)^2$$

Substituting the metric equation into the geodesic equation gives the following equations of motion.

$$\frac{d^2r}{d\tau^2} = -\frac{\mu}{r^2} + \left(r - \frac{3\mu}{c^2}\right)\left(\frac{d\phi}{d\tau}\right)^2 \tag{1.49}$$

$$\frac{d^2\phi}{d\tau^2} = -\frac{2}{r}\frac{dr}{d\tau}\frac{d\phi}{d\tau} \tag{1.50}$$

$$\left(\frac{dt}{d\tau}\right)^2 = \left(1-\frac{2m}{r}\right)^{-1} + \frac{1}{c^2}\left(1-\frac{2m}{r}\right)^{-2}\left(\frac{dr}{d\tau}\right)^2 + \frac{r^2}{c^2}\left(1-\frac{2m}{r}\right)^{-1}\left(\frac{d\phi}{d\tau}\right)^2 \tag{1.51}$$

The trajectory of a photon differs from that of a particle or spacecraft moving at the speed of light even in the limit of very small mass for the spacecraft. The difference arises because a photon has zero rest mass and thus there is no force of gravity acting on the photon that gives rise to Newtonian acceleration. The photon follows the contour of curved space and the resulting path is the called the null geodesic. Consider the metric associated with a particle traveling at the speed of light

$$ds^2 = 0 = -\frac{1}{c^2}\left\{(1-\frac{2m}{r})^{-1}\,dr^2 + r^2\,d\theta^2 + r^2\sin^2\theta\,d\phi^2\right\} + (1-\frac{2m}{r})\,dt^2$$

Since ds is zero, the geodesic equation degenerates to indeterminate forms that must be evaluated in the limit as ds goes to zero. The indeterminate form ds/ds, which has the value of 1 for a spacecraft, has the value 0 for a photon in the limit as ds approaches zero. We resolve the problem of ds approaching zero in the geodesic equation by introducing the affine parameter τ that acts like a clock on the photon. We know from special relativity that an observer's clock on the photon will not register any passage of time as the photon moves from point A to point B. However, we may imagine a universe filled with clocks all running at a rate dependent on the local gravity field. The proper time associated with a photon is simply the integral of these rates along the path of the photon. The difference of the affine parameter (τ) between two points times the speed of light is the distance that one would measure with a meter stick along the path of the photon. The equations of motion for a photon are given by

$$\frac{d^2r}{d\tau^2} = \left(r-\frac{3\mu}{c^2}\right)\left(\frac{d\phi}{d\tau}\right)^2 \tag{1.52}$$

$$\frac{d^2\phi}{d\tau^2} = -\frac{2}{r}\frac{dr}{d\tau}\frac{d\phi}{d\tau} \tag{1.53}$$

$$\left(\frac{dt}{d\tau}\right)^2 = \frac{1}{c^2}\left(1-\frac{2m}{r}\right)^{-2}\left(\frac{dr}{d\tau}\right)^2 + \frac{r^2}{c^2}\left(1-\frac{2m}{r}\right)^{-1}\left(\frac{d\phi}{d\tau}\right)^2 \tag{1.54}$$

1.8.3 Isotropic Schwarzschild Coordinates

In the Newtonian world, before general relativity, the trajectories of the planets were observed through telescopes and the data fit to a model of the solar system based on Newton's equations of motion. From this model, the gravitational constant of the sun and the planetary ephemerides were estimated to an accuracy consistent with the measurement and model errors. With the introduction of general relativity to the model, the data was refit and a new set of constants and planetary ephemerides determined. However, since the relativistic effects are small, the differences between the numerical values associated with the curved space coordinates and the classical coordinates are also small. This small difference often results in confusion of the two coordinate systems.

In order to make the classical system more nearly coincide with the relativistic system, a coordinate transformation or change of variable was devised to make the local curved space coordinates come into alignment with Euclidean coordinates. This transformation makes the relativistic coordinates look more classical but does not really change anything. The transformed coordinate system is called isotropic Schwarzschild coordinates and the transformation is given by

$$r = \left(1 + \frac{\mu}{2c^2\bar{r}}\right)^2 \bar{r}$$

$$\phi = \bar{\phi}$$

where $\bar{r}$ and $\bar{\phi}$ are the isotropic coordinates. In order to obtain the isotropic form of the equations of motion, we simply substitute the above equation for r into the exact Schwarzschild equations. The exact isotropic Schwarzschild line element is given by

$$d\bar{s}^2 = \frac{\left(1 - \frac{\mu}{2c^2\bar{r}}\right)^2}{\left(1 + \frac{\mu}{2c^2\bar{r}}\right)^2} dt^2 - \frac{1}{c^2}\left(1 + \frac{\mu}{2c^2\bar{r}}\right)^4 \left(d\bar{r}^2 + r^2 d\bar{\phi}^2\right)$$

and this is approximated by

$$d\bar{s}^2 = \left(1 - \frac{2\mu}{c^2\bar{r}}\right) dt^2 - \frac{1}{c^2}\left(1 + \frac{2\mu}{c^2\bar{r}}\right)\left(d\bar{r}^2 + r^2 d\bar{\phi}^2\right)$$

The exact isotropic Schwarzschild equations of motion for a spacecraft become

$$\frac{d^2\bar{r}}{d\tau^2} = -\frac{\mu}{\bar{r}^2}\left(1 + \frac{\mu}{2c^2\bar{r}}\right)^{-4} + \left(1 - \frac{\mu^2}{4c^4\bar{r}^2}\right)^{-1}$$
$$\times \left\{\frac{\mu^3}{2c^4\bar{r}^5}\left(1 + \frac{\mu}{2c^2\bar{r}}\right)^{-4}\left(\frac{d\bar{r}}{d\tau}\right)^2 + \left[\left(1 + \frac{\mu}{2c^2\bar{r}}\right)^2 \bar{r} - \frac{3\mu}{c^2}\right]\left(\frac{d\phi}{d\tau}\right)^2\right\}$$

$$\frac{d^2\bar{\phi}}{d\tau^2} = -\frac{\left(1 - \dfrac{\mu^2}{4c^4\bar{r}^2}\right)}{\left(1 + \dfrac{\mu}{2c^2\bar{r}}\right)^2} \frac{2}{\bar{r}} \frac{d\bar{r}}{d\tau} \frac{d\bar{\phi}}{d\tau}$$

$$\frac{d^2\bar{t}}{d\tau^2} = \frac{\left(1 + \dfrac{\mu}{2c^2\bar{r}}\right)^2}{\left(1 - \dfrac{\mu}{2c^2\bar{r}}\right)^2} + \frac{1}{c^2} \frac{\left(1 + \dfrac{\mu}{2c^2\bar{r}}\right)^6}{\left(1 - \dfrac{\mu}{2c^2\bar{r}}\right)^2} \left[\left(\frac{d\bar{r}}{d\tau}\right)^2 + \bar{r}^2 \left(\frac{d\bar{\phi}}{d\tau}\right)^2\right]$$

and these may be approximated by

$$\frac{d^2\bar{r}}{d\tau^2} = -\frac{\mu}{\bar{r}^2}\left(1 - \frac{2\mu}{c^2\bar{r}}\right) + \left(\bar{r} - \frac{2\mu}{c^2}\right)\left(\frac{d\phi}{d\tau}\right)^2 \tag{1.55}$$

$$\frac{d^2\bar{\phi}}{d\tau^2} = -\left(1 - \frac{\mu}{c^2\bar{r}}\right) \frac{2}{\bar{r}} \frac{d\bar{r}}{d\tau} \frac{d\bar{\phi}}{d\tau} \tag{1.56}$$

$$\frac{d^2\bar{t}}{d\tau^2} = 1 + \frac{2\mu}{c^2\bar{r}} + \frac{1}{c^2}\left(1 + \frac{4\mu}{c^2\bar{r}}\right)\left[\left(\frac{d\bar{r}}{d\tau}\right)^2 + \bar{r}^2\left(\frac{d\bar{\phi}}{d\tau}\right)^2\right] \tag{1.57}$$

The exact isotropic Schwarzschild equations of motion for a photon become

$$\frac{d^2\bar{r}}{d\tau^2} = \left(1 - \frac{\mu^2}{4c^4\bar{r}^2}\right)^{-1} \left\{\frac{-\mu^2}{2c^4\bar{r}^3}\left(\frac{d\bar{r}}{d\tau}\right)^2 + \left[\left(1 + \frac{\mu}{2c^2\bar{r}}\right)^2 \bar{r} - \frac{3\mu}{c^2}\right]\left(\frac{d\phi}{d\tau}\right)^2\right\}$$

$$\frac{d^2\bar{\phi}}{d\tau^2} = -\frac{\left(1 - \dfrac{\mu^2}{4c^4\bar{r}^2}\right)}{\left(1 + \dfrac{\mu}{2c^2\bar{r}}\right)^2} \frac{2}{\bar{r}} \frac{d\bar{r}}{d\tau} \frac{d\bar{\phi}}{d\tau}$$

$$\frac{d^2\bar{t}}{d\tau^2} = \frac{1}{c^2} \frac{\left(1 + \frac{\mu}{2c^2\bar{r}}\right)^6}{\left(1 - \dfrac{\mu}{2c^2\bar{r}}\right)^2} \left[\left(\frac{d\bar{r}}{d\tau}\right)^2 + \bar{r}^2 \left(\frac{d\bar{\phi}}{d\tau}\right)^2\right]$$

and these may be approximated by

$$\frac{d^2\bar{r}}{d\tau^2} = \left(\bar{r} - \frac{2\mu}{c^2}\right)\left(\frac{d\phi}{d\tau}\right)^2 \tag{1.58}$$

$$\frac{d^2\bar{\phi}}{d\tau^2} = -\left(1 - \frac{\mu}{c^2\bar{r}}\right) \frac{2}{\bar{r}} \frac{d\bar{r}}{d\tau} \frac{d\bar{\phi}}{d\tau} \tag{1.59}$$

$$\frac{d^2\bar{t}}{d\tau^2} = \frac{1}{c^2}\left(1 + \frac{4\mu}{c^2\bar{r}}\right)\left[\left(\frac{d\bar{r}}{d\tau}\right)^2 + \bar{r}^2\left(\frac{d\bar{\phi}}{d\tau}\right)^2\right] \tag{1.60}$$

1.8.4 Mercury Perihelion Shift

Integration of the classical equations of motion for the orbit of Mercury reveals a shift in perihelion that cannot be accounted for with Newtonian theory. For navigation, it is necessary to modify the equations of motion to account for perihelion precession which is caused by the relativistic curvature of space near the sun. This may be accomplished by use of a well-known formula or numerical integration of the relativistic equations of motion. The results obtained by numerical integration of the relativistic equations of motion may be compared with this formula. The well-known formula is simply

$$\delta\phi_0 = \frac{6\pi\mu_s}{c^2 a(1-e^2)}$$

where μ_s is the gravitational constant of the sun, a is the semimajor axis of Mercury's orbit, e is the orbital eccentricity, and c is the speed of light.

A simple derivation of the precession of Mercury's periapsis may be obtained by assuming that all the additional potential energy from General Relativity goes into increasing the period of the orbit. The addition of the General Relativity acceleration does not change the mean motion. After one revolution of the classical orbit, the perturbed orbit and the classical orbit have nearly the same angular orientation. At periapsis on the classical orbit, the perturbed orbit is descending for an additional δP to its periapsis. The precession is thus given by

$$\delta\phi_0 = 2\pi\frac{\delta P}{P}$$

$$\delta P = \frac{3P}{2a}\delta a$$

$$\delta a = \frac{a^2}{\mu}\delta C_3$$

and

$$\delta\phi = \frac{2\pi}{P}\frac{3P}{2a}\frac{a^2}{\mu}\delta C_3 = \frac{3\pi a}{\mu}\delta C_3$$

From the Schwarzschild isotropic equations of motion (Eq. 1.55), the radial acceleration is given by

$$\frac{d^2\bar{r}}{d\tau^2} = -\frac{\mu}{\bar{r}^2}\left(1-\frac{2\mu}{c^2\bar{r}}\right)$$

Integrating the acceleration from $\bar{r}$ to infinity yields the potential energy and the General Relativity contribution is

$$\delta E_r = \frac{\mu^2}{c^2 \bar{r}^2}$$

If the average radius $\langle \bar{r}^2 \rangle$ is approximated by $b^2 = a^2(1-e^2)$, the energy addition is

$$\delta C_3 = \frac{2\mu^2}{c^2 a^2 (1-e^2)} = 2\delta E_r$$

Collecting terms, the Mercury precession is approximated by

$$\delta\phi_0 = \frac{2\pi}{P} \frac{3P}{2a} \frac{a^2}{\mu} \delta C_3 = \frac{3\pi a}{\mu} \frac{2\mu^2}{c^2 a^2 (1-e^2)} = \frac{6\pi\mu}{c^2 a (1-e^2)}$$

The equations of motion are integrated with the initial conditions computed from the state vector of Mercury at perihelion. After one complete revolution of Mercury about the sun, the integrated results are transformed to osculating orbit elements and the argument of perihelion is computed. In order to remove the integration error, the Newtonian equations of motion are integrated by the same numerical integrator in parallel with the relativistic equations of motion. The arguments of perihelion are differenced and compared with the formula. The same integration is repeated, only this time the isotropic form of the Schwarzschild equations of motion is compared. The results are displayed below.

Mercury perihelion shift	
Perihelion shift formula	502.527×10^{-9} rad
Exact Schwarzschild integration	502.559×10^{-9} rad
Isotropic Schwarzschild integration	502.267×10^{-9} rad

The above results indicate that the formula for perihelion shift is quite accurate. The difference of 3×10^{-11} rad between the formula and the exact Schwarzschild integration may be attributed to the formula or perhaps integration error. The difference between the formula and the isotropic Schwarzschild integration is also small (26×10^{-11} rad). This difference may also be attributed to integration error but may be the truncation error associated with the isotropic metric.

1.8.5 Radar Delay

The transit time of a photon or electromagnetic wave between two points in space is a measurement that is used to determine the orbits of the planets and spacecraft for the purposes of navigation and science. Both the navigation of a spacecraft and science experiments, particularly associated with General Relativity, require precise measurements of the transit time. Since the Deep Space tracking stations can measure times to within 0.1 ns or about 3 cm, it is necessary to model the transit time to this accuracy.

The transit time of a photon or electromagnetic wave between two points in space is often referred to as the radar delay. This terminology originated with radar where a radio wave is transmitted and the delay in the reception of the reflected return is measured to determine the range. The time delay included that associated with transmission media and the path length. Individual delay terms from the troposphere, ionosphere, and solar plasma are identified and used to calibrate the measured delay. For planetary spacecraft, the path length is computed from the theory of General Relativity. For a round trip travel time, the additional delay attributable to the curved space of General Relativity, over what would be computed assuming flat space, can amount to approximately 250 μs.

$$ds^2 = \frac{\left(1 - \frac{\mu}{2c^2 r}\right)^2}{\left(1 + \frac{\mu}{2c^2 r}\right)^2} c^2 dt^2 - \left(1 + \frac{\mu}{2c^2 r}\right)^4 \left(dr^2 + r^2 d\phi^2 + r^2 \sin^2\theta d\theta^2\right)$$

For a photon, $ds^2 = 0$ and the equation to be integrated for the elapsed coordinate time (t) is obtained by transforming to Cartesian coordinates and solving for dt.

$$dt = \frac{1}{c} \frac{\left(1 + \frac{\mu}{2c^2 r}\right)^3}{\left(1 - \frac{\mu}{2c^2 r}\right)} \left(dx^2 + dy^2 + dz^2\right)^{\frac{1}{2}}$$

Expanding in a Taylor series and retaining terms of order c^{-5},

$$dt = \frac{1}{c}\left(1 + \frac{2\mu}{c^2 r} + \frac{7}{4}\frac{\mu^2}{c^4 r^2}\right)\left[dx^2 + dy^2 + dz^2\right]^{\frac{1}{2}} \tag{1.61}$$

The photon trajectory geometry is shown in Fig. 1.4. The motion is constrained to the $y - z$ plane and targeted from y_1, z_1 to y_2, z_2 such that the photon arrives at the same y coordinate which is taken to be R. For this geometry, the x coordinate is zero and the y coordinate variation is much smaller than the z coordinate variation. Since for this problem $\frac{dy}{dz} \sim 10^{-4}$, the line element differentials may be expanded as a Taylor series,

$$\left(dx^2 + dy^2 + dz^2\right)^{\frac{1}{2}} \approx dz + \frac{1}{2}\frac{dy^2}{dz} + \mathcal{O}\left(\frac{dy^4}{dz^3}\right) \tag{1.62}$$

Changing the y variable of integration to z and inserting Eq. (1.62) into Eq. (1.61),

$$dt = \frac{1}{c}\left(1 + \frac{2\mu}{c^2 r} + \frac{7}{4}\frac{\mu^2}{c^4 r^2}\right)\left(dz + \frac{1}{2}\frac{dy^2}{dz^2}dz + \mathcal{O}(\frac{dy^4}{dz^3})\right) \tag{1.63}$$

Fully expanded, there are nine terms in Eq. (1.63) and four of them are of order $1/c^5$ or greater. Consider a photon grazing the surface of the Sun. A maximum error of about 10 cm or 0.3 ns is desired. To achieve this accuracy, numerical integration of the equation of geodesics reveals that only four of the terms in Eq. (1.63) need be retained and these are

$$t_2 - t_1 = \frac{1}{c}\int_{z_1}^{z_2}\left[1 + \frac{2\mu}{c^2 r} + \frac{1}{2}\frac{dy^2}{dz^2} + \frac{7}{4}\frac{\mu^2}{c^4 r^2}\right] dz \tag{1.64}$$

In carrying out the integration, care should be taken in geometrically interpreting the results. A "straight line" in curved space geometry, the shortest measured distance between two points, is the photon trajectory and not the dashed line shown in Fig. 1.4. Consider the first term of Eq. (1.64)

$$\Delta t_f = \frac{1}{c}\int_{z_1}^{z_2} dz = \frac{1}{c}(z_2 - z_1) \tag{1.65}$$

This is called the flat space term. If the end points were in flat space, Δt_f would be the time a photon travels from point 1 in Fig. 1.4 to point 2. In curved space, there is no such thing as a straight line that connects these two points. The

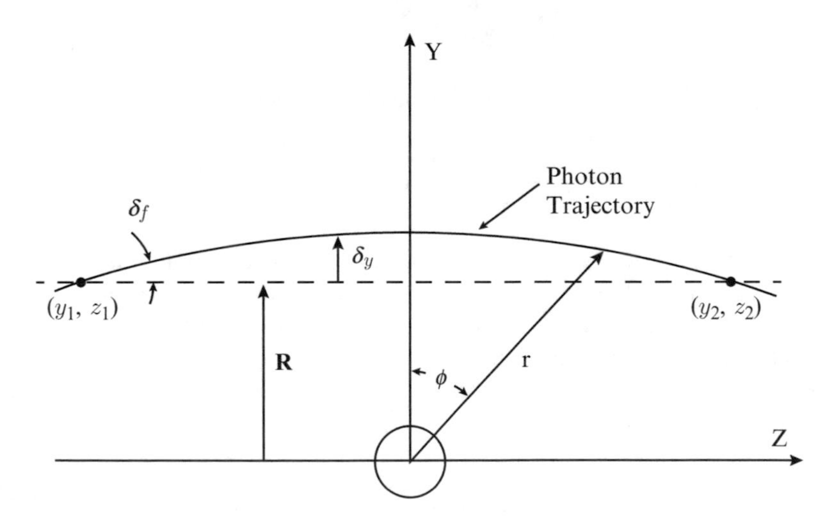

Fig. 1.4 Photon trajectory geometry

real interpretation of the term given by Eq. (1.65) is the mathematical result of performing the integration on the first term of Eq. (1.64).

The second term of Eq. (1.64) is called the logarithmic term for reasons that will become obvious.

$$\Delta t'_{log} = \frac{2\mu}{c^3} \int_{z_1}^{z_2} \frac{dz}{r}$$

Integration requires an equation for r as a function of z. An iterative solution may be obtained by assuming a solution for r and integrating to obtain a first approximation for t and y as a function of z. This solution is inserted into the remainder term, the difference between the assumed and actual function, and a second iterated solution may be obtained for t and y. This method of successive approximations is continued until the required accuracy is achieved. As a starting function, "straight line" motion is assumed. Making use of the approximation that

$$r \approx \sqrt{z^2 + R^2}$$

$$\Delta t'_{log} = \frac{2\mu}{c^3} \int_{z_1}^{z_2} \left[\frac{1}{\sqrt{z^2 + R^2}} + \left(\frac{1}{r} - \frac{1}{\sqrt{z^2 + R^2}} \right) \right] dz = \Delta t_{log} + \Delta t_{rr} \tag{1.66}$$

The first term of Eq. (1.66) integrates to the well-known equation for the time delay.

$$\Delta t_{log} = \frac{2\mu}{c^3} \ln \left[\frac{z_2 + \sqrt{z_2^2 + R^2}}{z_1 + \sqrt{z_1^2 + R^2}} \right] \tag{1.67}$$

The second term of Eq. (1.66), which will be referred to as the radial remainder term (Δt_{rr}), requires a more accurate equation for r to be evaluated.

In order to evaluate the terms associated with bending of the trajectory, an equation for y as a function of z is needed. The y coordinate is associated with the bending of the photon trajectory. Consider two photons in the plane of motion separated by ΔR. The plane containing these two photons and perpendicular to the velocity vector is the plane of the wave front. The bending is simply the distance one photon leads the other divided by their separation.

$$\delta = \frac{c\Delta t_d}{\Delta R}$$

In the limit as ΔR approaches zero, the equation for bending is

$$\delta = c \frac{dt_d}{dR}$$

The equation for the delay is taken to be the logarithmic term given by Eq. (1.67) and for simplicity the bending is computed starting at closest approach ($z_2 = 0$) to the origin.

$$t_d = \frac{2\mu}{c^3} \ln \left[\frac{z + \sqrt{z^2 + R^2}}{R} \right]$$

Taking the derivative with respect to R,

$$\delta = c\,\frac{dt_d}{dR} = -\frac{2\mu}{c^2 R} \frac{z}{\sqrt{z^2 + R^2}}$$

Therefore, the accumulated bending from z_1 to z, expressed as differentials, is given by

$$\frac{dy}{dz} = \delta_f - \frac{2\mu}{c^2 R} \left(\frac{z}{\sqrt{z^2 + R^2}} - \frac{z_1}{\sqrt{z_1^2 + R^2}} \right) \tag{1.68}$$

where δ_f is the initial angle between the photon velocity vector and the horizontal line shown in Fig. 1.4. Referring to Fig. 1.4, the y component of the photon is

$$y = R + \delta_y$$

$$\delta_y(z) = \int_{z_1}^{z} \left(\delta_f - \frac{2\mu}{c^2 R} \left(\frac{z'}{\sqrt{z'^2 + R^2}} - \frac{z_1}{\sqrt{z_1^2 + R^2}} \right) \right) dz'$$

and

$$\delta_y = \delta_f (z - z_1) - \frac{2\mu}{c^2 R} \left(\sqrt{z^2 + R^2} - \frac{z z_1 + R^2}{\sqrt{z_1^2 + R^2}} \right) \tag{1.69}$$

The angle δ_f may be determined by evaluating the bending over the interval from z_1 to z_2. The coordinates are rotated to target the photon to the point $z = z_2$, where $\delta_y = 0$ and the constant gravitational aberration angle δ_f was determined as

$$\delta_f = \frac{1}{z_2 - z_1} \frac{2\mu}{c^2 R} \left(\sqrt{z_2^2 + R^2} - \frac{z_2 z_1 + R^2}{\sqrt{z_1^2 + R^2}} \right) \tag{1.70}$$

The angle δ_f simply rotates the coordinates of Fig. 1.4 such that y_1 and y_2 have the same value R.

The geometrical part of the radial remainder term may be approximated by making use of

$$\frac{1}{r} - \frac{1}{\sqrt{R^2+z^2}} = \frac{1}{\sqrt{(R+\delta_y)^2+z^2}} - \frac{1}{\sqrt{R^2+z^2}} \approx \frac{-R\delta_y}{(R^2+z^2)^{\frac{3}{2}}}$$

The complete radial remainder term (Δt_{rr}) is then given by

$$\Delta t_{rr} = -\frac{2\mu}{c^3}\int_{z_1}^{z_2} \frac{R}{(R^2+z^2)^{\frac{3}{2}}}\left[\delta_f(z-z_1) - \frac{2\mu}{c^2 R}\left(\sqrt{z^2+R^2} - \frac{zz_1+R^2}{\sqrt{z_1^2+R^2}}\right)\right] dz$$

$$\Delta t_{rr} = \frac{2\mu}{c^3 R}\left\{\delta_f\left[\frac{z_1 z_2+R^2}{\sqrt{z_2^2+R^2}} - \sqrt{z_1^2+R^2}\right]\right.$$
$$\left. - \frac{2\mu}{c^2}\left[\arctan\left(\frac{z_1}{R}\right) - \arctan\left(\frac{z_2}{R}\right) + \frac{R(z_2-z_1)}{\sqrt{z_1^2+R^2}\sqrt{z_2^2+R^2}}\right]\right\} \tag{1.71}$$

The third term of Eq. (1.64) is the direct contribution of the trajectory bending to the time delay. This term is referred to as the bending term and is given by

$$\Delta t_b = \frac{1}{2c}\int_{z_1}^{z_2}\left(\frac{dy}{dz}\right)^2 dz$$

Substituting Eq. (1.69) for the slope into the above equation gives

$$\Delta t_b = \frac{1}{2c}\int_{z_1}^{z_2}\left(\delta_f - \frac{2\mu}{c^2 R}\left(\frac{z}{\sqrt{z^2+R^2}} - \frac{z_1}{\sqrt{z_1^2+R^2}}\right)\right)^2 dz$$

Carrying out the integration

$$\Delta t_b = \frac{1}{2c}\left\{\delta_f^2(z_2-z_1) - \frac{4\mu}{c^2 R}\delta_f\left[\sqrt{z_2^2+R^2} - \frac{z_1 z_2+R^2}{\sqrt{z_1^2+R^2}}\right]\right.$$
$$+ \frac{4\mu^2}{c^4 R^2}\left[\frac{R^2(z_1+z_2)+2z_1^2 z_2}{R^2+z_1^2} - 2z_1\sqrt{\frac{z_2^2+R^2}{z_1^2+R^2}}\right.$$
$$\left.\left. + R\left[\arctan\left(\frac{z_1}{R}\right) - \arctan\left(\frac{z_2}{R}\right)\right]\right]\right\} \tag{1.72}$$

The fourth and final term of Eq. (1.64) is the c^5 approximation to the error in the metric. This is a small term and contributes less than a nanosecond to the delay. The equation is given by

$$\Delta t_m = \frac{7}{4}\frac{\mu^2}{c^5}\int_{z_1}^{z_2}\frac{1}{r^2}dz \approx \frac{7}{4}\frac{\mu^2}{c^5}\int_{z_1}^{z_2}\frac{1}{R^2+z^2}\,dz$$

Carrying out the integration

$$\Delta t_m \approx \frac{7}{4}\frac{\mu^2}{c^5 R}\left[\arctan\left(\frac{z_2}{R}\right) - \arctan\left(\frac{z_1}{R}\right)\right] \tag{1.73}$$

The complete equation for the coordinate time delay of a photon moving from (y_1, z_1) to (y_2, z_2) is obtained by summing all the individual terms and

$$t_2 - t_1 = \Delta t_f + \Delta t_{log} + \Delta t_{rr} + \Delta t_b + \Delta t_m \tag{1.74}$$

Before evaluating the individual terms of Eq. (1.74), the parameters used in the individual terms must be determined unambiguously from the end points of the photon trajectory. If two arbitrary end points in the $y-z$ plane are defined by (y_1', z_1') and (y_2', z_2'), the vectors from the origin to these points are given by

$$\mathbf{r}_1 = (0, y_1', z_1') \qquad \text{and} \qquad \mathbf{r}_2 = (0, y_2', z_2')$$

and the vector from point 1 to point 2 is

$$\mathbf{r}_{12} = (0, y_2' - y_1', z_2' - z_1')$$

The angles between the vectors $\mathbf{r}_1$ and $\mathbf{r}_2$ and the vector $\mathbf{r}_{12}$ are computed from the dot products.

$$\phi_1 = \arccos\left(\frac{\mathbf{r}_1 \cdot \mathbf{r}_{12}}{r_2 r_{12}}\right), \qquad \phi_2 = \arccos\left(\frac{\mathbf{r}_2 \cdot \mathbf{r}_{12}}{r_2 r_{12}}\right)$$

The parameters needed in Eq. (1.64), with the coordinates rotated as shown in Fig. 1.4 are then given by

$$R = r_1 \sin\phi_1 = r_2 \sin\phi_2$$

$$z_1 = r_1 \cos\phi_1, \qquad z_2 = r_2 \cos\phi_2$$

and the angle δ_f is given by Eq. (1.70). The fully expanded equation for the transit time is given by,

$$
\begin{aligned}
t_2 - t_1 \approx\ & \frac{1}{c}(z_2 - z_1) + \frac{2\mu}{c^3}\ln\left[\frac{z_2 + \sqrt{z_2^2 + R^2}}{z_1 + \sqrt{z_1^2 + R^2}}\right] \\
& + \frac{2\mu}{c^3 R}\left\{\delta_f\left[\frac{z_1 z_2 + R^2}{\sqrt{z_2^2 + R^2}} - \sqrt{z_1^2 + R^2}\right]\right. \\
& \left. - \frac{2\mu}{c^2}\left[\arctan\left(\frac{z_1}{R}\right) - \arctan\left(\frac{z_2}{R}\right) + \frac{R(z_2 - z_1)}{\sqrt{z_1^2 + R^2}\sqrt{z_2^2 + R^2}}\right]\right\} \\
& + \frac{1}{2c}\left\{\delta_f^2(z_2 - z_1) - \frac{4\mu}{c^2 R}\delta_f\left[\sqrt{z_2^2 + R^2} - \frac{z_1 z_2 + R^2}{\sqrt{z_1^2 + R^2}}\right]\right. \\
& + \frac{4\mu^2}{c^4 R^2}\left[\frac{R^2(z_1 + z_2) + 2z_1^2 z_2}{R^2 + z_1^2} - 2z_1\sqrt{\frac{z_2^2 + R^2}{z_1^2 + R^2}}\right. \\
& \left.\left. + R\left[\arctan\left(\frac{z_1}{R}\right) - \arctan\left(\frac{z_2}{R}\right)\right]\right]\right\} \\
& + \frac{7}{4}\frac{\mu^2}{c^5 R}\left[\arctan\left(\frac{z_2}{R}\right) - \arctan\left(\frac{z_1}{R}\right)\right]
\end{aligned}
$$

After simplification this equation takes the following form

$$
\begin{aligned}
t_2 - t_1 \approx\ & \frac{1}{c}(z_2 - z_1)(1 + \frac{1}{2}\delta_f^2) + \frac{2\mu}{c^3}\ln\left[\frac{z_2 + \sqrt{z_2^2 + R^2}}{z_1 + \sqrt{z_1^2 + R^2}}\right] \\
& + \frac{2\mu}{c^3 R}\delta_f\left[\frac{(z_1 z_2 - z_2^2)\sqrt{z_1^2 + R^2} + (z_1 z_2 - z_1^2)\sqrt{z_2^2 + R^2}}{\sqrt{z_1^2 + R^2}\sqrt{z_2^2 + R^2}}\right] \\
& + \frac{2\mu^2}{c^5 R^2}\left[\frac{R^2(z_1 + z_2) + 2z_1^2 z_2}{z_1^2 + R^2} - \frac{2z_2(z_1 z_2 + R^2)}{\sqrt{z_1^2 + R^2}\sqrt{z_2^2 + R^2}}\right] \\
& + \frac{15}{4}\frac{\mu^2}{c^5 R}\left[\arctan\left(\frac{z_2}{R}\right) - \arctan\left(\frac{z_1}{R}\right)\right] \\
\delta_f \approx\ & \frac{1}{z_2 - z_1}\frac{2\mu}{c^2 R}\left(\sqrt{z_2^2 + R^2} - \frac{z_2 z_1 + R^2}{\sqrt{z_1^2 + R^2}}\right)
\end{aligned}
\tag{1.75}
$$

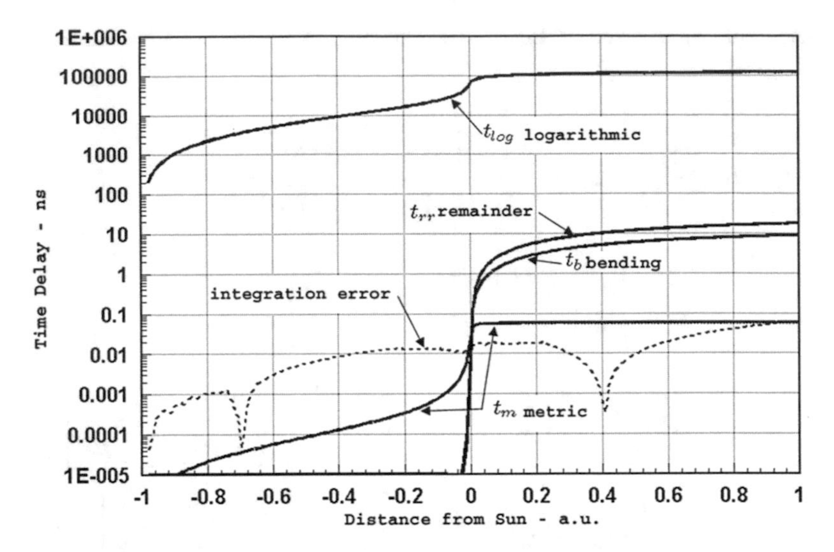

Fig. 1.5 Time delay for solar graze

Equation (1.75) is the time delay associated with a photon or electromagnetic wave that passes through the gravitational field of a massive spherical body. The time delay is a function of only the gravitational constant of the massive body and the parameters z_1, z_2, and R which may be computed directly from the isotropic Schwarzschild coordinates of the end points.

In order to determine the veracity of Eq. (1.75), a comparison with the time delay computed from numerical integration of the geodesic equations of motion was made and the result plotted in Fig. 1.5. In carrying out the numerical integration, a photon was initialized with a z coordinate of $-149{,}000{,}000$ km and y coordinate of 696,000 km. The y component of velocity was set to zero and the z component to c. The x coordinates of position and velocity were set to zero. Thus, the photon is initialized with a velocity magnitude equal to the speed of light and parallel to the z axis about 1 A.U. from the sun and on a flight path that would graze the surface of the Sun if there were no bending due to General Relativity. The polar coordinates of the initial conditions were used to initialize the equations of motion and these were integrated by a fourth-order Runge-Kutta integrator with fifth-order error control. The integration was stopped at various times along the flight path and Eq. (1.75) was evaluated. The required parameters were computed from the initial coordinates and the integrated coordinates at the time of the evaluation.

Also shown in Fig. 1.5 are some of the individual groupings of terms from Eq. (1.64). The linear term has been omitted since this term would require an additional 6 cycles of logarithmic scale. The dashed curve is the difference between the time delay computed by Eq. (1.75) and the results of numerical integration. This difference is attributed to error in the numerical integration algorithm. This was verified by setting the mass of the Sun to zero and integrating straight line motion in the same coordinate system. Unfortunately, the integration error masked the error in the metric. Therefore, Eq. (1.75) could only be verified to about 0.05 ns which is about the same level of error as the error in the metric. The integration error of about $10^{-14}\ t$ is consistent with the error obtained integrating spacecraft orbits for navigation. Observe that the radial remainder term and bending term cause errors on the order of 10 ns or 37 cm.

1.8.6 Light Deflection

Light deflection is the bending of a photon or radio wave trajectory as it passes by a massive object. An experiment performed during a solar eclipse in 1919 measured the deflection of star light and was the first confirmation of General Relativity theory. For this comparison, we integrate the equations of motion for a photon and compare it with an analytic formula. The analytic formula is for a photon arriving at the Earth from infinity. This formula has been adapted to provide a continuous measure of the bending between any two points and is given by

$$\delta\phi = \frac{2\mu}{c^2 R}\{(\cos(90 + \phi_1) - \cos(90 + \phi_2)\}$$

where R is the closest approach to the sun, ϕ_1 is the angle from the y axis to the source, and ϕ_2 is the angle from the y axis to the receiver. The y axis is in the direction of closest approach as illustrated in Fig. 1.4. Einstein's formula for the total bending is simply

$$\delta\phi_{12} = \frac{4\mu}{c^2 R}$$

where $\phi_1 = -90°$ and $\phi_2 = 90°$.

Another formula for the bending is derived in Sect. 1.8.5 for the radar delay and is given by Eq. (1.69).

$$\delta_f = \frac{1}{z_2 - z_1}\frac{2\mu}{c^2 R}\left(\sqrt{z_2^2 + R^2} - \frac{z_2 z_1 + R^2}{\sqrt{z_1^2 + R^2}}\right)$$

If we take the limit as z_1 approaches minus infinity and z_2 approaches plus infinity, δ_f is one half of the Einstein bending formula. Since the total bending is the sum of the approach and departure bending, which are equal, the δ_f formula, when multiplied by two, is the Einstein formula.

Comparison of the Einstein formula with numerical integration of the isotropic Schwarzschild equations of motion is a little tricky because we must define what is meant by bending in curved space. The generally accepted definition is the angle between the local tangent of the photon trajectory and the straight line path that the photon would follow if the sun was removed. Thus, in isotropic Schwarzschild coordinates the deflection is given by

$$\delta\phi = \tan^{-1}\left(\frac{V_r \cos\phi - V_n \sin\phi}{V_r \cos\phi + V_n \sin\phi}\right)$$

where

$$V_r = \frac{dr}{d\tau}$$

$$V_n = r\frac{d\phi}{d\tau}$$

and the undeflected photon is assumed to move parallel to the z axis.

The equations of motion are initialized with the position and velocity of the photon. We place the photon far from the sun on a trajectory that will graze the surface of the sun. The initial state vector is given by

$$r_1 = 149,001,625\,\text{km}$$

$$\phi_1 = -89.73236^\circ$$

$$\frac{dr_1}{d\tau} = -299,789.729\,\text{km/s}$$

$$\frac{d\phi_1}{d\tau} = 9.3982872536 \times 10^{-6}\,\text{rad/s}$$

and the constants are

$$\mu = 1.327124399 \times 10^{11}\,\text{km}^3/\text{s}^2$$

$$c = 299792.458\,\text{km/s}$$

The equations of motion are integrated along a trajectory that grazes the sun and terminated at

$$\tau_2 = 954.901039554\,\text{s (affine parameter time)}$$

$$t_2 = 954.901158130\,\text{s (coordinate time)}$$

$$r_2 = 137,274,407\,\text{km}$$

$$\phi_2 = 89.70998749^\circ$$

$$\frac{dr_2}{d\tau} = 299,789.146\,\text{km/s}$$

$$\frac{d\phi_2}{d\tau} = 11.072650234 \times 10^{-6}\,\text{rad/s}$$

A comparison of the total bending obtained by numerical integration with the theoretical formula derived by Einstein gives

Total light deflection angle	
Einstein's formula	8.48622×10^{-6} rad
Exact Schwarzschild integration	8.48642×10^{-6} rad

1.9 Numerical Integration

Integration of a function on a computer makes use of the definition of a derivative and some results that are associated with proof of the fundamental theorem of calculus. The definition of an integral is a Riemann sum in the limit as the number of sums approaches infinity and the width of the interval approaches zero. A computer cannot deal with infinity so the Riemann sums must be finite. This does not impose a limit on accuracy since given a required accuracy (ϵ) the width of the Riemann sum intervals may be made small enough (δ) to achieve this accuracy. Given an epsilon, there is a delta which is a refrain used by mathematicians in proving theorems. The real limitation of accuracy on a computer is machine precision which may be overcome by computing in extended precision. Orbit determination software is written in double precision which was a problem when computers were expensive. Computation of the Doppler observable strains the limit of double precision (64 bits).

1.9.1 Fundamental Theorem of Calculus

The fundamental theorem of calculus involves performing two operations on an arbitrary function (f) and then showing that the resultant function is also f. The two operations are integration over an interval and differentiation. The integral is defined as the area under the curve defined by $f(s)$ and the s axis and the derivative is defined as the slope of $f(s)$. The function (f(s)) is continuous and has continuous first and second derivatives. A continuous function is defined here as one that can be drawn on graph paper without lifting the pencil and is the only kind of function needed for navigation of spacecraft.

$$\frac{d}{dx}\int_{\alpha*}^{x} f(s)\,ds = f(x)$$

Fig. 1.6 Plot of function

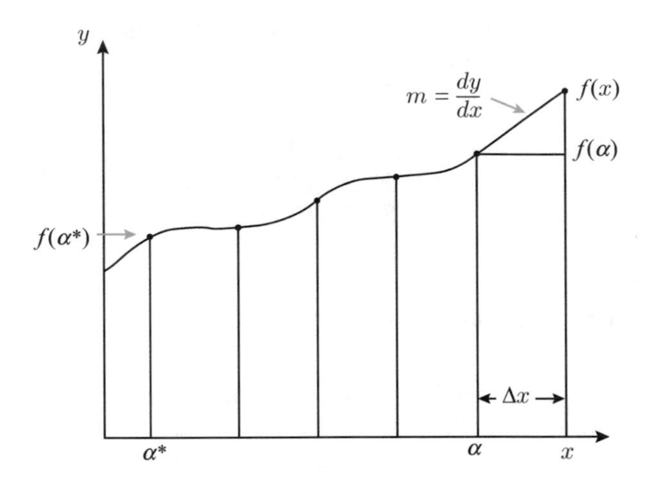

Figure 1.6 shows a plot of the geometry. An intermediate evaluation value α is defined that is within an infinitesimal interval of the value x. The reason for doing this is that the integral from α^* to α will be a constant (not a function of x) and have a derivative of zero.
In the limit as Δx approaches zero

$$f(s) = f(\alpha) + m\ (s - \alpha)$$

where

$$m = \frac{df(s)}{ds} \approx \frac{f(x) - f(\alpha)}{x - \alpha}$$

The integral from α to x is thus the area under that part of the curve which is the slim addition of width Δx. The big area from α^* to α can be ignored because this area is not a function of x and will have a derivative of zero with respect to x.

$$\int_{\alpha}^{x} f(s)\ ds = f(\alpha)\ (x - \alpha) + \frac{1}{2}\ m\ (x - \alpha)^2$$

From the definition of the derivative, we have in the limit as δx goes to zero.

$$\frac{d}{dx}\left[\int_{\alpha}^{x} f(s)\ ds\right] =$$

$$\lim_{\delta x \to 0} \frac{f(\alpha)(x-\alpha) + \frac{1}{2}\ m(x-\alpha)^2 - f(\alpha)(x-\alpha-\delta x) - \frac{1}{2}\ m(x-\alpha-\delta x)^2}{\delta x}$$

$$\frac{d}{dx}\left[\int_{\alpha}^{x} f(s)\ ds\right] = f(\alpha) + m(x - \alpha) = f(x)$$

The proof would be complete if the slope or straight line in Fig. 1.6 was the same as the curve. The problem is the small area between the curve and the slope. In the limit as Δx approaches zero, this small area vanishes. However, the triangle which defines the change in the function also vanishes. We have to show that the small area vanishes faster than the triangle or the ratio of the small area to the triangle approaches zero. The integral is the infinite sum of the infinitely small triangles. This is called a squeeze by mathematicians. If we assume that

$$f(s) = f(\alpha) + m\ (s - \alpha) + c_2(s - \alpha)^2 + \cdots c_n(s - \alpha)^n$$

and process this function as described above, we obtain

$$f(x) = f(\alpha) + m\ (x - \alpha) + c_2(x - \alpha)^2 + \cdots c_n(x - \alpha)^n$$

The change in $f(x)$ over the interval from α to x is

$$\Delta f(x) = m\ (x - \alpha) + c_2(x - \alpha)^2 + c_n \cdots (x - \alpha)^n$$

Dividing by $(x - \alpha)$

$$\frac{\Delta f(x)}{(x - \alpha)} = m + c_2(x - \alpha) + \cdots c_n(x - \alpha)^{n-1}$$

In the limit as $(x - \alpha)$ approaches zero,

$$f(x) = f(\alpha) + m(x - \alpha)$$

This proof is a bit circular in that the fundamental theorem of calculus is probably needed to prove that the function can be represented by a Taylor series over a small interval. Tom Apostle provides two proofs of this theorem in his book on calculus. A third proof is given in *Mathematical Analysis* and this is probably the best proof. All of his proofs involve elegant squeezes. Newton's proof is similar to the one given here but would probably not be accepted today by mathematicians. Newton and Leibnitz stated the theorem and thus get all the credit. The proof is secondary.

1.9.2 Runge-Kutta Numerical Integration

Numerical integration uses many of the ideas associated with proving the fundamental theorem of calculus. The difference is that the fundamental theorem of calculus is an exact result and numerical integration is an approximation that is exact in the limit as the integration interval goes to zero which can never be achieved on a computer. Numerical integration is in a sense more difficult because the result is

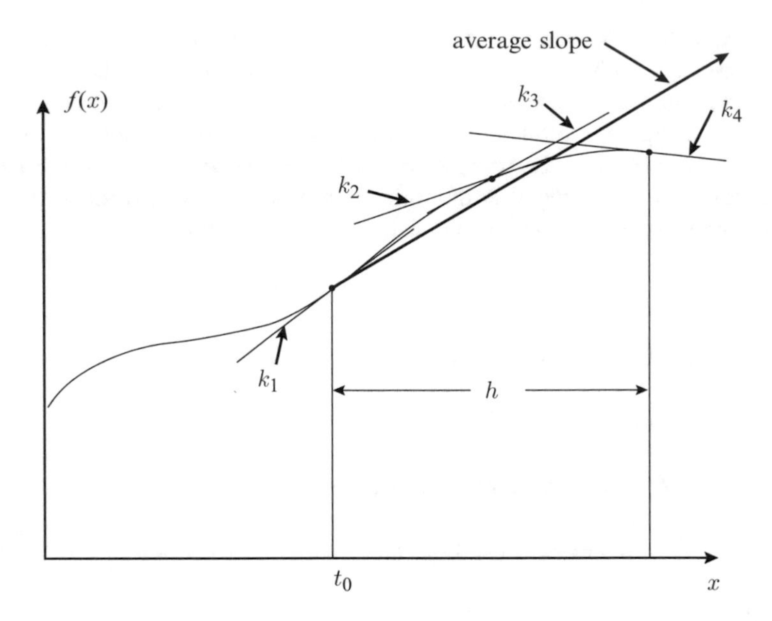

Fig. 1.7 Fourth-order Runge-Kutta

unknown and the integration uses the result. Therefore, numerical integration is a boot strapping process. Given,

$$\frac{dx}{dt} = f(x, t)$$

numerical integration involves finding $f(x, t)$ knowing only the derivative. We first obtain the following derivatives over the interval from t_0 to $t_0 + h$. The geometry is illustrated in Fig. 1.7.

$$k_1 = f(x(t_0),\ t_0)$$

$$k_2 = f\left(x(t_0) + k_1\frac{h}{2},\ t_0 + \frac{h}{2}\right)$$

$$k_3 = f\left(x(t_0) + k_2\frac{h}{2},\ t_0 + \frac{h}{2}\right)$$

$$k_4 = f\left(x(t_0) + k_3\, h,\ t_0 + h\right)$$

The average slope over the interval is a weighted average and $x(t_0 + h)$ is thus,

$$x(t_0 + h) = x(t_0) + \frac{k_1 + 2k_2 + 2k_3 + k_4}{6}\, h$$

Euler integration involves only one derivative evaluation. The derivative evaluated at the left side of the integration interval and extended across the interval is given by,

$$x(t_0 + h) = x(t_0) + f\,(x(t_0),\; t_0)\; h$$

An orbit was integrated around one full revolution of the sun to test various numerical integrator accuracies. The orbit had a semi-major axis of 149.4×10^6 km, an eccentricity of 0.8, and orientation angles in the plane of the ecliptic. The gravitational constant for the sun was $0.132712440017987 \times 10^{12}\ \text{km}^3/\text{s}^2$ which is far more digits than required for the test. The orbit period is about 1 year and periapsis is inside the Earth's orbit. The integration was started 11,805,133.8 s from periapsis to avoid the symmetry associated with starting at periapsis. The results are tabulated below for the Euler integration error. The number of steps is tabulated along with the position error, energy error, and momentum error. The integration step size is the period of the orbit divided by the number of steps. Thus, 365 steps would be about one day. The dimensionless errors are obtained by dividing the actual error by the parameter nominal value. Thus, 10^{-2} would be a one percent error. The exponents indicate the number of decimal places of accuracy.

Euler integration error

Steps	Δ x	Δ y	Energy	Momentum
36,500	10^{-1}	10^{-1}	10^{-1}	10^{-1}
365,000	10^{-2}	10^{-1}	10^{-2}	10^{-1}
3,650,000	10^{-3}	10^{-2}	10^{-3}	10^{-2}
36,500,000	10^{-4}	10^{-3}	10^{-4}	10^{-3}

The Euler integration does not do very well. It takes a one second step size to get four-decimal place accuracy. However, a 100 s step size (step = 365,000) may be accurate enough for some applications and only requires a few lines of computer code to implement.

The classical fourth-order Runge-Kutta algorithm results are shown below. An 864 s step size yields 11 decimal place accuracy which is sufficient for planetary spacecraft navigation. This algorithm was used for about 10 years by the author without any need for more accuracy. It was used in the Mars Orbit Insertion software for the Viking mission to Mars in 1976. The original coding was to replace an analog computer integrator with digital computer code.

Fourth-order Runge-Kutta integration error				
Steps	Δ x	Δ y	Energy	Momentum
365	10^{-3}	10^{-2}	10^{-3}	10^{-3}
3650	10^{-9}	10^{-6}	10^{-8}	10^{-8}
36,500	10^{-11}	10^{-11}	10^{-13}	10^{-12}
365,000	10^{-10}	10^{-9}	10^{-12}	10^{-12}

The integrator used today is the fifth-order Runge-Kutta-Fehlberg integrator with sixth degree error control. This integrator achieves 13-decimal place accuracy for an 865 s step size and this integrator was used for the Near Earth Asteroid Rendezvous software. It was necessary to tailor the error control for integrating spacecraft by planets. The variable step size had a tendency to miss planets or asteroids and flyby without reducing the step size enough.

Fifth-order Runge-Kutta-Fehlberg integration error				
Steps	Δ x	Δ y	Energy	Momentum
365	10^{-2}	10^{-1}	10^{-4}	10^{-3}
3650	10^{-9}	10^{-7}	10^{-8}	10^{-8}
36,500	10^{-13}	10^{-11}	10^{-15}	10^{-15}
365,000	10^{-13}	10^{-11}	10^{-13}	10^{-13}

The Fehlberg integrator is accurate enough to predict solar eclipses at the time of Alexander the Great.

1.10 Summary

The equations of motion have been developed for motion that is relative to navigation. Flight operations software is dominated by numerical integration of Newton's equations of motion. Energy and momentum are useful concepts for understanding motion but there are very few equations in operational software that explicitly acknowledge their existence. One notable example is the momentum and energy supplied by springs as a probe separates from a spacecraft. Energy and momentum are more an artifact of Newton's second law. The equations of motion describing particle collisions and the kinetic theory of gasses acknowledge energy and momentum, but operational software does not explicitly integrate these equations. Therefore, a knowledge of Newton's equations of motion is just about all one needs to navigate spacecraft.

One minor exception is the Theory of General Relativity. Clocks, the path of photons near the Sun, and the precession of Mercury's orbit are affected by general relativity and enter as corrections or calibrations of the data or equations of motion. Since these corrections can be seen in the data, they are included in the flight operations software. However, sufficiently accurate navigation could be performed ignoring General Relativity. Navigators, who conduct mission operations, generally regard General Relativity as a curiosity.

Since the computer is used in the conduct of flight operations, understanding of the mathematics of algorithms contained in computer software is needed. Numerical integration algorithms and finite difference partial derivatives or difference equations are examples. Analytic partial derivatives, that are in the provence of calculus text books, are also needed for orbit determination.

Exercises

1.1 A basketball is dropped on the floor from a height of 1.5 m and rebounds to a height of 1.5 m neglecting energy loss. A golf ball also rebounds to the same height. The golf ball is now held above the basketball and they are both dropped together. The basketball rebounds and hits the golf ball. How high will the golf ball go? The basketball weighs 560 g and the golf ball weighs 45 g. The acceleration is assumed to be uniform at 9.82 m/s^2.

1.2 The partial derivatives of the gravitational acceleration with respect to position is needed for the variational equations. If the gravitational acceleration is given by

$$\mathbf{a} = \frac{-\mu}{r^3}\mathbf{r}$$

determine $\frac{\partial \mathbf{a}}{\partial \mathbf{r}}$ which is a 3×3 matrix.

1.3 The spin vector of the Earth precesses around the Ecliptic pole with a period of about 26,000 years. The applied moment is from the Sun and Moon gravity gradients. The applied moment is given by applying the result from Exercise 1.2.

$$\mathbf{M} = m_e \frac{\partial \mathbf{a}}{\partial \mathbf{r}} \Delta r \Delta r_l \begin{bmatrix} \sin(\alpha) \\ \cos(\alpha) \\ 0 \end{bmatrix}$$

where m_e is the mass of the Earth, Δr is the separation of the dipole associated with oblateness, and Δr_l is the effective moment arm. Assume that the product $m_e\ \Delta r\ \Delta r_l$ is given by

$$m_e \Delta r\ \Delta r_l = I_{zz} - (I_{xx} + I_{yy})/2.$$

Determine an equation for the applied moment.

1.4 For the purpose of computing gravity gradient torques, a body may be represented by dipole point masses distributed on the body fixed coordinate axes.

Since $I_{zz} = m\, k_z^2$, two point masses may be located on the z axis at plus and minus k_z. The torque may be calculated from the I_{xx} and I_{zz} point masses by projecting the positions of the point masses on to the moment arm and sun vector directions. Show that

$$\sin\epsilon\cos\epsilon\ \Delta r \Delta r_l = \sin\epsilon\cos\epsilon\ (I_{zz} - I_{xx})$$

where ϵ is the complement of the angle between the sun vector and body fixed z axis.

1.5 The precession contributed by the Sun is given by

$$\dot{\alpha} = \frac{M}{2\omega_e I_{zz}}$$

The factor of two is needed to attenuate the vector magnitude by the average of sine function squared associated with the orbital motion and separation of precession from nutation. Determine the precession rate for the sun contribution where $(I_{xx} - I_{yy})/I_{zz} = 3.27376 \times 10^{-3}$, $\omega_e = 7.292 \times 10^{-5}\,\mathrm{s}^{-1}$, $GM = 0.132712 \times 10^{12}\,\mathrm{km}^3\mathrm{s}^{-2}$, $\mathrm{r} = 0.149577 \times 10^9\,\mathrm{km}$, $\epsilon = 23.439°$.

1.6 Show that the partial derivative of angular acceleration with respect to the inertia tensor is given by

$$\frac{\partial f_{\dot{\Omega}}}{\partial I_e} = -I^{-1}\begin{bmatrix} 0 & -\omega_z & \omega_y \\ \omega_z & 0 & -\omega_x \\ -\omega_y & \omega_x & 0 \end{bmatrix}\begin{bmatrix} \omega_x & 0 & 0 & \omega_y & \omega_z & 0 \\ 0 & \omega_y & 0 & \omega_x & 0 & \omega_z \\ 0 & 0 & \omega_z & 0 & \omega_x & \omega_y \end{bmatrix}$$

$$-I^{-1}\begin{bmatrix} \dot{\omega}_x & 0 & 0 & \dot{\omega}_y & \dot{\omega}_z & 0 \\ 0 & \dot{\omega}_y & 0 & \dot{\omega}_x & 0 & \dot{\omega}_z \\ 0 & 0 & \dot{\omega}_z & 0 & \dot{\omega}_x & \dot{\omega}_y \end{bmatrix} \tag{1.76}$$

The elements of the inertia tensor are contained in a column matrix of dimension six. Unless highly skilled in tensor algebra, this problem can be solved by expanding the vectors and matrices as equations in terms of their elements, differentiating, and then reassembling into the above matrices. The result in this form can be easily programmed on a computer.

1.7 Determine the root-mean-square velocity of a nitrogen gas molecule (M = 14) at room temperature (72°F).

1.8 A sample of air at standard temperature and pressure (0°C, $1.01325\,\mathrm{nt/m^2}$) occupies a volume of $22{,}421\,\mathrm{cm}^3$. A mole of any gas at standard temperature and pressure will occupy a volume of $22{,}421\,\mathrm{cm}^3$ and contain Avogadro's number of molecules (6.022×10^{23}). Assume the air is an ideal gas with molecular weight of 29 and the diameter of an air molecule is 3 Å. Determine the mean free path length of a molecule between collisions and the number of collisions per second.

1.9 On Jan 1, 65,000,000 12:00:00 BC, a tyrannosaurus rex sets his atomic watch at high noon. At the same time, a photon is emitted from a distant galaxy and sets his identical watch to the same time. On Jan 1, 2017 12:00:00, the photon hits t rex. Both watches gain 3×10^{-16} seconds for each second of elapsed time, the same accuracy as the atomic clocks used by the DSN. What time will both watches read?

1.10 Perform Euler integration of the sine function from 0 to 90° assuming an integration step size of 30° and evaluating the function at the right side of the interval. Repeat evaluating the function in the middle of the interval. Repeat again only assume an integration step size of 10°.

Bibliography

Bate, R. R., Mueller, D. D., and White, J. E., *Fundamentals of Astrodynamics*, Dover Publications, New York, 1971.

Davies, M. E., V. K. Abalakin, M. Bursa, T. Lederle, J. H. Lieske, R. H. Rapp, P. K. Seidelmann, A. T. Sinclair, V. G. Teifel, Y. S. Tjuflin 1986. Report of the IAU/IAG/COSPAR Working Group on Cartographic Coordinates and Rotational Elements of the Planets and Satellites: *Celest. Mech.* 39, 103–113., 1985.

Greenwood D. T. 1965, *Principles of Dynamics*, Prentice-Hall Inc., Englewood Cliffs, NJ.

Hellings, R. W., "Relativistic Effects in Astronomical Timing Measurements," The Astronomical Journal, vol. 91, no. 3, pp. 650–659, March 1986.

Ichinose, S. and Y. Kaminaga, "Inevitable ambiguity in perturbation around flat space-time," *Physical Review*, Vol 40, N 12, pp 3997–4010, 15 December 1989.

Lass, H., *Vector and Tensor Analysis*, McGraw-Hill, New York, 1950.

Llanos, P. J., Miller, J. K. and Hintz, G. R., "Trajectory Dynamics of Gas Molecules and Galaxy Formation", Paper AAS 13-863. AAS/AIAA 2013 Astrodynamics Specialist Conference, Hilton Head, SC, January 2013.

Lieske, J. H., "Improved Ephemerides of the Galilean Satellites", Astronomy Astrophysics, Vol. 82, pp. 340–348, 1980.

McVittie, G. C., *General Relativity and Cosmology*, The University of Illinois Press, Urbana, 1965.

Miller, J, K. and S. G. Turyshev, "The Trajectory of a Photon", AAS03-255, 13th AAS/AIAA Space Flight Mechanics Meeting, Ponce, Puerto Rico, February 9, 2003.)

Misner, C. W., Thorne, K. S., Wheeler, J. A., *Gravitation*, W. H. Freeman, New York, 1972.

Moyer, T. D., "Transformation from Proper Time on Earth to Coordinate Time in Solar System Barycentric Space-Time Frames of Reference: Parts 1 and 2," Celestial Mechanics, vol. 23, pp. 33–08, January 1981.

Moyer, T. D., Mathematical Formulation of the Double-Precision Orbit Determination Program, JPL Technical Report 32–1527, Jet Propulsion Laboratory, Pasadena, California, May 15, 1971.

Richter, G. W. and Matzner, R. A., "Second-order contributions to relativistic time delay in the parameterized post-Newtonian formalism," *Physical Review*, Vol. 28, N 12, 15 December 1983.

Sokolnikoff, I. S., *Tensor Analysis Theory and Applications to Geometry and Mechanics of Continua*, John Wiley and Sons, New York, 1964.

Standish, E. M., E. M. Keesey, and X. X. Newhall, "JPL Development Ephemeris Number 96," Technical Report 32–1603, Jet Propulsion Laboratory internal document, Pasadena, California, 1976.

Van De Kamp, P., *Elements of Astromechanics*, W. H. Freeman and Company, San Francisco, 1964.

Weeks, C. J., J. K. Miller, B. G. Williams, "Calibration of Radiometric Data for Relativity and Solar Plasma During a Solar Conjunction", *Journal of the Astronautical Sciences*, Vol. 49, No. 4, pp. 615–628, October-December, 2001.

Chapter 2
Force Models

The acceleration of a spacecraft is proportional to the vector sum of all the forces acting on the spacecraft. Each component of the resultant force is computed by individual force models. The required accuracy of force models is dependent on the magnitude of the force and the observability of the force in orbit determination software. By far the most important force model is gravity. Gravity force models are formulated as acceleration but this is only a matter of convenience because the mass of the spacecraft factors out of the equations of motion. Force models are generally independent of motion. Even though solar pressure and rocket thrust involve motion of molecules and photons, the force on the spacecraft does not depend on its motion. A notable exception is atmospheric drag forces that are dependent on the velocity of the spacecraft relative to the atmosphere.

2.1 Rocket Equation

The acceleration of a spacecraft from a rocket engine is accomplished by ejecting gas molecules at high velocity. The energy source can be gas stored in a container under pressure. Cold gas rocket thrusters were used by early spacecraft for attitude control. In order to attain the high thrust and high efficiency of modern rocket engines, the gas molecules are heated to a very high temperature by burning rocket fuel in a thrust chamber that directs the gas in a steady stream opposite to the direction of acceleration. The acceleration of a spacecraft subject to the thrust T is given by,

$$\frac{dv}{dt} = \frac{T}{m_{sc}}$$

J. Miller, *Planetary Spacecraft Navigation*, Space Technology Library 37,
https://doi.org/10.1007/978-3-319-78916-3_2

If the rocket exhaust is throttled to flow at a constant velocity (u), the thrust is given by the rate of change of momentum as described above for particle impacts.

$$T = u\,\frac{dm}{dt}$$

The propellant is assumed to be an integral part of the spacecraft and the equation for the velocity change of the total mass including spacecraft and propellant is

$$dv = u\,\frac{dm}{m}$$

Integrating gives what is known as the rocket equation,

$$\Delta v = u \ln \frac{m_0}{m} \tag{2.1}$$

where Δv is the change in velocity of the spacecraft and m_0 is the initial mass. The propellant burned is the difference between m and m_0. The simplicity of the rocket equation belies its usefulness. Interpretation of the parameters of the rocket equation provides considerable insight into the design of rocket engines. The key parameter is the rocket exhaust velocity. Historically, the designers of rocket engines have attempted to maximize the exhaust velocity and have devised a parameter called the specific impulse (I_{sp}) to provide a measure of efficiency. The specific impulse is defined by,

$$I_{sp} = \frac{u}{g_0}$$

The constant g_0 is the acceleration of Earth's gravity at sea level and appears to be a relic from the time when pound mass was used instead of the slug which is the current unit of mass in the English system. The unit of specific impulse is a second and provides a measure of the time a particle would fall near the surface of the Earth to attain the speed of the rocket exhaust which is a fairly meaningless concept. In the modern era, the I_{sp} is a measure of the overall efficiency of the rocket engine and has been incorporated into the rocket equation.

$$\Delta v = g_0 I_{sp} \ln \frac{m_0}{m} \tag{2.2}$$

The specific impulse that is quoted for rocket engines is not directly proportional to the exhaust velocity. It factors in inefficiencies associated with the flow of the exhaust gas and is adjusted to give the right performance when used in the rocket equation. The specific impulse associated with a cold gas rocket engine or with discreet venting from a spacecraft may be computed from the ideal gas law (Eq. 1.46).

$$I_{sp} = \frac{1}{g_0}\sqrt{\frac{3RT}{M}} \tag{2.3}$$

Observe that the specific impulse and hence the overall efficiency of this type of rocket engine is only a function of the temperature and molecular weight of the gas that is expelled. For this reason, rocket engine designers tend to favor hydrogen as a fuel because of its low-molecular weight. Solid rocket motors use propellant with a relatively high-molecular weight and attain specific impulses in the range of 200 s. Liquid propellant rocket motors that use hydrogen as a fuel attain specific impulses in the high 300's. The temperature of the fuel is also a major factor. Cold gas systems attain specific impulses in the 50 s range. In order to attain really high temperatures for the gas molecules, the gas may be ionized and accelerated with an electric field. Ion drive engines can obtain specific impulses of several thousand.

2.2 Aerodynamic Forces

A spacecraft moving through a planetary atmosphere or through the tail of a comet experiences aerodynamic forces opposite to the direction of motion called drag and normal to the direction of motion called lift or side slip. The drag force is generally beneficial to spacecraft since it may be used to remove kinetic energy from the spacecraft. During descent to the planet's surface, the drag force acts in a direction to aid thrusters used for braking and in orbit the drag force may be used to circularize the orbit. Consider a spacecraft that may be approximated by a flat plate oriented perpendicular to the velocity vector relative to an atmosphere. Since the moving spacecraft encounters atmospheric molecules at rest, a force must be applied by the spacecraft to accelerate the atmosphere molecules from rest to the velocity of the spacecraft. This force is given by,

$$F = m\,\frac{dv}{dt}$$

The rate of change of energy or power that must be supplied by the spacecraft to accelerate the gas molecules of mass m is given by,

$$F\,v = mv\,\frac{dv}{dt}$$

Over an incremental distance (dx), the work (dE) done on the atmosphere is

$$dE = \int_{x}^{x+\delta x} F\,dx = F\delta x = m\int_{0}^{v} v\,dv = m\frac{v^2}{2}$$

The mass of the atmosphere swept out is $\rho A\delta x$ where A is the cross section area of the spacecraft and ρ is the atmospheric density. Thus, we have

$$F = A\rho\frac{v^2}{2}$$

for the force on the spacecraft. The force distributed over the spacecraft is the pressure (q) times the area ($F = qA$). The pressure exerted on the spacecraft is called dynamic pressure and is given by,

$$q = \frac{1}{2}\rho v^2$$

The drag force is given by,

$$F_d = C_d q A \tag{2.4}$$

where the drag coefficient C_d is a parameter computed by aerodynamicists to account for the flow of the atmosphere around the spacecraft and has a maximum value of one. As the spacecraft moves through the atmosphere, the gas molecules that are accelerated to the velocity of the spacecraft build up in front of the spacecraft and form a wedge that deflects some of the molecules encountered to flow around the spacecraft. As a result, not all of the molecules are accelerated to the velocity of the spacecraft and the drag coefficient is less than one. If the spacecraft is tilted by some angle, aerodynamic forces are generated that are normal to the velocity vector called lift. For spacecraft navigation in an atmosphere, an aerodynamic model is required to enable the computation of all the force components as a function of q for all the permissible attitudes into which the spacecraft may be maneuvered. The aerodynamic coefficients are computed by an aerodynamicist or reduced from wind tunnel data and supplied to navigation enabling the trajectory of the spacecraft to be computed.

2.3 Solar Pressure

Photons emanating from the sun impinge on the spacecraft resulting in a force that accelerates the spacecraft. The force results from the change in momentum as the photon decelerates from the speed of light (c) to rest with respect to the spacecraft and is converted to heat assuming that the spacecraft is a black body. The force is related to the change in linear momentum given by,

$$F = \frac{d(mc)}{dt} \tag{2.5}$$

The incremental energy required to decelerate the photon is

$$dE = F dx$$

and the power supplied to the spacecraft is obtained by dividing by dt.

$$\frac{dE}{dt} = Fc \tag{2.6}$$

Over the time interval dt, the photon moves an infinitesimal amount dx. The mathematics are a bit oversimplified, but if Eq. (2.5) is substituted into Eq. (2.6) we get $E = mc^2$ and this result is consistent with Special Relativity. The force of the photons on the spacecraft creates a pressure over the exposed surface area (A) and the net force is obtained by integrating all the photons over the area (A) of the spacecraft.

$$F = \frac{1}{c}\frac{dE}{dt}$$

$$\frac{dE}{dt} = I\ A$$

The power supplied to the spacecraft per unit area (I) may be computed from the solar intensity measured at Earth (I_e) and scaled by the inverse square of the distance from the sun.

$$I = \left(\frac{R_e}{R_s}\right)^2 I_e \tag{2.7}$$

Collecting terms and solving for the force on the spacecraft gives

$$F = \frac{K\ A}{R_s^2}$$

$$K = \frac{1}{c} R_e^2 I_e \tag{2.8}$$

where

$$I_e = 1,353\ \mathrm{w/m^2},\ \mathrm{c} = 2,999,793.458\ \mathrm{m/s^2},\ \text{and}\ \mathrm{R_e} = 149.4 \times 10^9\ \mathrm{m}$$

and

$$K = 1.01 \times 10^{17}\ \mathrm{kg\,m/s^2} = 1.01 \times 10^8\ \mathrm{kg\,km^3/m^2\,s^2}$$

K is given in both MKS units and mixed units since area is generally given in m^2 and distance in km. The solar pressure model used for navigation is more complicated than the simple flat plate black body model described above suggests. The above result for the force assumes that all the photons are absorbed by the spacecraft and are directed radially away from the Sun. Some of the photons will be reflected from the spacecraft which will increase the solar pressure. If the spacecraft were a perfect mirror, the force would be doubled. The incident and reflected momentum exchange would be in the same direction. The reflected solar energy is composed of specular and diffuse radiation. For specular reflection, the angle of incidence is equal to the angle of reflection and for diffuse radiation, the energy is scattered by the cosine of the sun angle. The solar pressure model used for

Table 2.1 Solar pressure model parameters

	Front side	Cylindrical side	Back side	Antenna
Area	$8.92\,m^2$	$2.22\,m^2$	$11.25\,m^2$	$2.33\,m^2$
γ	0.165	0.039	0.750	0.750
β	0.742	0.101	0.100	0.107
ϵ	−0.112	0.400	0.400	0.398

the Near Earth Asteroid Rendezvous (NEAR) mission had three components. The specular radiation component is γ, the diffuse radiation component is β, and the third component ϵ accounts for thermal reradiation. Table 2.1 gives the values of these coefficients for each part of the spacecraft. A separate set of coefficients was specified for the front side, cylindrical side, back side, and antenna.

2.4 Gravity Models

The gravitational acceleration (**A**) of a point mass by a body may be obtained by integrating Newton's inverse square law of gravity $\mathbf{a}(x, y, z)$ over the body and is given by,

$$\mathbf{A} = \int_V \mathbf{a}(x, y, z)\, dx\, dy\, dz = G\rho_i \sum_{i=0}^{\infty} \frac{\mathbf{r_i} - \mathbf{r}'}{|\mathbf{r_i} - \mathbf{r}'|^3}\, r_i^2 \cos\phi_i\, dr_i\, d\phi_i\, d\lambda_i \tag{2.9}$$

where G is the gravitational constant and ρ is the density. Equation (2.9) is a mathematical statement of Newton's law of gravity and requires no proof. The acceleration may be replaced by force per unit mass. Thus, a problem in dynamics is replaced by a problem in statics which greatly simplifies the physics. We no longer need to be concerned with the concept of energy and momentum. The vector $\mathbf{r}$ is from the center of the coordinate system to the mass element and the vector $\mathbf{r}'$ is to the spacecraft. The geometry is illustrated in Fig. 2.1. The right side of the above equation is simply the definition of a volume integral transformed to spherical coordinates. In the limit as the index (i) goes to infinity, the size of the volume elements goes to zero. The volume elements may be replaced by mass elements defined by,

$$dm_i = \rho\, r_i^2 \cos\phi_i\, dr_i\, d\phi_i\, d\lambda_i$$

The sum of the mass elements is the total mass of the body. If ρ equals one, the sum is the volume of the body. The density on the surface (ρ) may be defined by an expansion of Legendre polynomials and associated functions as

$$\rho = \sum_{n=0}^{\infty} \sum_{m=0}^{n} P_n^m(\sin\phi)[A_{nm}\ \cos m\lambda + B_{nm}\ \sin m\lambda] \tag{2.10}$$

Alternatively, the density could be given in a table as a function of latitude and longitude.

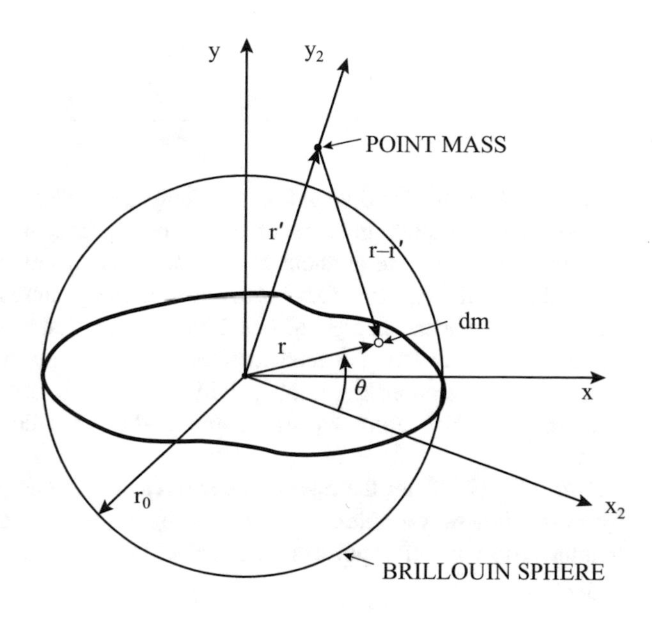

Fig. 2.1 Planet or asteroid body gravity geometry

2.4.1 Harmonic Expansion Model

The gravitational acceleration given by Eq. (2.9) may be recast as the sum of inverse square relationships for each mass element,

$$\mathbf{A} = \sum_{i=0}^{\infty} \frac{-\mu_i(\mathbf{r}' - \mathbf{r}_i)}{|\mathbf{r}' - \mathbf{r}_i|^3} \tag{2.11}$$

where $\mu_i = Gm_i$. Integrating the total acceleration term by term, we obtain the following,

$$U = \sum_{i=0}^{\infty} \frac{\mu_i}{|\mathbf{r}' - \mathbf{r}_i|} \tag{2.12}$$

If we take the derivatives of **A** term by term with respect to x', y', and z' and sum them, we obtain after a little algebra,

$$\nabla \cdot \mathbf{A} = \sum_{i=0}^{\infty} \frac{3\mu_i}{|\mathbf{r}' - \mathbf{r}_i|^3} - \frac{3\mu_i}{|\mathbf{r}' - \mathbf{r}_i|^5} \left[(x' - x_i)^2 + (y' - y_i)^2 + (z' - z_i)^2 \right] = 0 \tag{2.13}$$

Since the right side of Eq. (2.11) is the gradient of the right side of Eq. (2.12) and the right side of Eq. (2.13) are the derivatives of the right side of Eq. (2.11) summed, the same relationships hold for the left sides and

$$\mathbf{A} = \nabla U \tag{2.14}$$

$$\nabla^2 U = \frac{\partial^2 U}{\partial x'^2} + \frac{\partial^2 U}{\partial y'^2} + \frac{\partial^2 U}{\partial z'^2} = 0 \tag{2.15}$$

Equation (2.15) for the divergence of U may be solved for U. For U to be unique, the boundary condition must be satisfied. The fundamental theorem of calculus will then guarantee that the gradient of U will be the acceleration. For heat flow, the above differential equation for U is the flow of energy across the faces of the volume elements. For electric charge, Michael Faraday, the father of electrical engineering, postulated a flux. The real mathematical basis for the divergence theorem is the inverse square relationship. Heat, gravity, and electric charge are all inverse square and obey Newton's action equals reaction and acceleration equals force divided by mass.

Since Eq. (2.15) for the divergence is linear and homogeneous, it may be solved by separation of variables. It is interesting to note that Legendre's equation is obtained from the offset point mass given in Eq. (2.11). The solution for the potential is given by,

$$U = \frac{\mu}{r}\left\{C_{00} + \sum_{n=1}^{\infty}\left(\frac{r_o}{r}\right)^n \sum_{m=0}^{n} P_n^m(\sin\phi)[C_{nm}\ \cos m\lambda + S_{nm}\ \sin m\lambda]\right\} \tag{2.16}$$

The acceleration is obtained by taking the gradient of U, which involves first derivatives of Legendre polynomials that are generally obtained by recursion relationships. An explicit formula for the Legendre polynomials and associated functions that do not involve recursion relationships is given by Heiskanen and Moritz. For orbit determination, the partial derivatives of acceleration with respect to state and gravity coefficients are needed for the variational equations. These partials are difficult to derive because they involve second derivatives of Legendre polynomials and transformations back and forth between spherical and Cartesian coordinates.

2.4.2 Point Mass Model

The next gravity model to be considered is the point mass model. In Europe, this model is called the punctual model. In German, punc means point as we know from the radios in Mercedes Benz manufactured by Blau Punc (blue dot). Equation (2.9) may be modified by replacing the spherical volume element with the Cartesian volume element.

$$\mathbf{A} = G\sum_{i=0}^{\infty}\rho_i \frac{\mathbf{r}_i - \mathbf{r}'}{|\mathbf{r}_i - \mathbf{r}'|^3}\, dxdydz \tag{2.17}$$

On a computer, the body is divided into a large number of cubes or parallelepipeds that define each volume element of size dx by dy by dz. The volume elements do not have to be of the same size, but they must sum to the correct volume. No overlapping is allowed. An easy way to do this summing on a computer is to circumscribe the body with a parallelepiped and then use a triple do loop to access all the volume elements of the parallelepiped. Volume elements outside the body are discarded or assigned a density of zero. This is the same technique used by Michelangelo. He simply procured a large block of marble and chipped away the part that was not David's. Each volume element is assigned a density. One way of determining ρ is to compute the density as a function of latitude and longitude from the above harmonic expansion. The density would thus be constant along any radius vector. This is the density distribution that will be exploited by a later gravity model.

The point mass model is singular when $\mathbf{r}' - \mathbf{r}_i$ is equal to zero. This singularity can only occur when the point where the gravitational acceleration is being computed ($\mathbf{r}'$) is below the surface of the body. Of course, the actual acceleration cannot reach infinity, because there is no such thing as infinite density. When $\mathbf{r}'$ is inside a mass element, the mass element can be replaced with a sphere of equal volume and density. The geometry is illustrated in Fig. 2.2. The radius of the sphere is given by,

$$R_o = \left[\frac{3\,dx\,dy\,dz}{4\pi}\right]^{\frac{1}{3}}$$

The acceleration inside a sphere is zero at the center of gravity and the magnitude is linear from the center to the surface where the acceleration is Gm over R_o^2. The acceleration inside a little cave inside the mass element is simply,

$$\mathbf{a}_i = G\frac{4\pi}{3}\rho_i(\mathbf{r}_i - \mathbf{r}')$$

2.4.3 Pyramid Model

The acceleration of a point mass or spacecraft is determined by summing or integrating the acceleration contribution of each mass element (dm) in a gravitationally attractive body. The mass of the spacecraft is assumed to be too small to affect the gravity field of the attractive body. The geometry is illustrated in Fig. 2.1. The acceleration is

$$\mathbf{A} = G\int_V \frac{\mathbf{r} - \mathbf{r}'}{|\mathbf{r} - \mathbf{r}'|^3}\,dm \tag{2.18}$$

$$dm = \rho\, r^2 \cos\phi\, dr\, d\phi\, d\lambda$$

Fig. 2.2 Internal mass element model

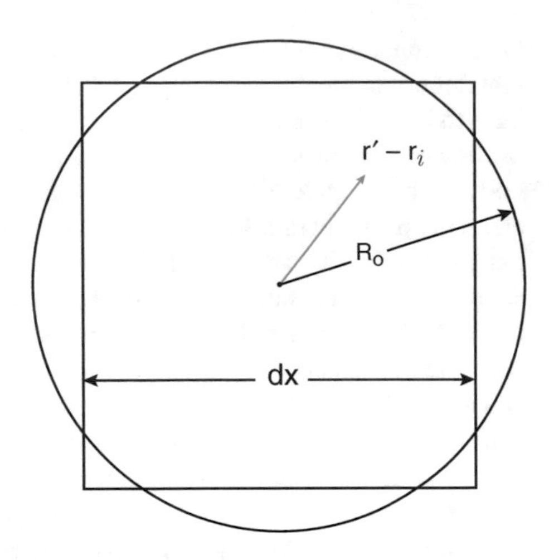

where V denotes the irregularly shaped body. The vector $\mathbf{r}$ is from the center of the coordinate system to the mass element and the vector $\mathbf{r}'$ is to the spacecraft. The density ρ is defined by an expansion of Legendre polynomials and associated functions as defined by Eq. (2.10). The components of the vectors $\mathbf{r}$ and $\mathbf{r}'$ in Cartesian coordinates are

$$\mathbf{r} = (r\cos\lambda\cos\phi,\, r\sin\lambda\cos\phi,\, r\sin\phi)$$
$$\mathbf{r}' = (x', y', z')$$

Substituting the vector components into Eq. (2.18) gives

$$\mathbf{A} = G\int_V \frac{(r\cos\lambda\cos\phi - x',\, r\sin\lambda\cos\phi - y',\, r\sin\phi - z')}{\left[(r\cos\lambda\cos\phi - x')^2 + (r\sin\lambda\cos\phi - y')^2 + (r\sin\phi - z')^2\right]^{\frac{3}{2}}}\, dm \tag{2.19}$$

Before integrating with respect to r, a rotation of coordinates may be defined that will simplify the integrand. The y_2 axis is placed through the location of the spacecraft ($\mathbf{r}'$) and the x_2 axis is placed in the plane defined by the mass element (dm) and $\mathbf{r}'$.

$$\begin{bmatrix} x_2 \\ y_2 \\ z_2 \end{bmatrix} = T \begin{bmatrix} x \\ y \\ z \end{bmatrix}$$

$$T = \begin{bmatrix} \dfrac{\mathbf{r}' \times (\mathbf{r} \times \mathbf{r}')}{|\mathbf{r}' \times (\mathbf{r} \times \mathbf{r}')|} \\ \\ \dfrac{\mathbf{r}'}{|\mathbf{r}'|} \\ \\ \dfrac{\mathbf{r} \times \mathbf{r}'}{|\mathbf{r} \times \mathbf{r}'|} \end{bmatrix}$$

The coordinates of the spacecraft $\mathbf{r}_2'$ and a mass element $\mathbf{r}_2$ in the rotated coordinate system are

$$\mathbf{r}_2' = (0,\ r',\ 0)$$
$$\mathbf{r}_2 = (r\cos\theta,\ r\sin\theta,\ 0)$$
$$\theta = 90 - \cos^{-1}\left(\frac{\mathbf{r}\cdot\mathbf{r}'}{rr'}\right)$$

and

$$\mathbf{A} = G\int_V T^T \frac{[r\cos\theta,\ r\sin\theta - r',\ 0]^T}{\left[(r^2\cos^2\theta + (r\sin\theta - r')^2\right]^{\frac{3}{2}}}\,dm$$

Replacing the mass element (dm) with $\rho\, r^2 \cos\phi dr d\phi d\lambda$ gives

$$\mathbf{A} = G\int_S \rho\, T^T \int_0^R \frac{[\cos\theta,\ \sin\theta - \frac{r'}{r},\ 0]^T}{\left((\cos^2\theta + (\sin\theta - \frac{r'}{r})^2\right)^{\frac{3}{2}}}\,dr\ \cos\phi\ d\phi\ d\lambda$$

where R is the radius of the body as a function of λ and ϕ. The density (ρ) and coordinate transformation (T) factor out of the r integration since they are only a function of latitude and longitude. Performing the r integration gives

$$\mathbf{A} = G\int_S \rho\, T^T \begin{bmatrix} a_{2x}(R) - a_{2x}(0) \\ a_{2y}(R) - a_{2y}(0) \\ 0 \end{bmatrix} \cos\phi\ d\phi\ d\lambda \tag{2.20}$$

$$a_{2x}(r) = \frac{r^2\cos^2\theta + r'\sin^2\theta(r\sin\theta - r') + \cos^2\theta(2r'^2 - 5rr'\sin\theta)}{\cos\theta\sqrt{r^2 + r'^2 - 2rr'\sin\theta}}$$

$$+ 3\cos\theta\sin\theta r' \ \ln\left[2r - 2r'\sin\theta + 2\sqrt{r^2 + r'^2 - 2rr'\sin\theta}\right]$$

$$a_{2y}(r) = \frac{rr'\cos^2\theta + r^2\sin\theta - 5rr'\sin\theta^2 + 3r'^2\sin\theta}{\sqrt{r^2 + r'^2 - 2rr'\sin\theta}}$$

$$- r'(1 - 3\sin^2\theta) \ \ln\left[2r - 2r'\sin\theta + 2\sqrt{r^2 + r'^2 - 2rr'\sin\theta}\right]$$

$$a_{2z} = 0$$

The gravitational acceleration computed assuming constant density will generally not yield sufficient accuracy for navigation of a spacecraft. Therefore, the density is varied as a function of latitude and longitude. The density is assumed to be uniform from the surface to the center of mass of the body. This assumption enables the density to be factored out of the r integration, simplifying the mathematics. The resultant mass distribution does not model reality inside the body but provides an exact model of the external gravity field. The integral given in Eq. (2.20) may be evaluated by tiling the unit sphere with area patches that are nearly square and sum to exactly 4π over the unit sphere. The total acceleration is obtained by evaluating the integrand of Eq. (2.20) at the center of each area patch, multiplying by the area and summing over all the area patches. Observe that the shape of the body (R) enters explicitly through the limit of integration with respect to r. The shape of the object may be obtained from a shape model or input directly from a table of radii as a function of latitude and longitude. The surface integral is not over the surface of the body, but rather over the unit sphere. The integration is performed by an algorithm developed for integration over the actual surface of a body and this algorithm is described below under Shape Model Gravity.

2.4.4 Polyhedral Model

The concept behind a polyhedral model is to pack polyhedrons inside a body and sum the gravitational acceleration from each polyhedra. An easy way to do this is to fill the body with cubes or parallelepipeds. This approach works fine for uniform density, provided that the cubes are small enough such that the error resulting from the cubes overlapping the surface of the body is small. Another approach that may be tried is to partition the equatorial plane into small squares and stack long thin parallelepipeds on the equatorial plane that reach the surface. A variable density may be accommodated by assigning a different density to each column or parallelepiped. This approach will not work because a sphere with a density gradient along the column length will always have the center of gravity on the equatorial plane because each column has uniform density and the columns are parallel to each other.

Another approach which does work is to pack tetrahedrons into the body. If the surface is defined by a triangular plate model, the base of the tetrahedron is on the surface and the apex is at the center of the reference coordinate system which is usually the center of mass. For a parallelepiped shape of the body, the body can be filled with 12 tetrahedrons by passing a diagonal through each of the six faces. This polyhedra model was implemented by Werner and the numerical comparisons with other models indicate that this model is exact for constant density. If the triangles are made small enough and the apex of all the polyhedra are at the center, this model would look like the pyramid model described above. Variable density could probably be accommodated by assigning density as a function of latitude and longitude as was done for the pyramid model. Since Werner's polyhedral model is exact for large tetrahedra, the final triangularly shaped model is exact. The pyramid model requires thousands of area patches to achieve the same accuracy but has other advantages when applied to the problem of orbit determination. The problem of variable density requires a large number of tetrahedrons and the pyramid model probably requires less computation per tetrahedron.

2.4.5 *Mass Distribution of an Irregularly Shaped Body*

A cross section of the pyramid gravity model is shown in Fig. 2.3. The cross section of Eros is used as an example viewed looking down on the North pole with longitude measured counterclockwise, or East, from the x axis. The cross hatched segments have uniform density from the surface defined by latitude and longitude to the center of mass. Each cross hatched segment may have a different density. In three dimensions, the cross hatched segments are actually pyramids with the apex at the center of figure and the base defined by rectangular patches whose sides are delimited by latitude and longitude. In the limit, as the size of the area patches approaches zero, the gravity potential on the surface is exact. If the surface gravity potential is exact, the resulting external gravity field in vacuum is exact. The internal mass distribution is of little interest for navigation since there is an infinity of mass distributions that will yield the same external field. The mass distribution of the pyramid model is selected because it is mathematically convenient.

Since any object may be replaced by an infinite cluster of point masses, verifying the pyramid model for a single point mass verifies the model for the object. The acceleration of a spacecraft that is outside the body is simply the sum of the accelerations from all the point masses. The geometry is illustrated in Fig. 2.4. The mass distribution is computed for two bodies. One has the shape of the asteroid Eros and the other is a sphere. Imagine that the two bodies are composed of a massless surface shell filled with cotton candy and a small lump of uranium hidden inside. The object is to compute a density distribution for the interior of the body that will have the same external gravity field as the lump of uranium. Another sphere is defined that circumscribes the bodies and the acceleration of the point mass is computed at sample points that cover the larger sphere. A typical sample point is

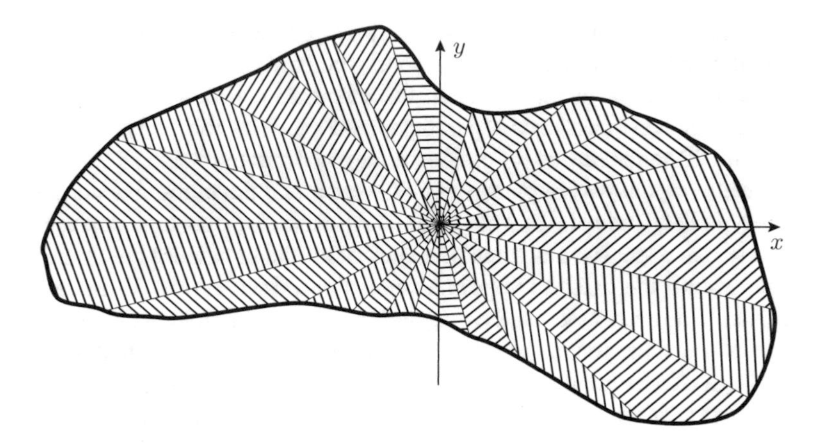

Fig. 2.3 Irregular body cross section

shown in the figure. The pyramid model acceleration is also computed for each sample point. As a first guess, the bodies are assumed to have uniform density and the mass is the same as the point mass. The magnitude of the acceleration is computed for the point mass and the body of interest. Using a square root information filter (SRIF), a least square solution is obtained for the density harmonic coefficients which give the magnitude of the acceleration of a spacecraft at the sample point. If the acceleration magnitudes of the point mass and gravity model are equal on the sampled sphere, the gradient of the potential is equal and a solution is obtained for the density harmonic coefficients.

The observable used for obtaining a solution for the density harmonic coefficients is defined by

$$a^2 = a_x^2 + a_y^2 + a_z^2$$

and the required partial derivatives are

$$\frac{\partial a}{\partial(A_{nm}, B_{nm})} = \frac{\partial a}{\partial(a_x, a_y, a_z)} \frac{\partial(a_x, a_y, a_z)}{\partial\rho} \frac{\partial\rho}{\partial(A_{nm}, B_{nm})} \tag{2.21}$$

where

$$\frac{\partial a}{\partial\rho} = \frac{a_x}{a}\frac{\partial a_x}{\partial\rho} + \frac{a_y}{a}\frac{\partial a_y}{\partial\rho} + \frac{a_z}{a}\frac{\partial a_z}{\partial\rho}$$

$$\frac{\partial(a_x, a_y, a_z)}{\partial\rho} = G\int_S T^T \begin{bmatrix} a_{2x}(R) - a_{2x}(0) \\ a_{2y}(R) - a_{2y}(0) \\ 0 \end{bmatrix} \cos\phi d\phi d\lambda$$

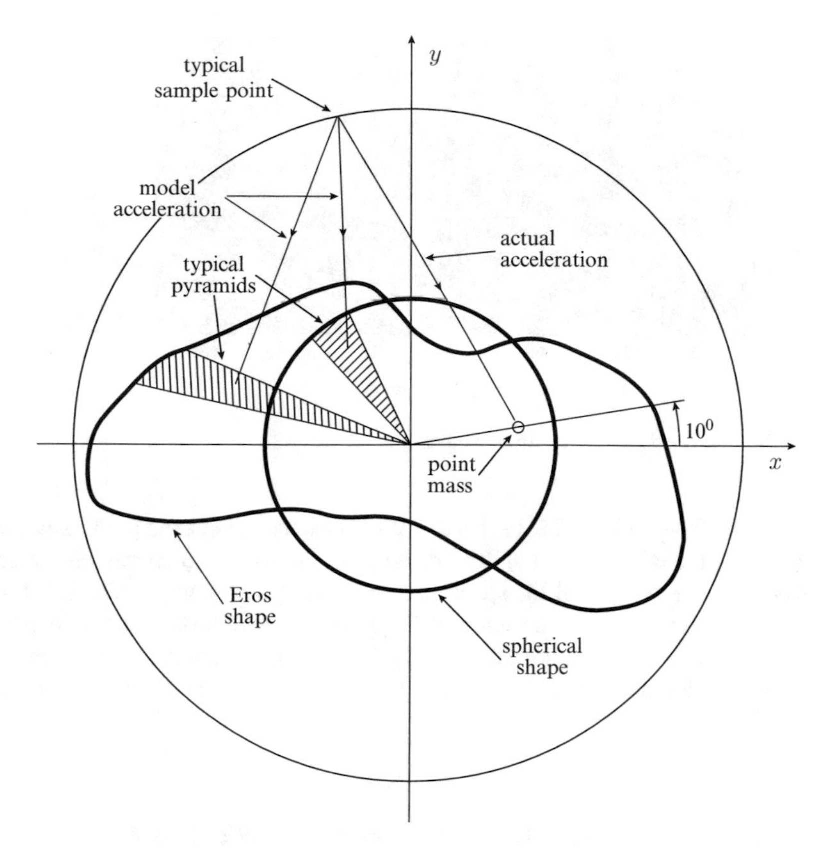

Fig. 2.4 Surface density of point mass

$$\frac{\partial \rho}{\partial A_{nm}} = P_n^m(\sin\phi)\cos m\lambda$$

$$\frac{\partial \rho}{\partial B_{nm}} = P_n^m(\sin\phi)\sin m\lambda$$

The above partial derivatives and the difference between the observed acceleration and computed acceleration are packed into a square root information matrix which is inverted after all the sample points have been processed to obtain the solution for the density harmonic coefficients.

A simple test to verify the pyramid theory involves computing the pyramid distribution for a point mass offset from the center of the coordinate system. The point mass is located at 10° longitude, 10° latitude, and 6 km from the center of the sphere. The radius of the sphere is 8.43259 km which is sized to equal the volume of Eros. The geometry is illustrated in Fig. 2.4. Accelerations of a spacecraft from the point mass and the pyramid model are computed at various sample points that cover a sphere of radius 18 km. A plot of the surface density is shown in Fig. 2.5 for the

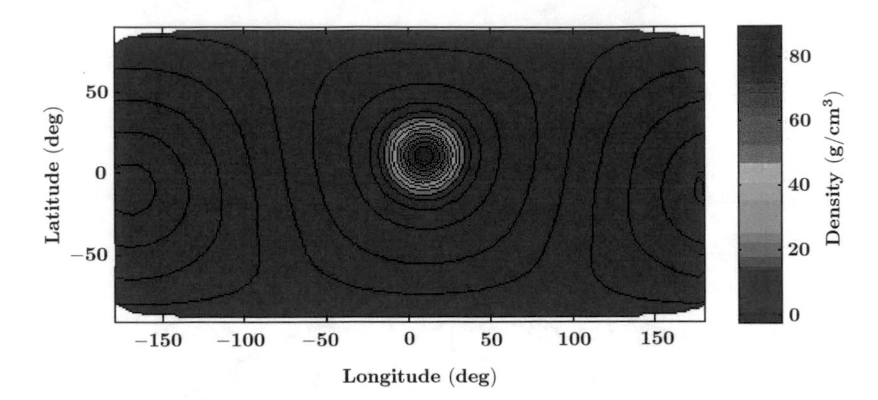

Fig. 2.5 Surface density on sphere from point mass

sphere. Since the gravity field is spherically symmetrical about the point mass, one would expect the contours of constant density on the sphere to be circles centered at the point on the surface of the sphere closest to the point mass. Figure 2.5 shows the maximum density at 10° latitude and 10° longitude and the contours of constant density are circles with a minimum density 180° from the maximum density. The spacecraft acceleration from the pyramid model and point mass are equal to very high precision.

2.4.6 Pyramid Gravity Model Comparison with Eros Harmonic Model

A comparison of the pyramid model with the gravity of Eros may be obtained by processing real data from an orbiting spacecraft in an orbit determination program. A high-quality set of data is available for the NEAR spacecraft orbiting the asteroid Eros. For 1 month, the spacecraft was in a 25-km polar orbit and nearly continuous Doppler data was obtained. Processing this data with the pyramid gravity model would require some modification of existing orbit determination software. However, we have a harmonic expansion of the gravity field obtained during Eros flight operations. This harmonic expansion closely replicated an independent gravity model obtained by integrating a laser altimetry derived shape model over the surface assuming constant density. This verification of the pyramid gravity model for constant density provided confidence in its use for Eros landing.

As a substitute for real data, the Eros harmonic expansion was used to compute acceleration at sample points on the 18-km sphere defined above. These acceleration data points were processed in a square root information filter to obtain surface density harmonic coefficients. Figure 2.6 shows the result of this simulation and provides insight into the possible nonuniformity of the Eros mass distribution.

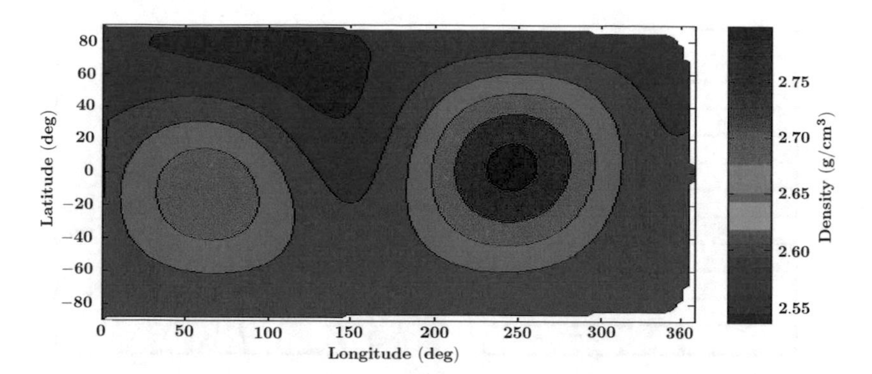

Fig. 2.6 Surface density of Eros from harmonic expansion

Inspection of Fig. 2.6 indicates that a large object may be buried at 260° longitude and 0° latitude. This conclusion is a bit premature since further analysis indicated that the Eros shape model is shifted about 100 m along the y axis. This apparent discrepancy was also observed when comparing the LIDAR shape model with the optical shape model derived independently by Peter Thomas. It should also be noted that this discrepancy is small compared to the mean density of Eros and did not affect the Eros landing trajectory.

2.4.7 Comparison of Gravity Model Mass Distributions

A mass distribution is defined for a parallelepiped or brick that encloses the asteroid Eros. The density is scaled such that the mass of the brick is equal to the mass of Eros. As a result, the mean density of the brick is less than the mean density of Eros. The brick is 32 km long, 18.48 km wide, and 11.28 km high. The density varies from 10% below the mean density at the bottom or south pole to 10% above the mean density at the top or north pole. The brick and Eros are thus layered with layers of constant density parallel to the equatorial plane.

A gravity field is computed for the brick, the cross section shown on the left side of Fig. 2.7, using the point mass model which is assumed to be exact. The gravitational acceleration is computed from the point mass model as a function of latitude and longitude distributed over a sphere of radius 18 km. At each latitude and longitude, the acceleration is also computed on the surface. These accelerations are treated as measurements and packed into a SRIF modified to provide a simple least square solution. The SRIF matrix is inverted to determine the coefficients of Legendre polynomials and associated functions that define the density of the pyramid gravity model as a function of latitude and longitude. A cross section of the pyramid model through the center of the coordinate system is shown on the right side of Fig. 2.7. Because of symmetry, all cross sections through the center will look the same.

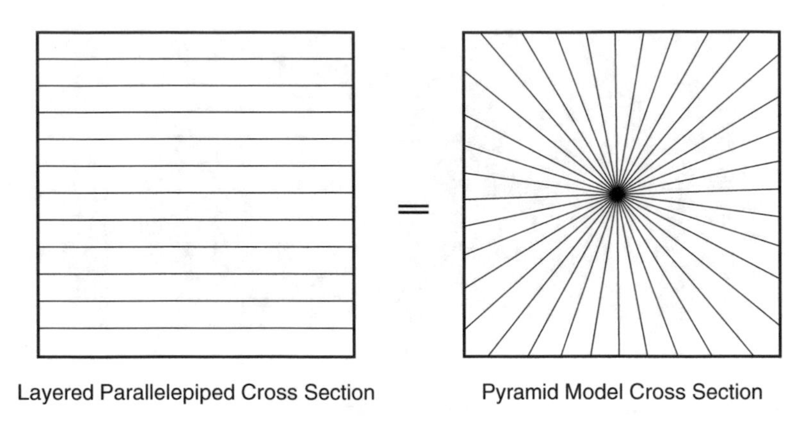

Fig. 2.7 Parallelepiped and pyramid gravity model mass distributions

The equal sign between the layered parallelepiped on the left and the radially distributed density on the right indicates that the external gravity field of these two parallelepipeds are the same. This counterintuitive result suggests that attempts to determine internal structure of planets or asteroids or the Moon by gravity measurements are highly suspicious. The first degree coefficients of a harmonic expansion provide the location of the center of mass, which gives some insight into mass distribution. The second degree coefficients give the ratios of inertia tensor elements and provides additional insight into mass distribution. However, a key piece of information, namely the trace of the inertia tensor, is missing from gravity measurements. The trace can be determined by observation of non-principal axis rotation as was attempted on the NEAR mission. Unfortunately, Eros and most other bodies are nearly in principal axis rotation. Direct comparison of the gravity coefficients determined by orbit determination tracking measurements and computed from a shape model from laser altimetry measurements revealed that the harmonic coefficients through degree six were very nearly the same. This result indicated that Eros was very nearly constant density and provided a valuable set of coefficients that were used as *a priori* for subsequent orbit solutions.

In the above discussion, the gravity fields of two bodies were computed using different gravity models. The pyramid model uses a universal mass distribution that will produce the same gravity field as for any actual mass distribution. This property of the pyramid model enables the production of two bodies with radically different internal mass distributions that have the same gravity field outside the bodies. With a 3D printer, one can actually go down into his or her basement and manufacture these bodies. However, it is easier to do this on a computer. We select the parallelepiped shape for a numerical demonstration. Eros shape models may not be readily available for those who may want to repeat this demonstration. The shape has a Cartesian coordinate system centered at the geometric center with the x axis along the length from -16 km to $+16$ km, the y axis along the width from -9.24 km to $+9.24$ km, and the z axis along the height from -5.64 km to $+5.64$ km.

The total acceleration is obtained by summing the acceleration associated with each mass element given by,

$$\mathbf{A}_i = G\rho_{avg}\, v_i \left(1 + \frac{r_{i\,z}}{r_{max\,z}}\delta\rho\right) \frac{\mathbf{r}_i - \mathbf{r}'}{|\mathbf{r}_i - \mathbf{r}'|^3}$$

The average density times G is $G\rho_{avg} = 6.5826 \times 10^{-8}$, the volume of the mass element is v_i, the z component of the mass element is $r_{i\,z}$, the maximum z component is $r_{max\,z} = 5.64$ km, and the density variation is $\delta\rho = 0.1$. The volume of the parallelepiped is 6670.54 km^3 and thus the GM of the body is $GM = G\rho_{avg} \times 6670.54 = 4.3909 \times 10^{-4}$ km^3/s^2 which is the GM of Eros. For a parallelepiped divided into 200×100×75 = 1,500,000 mass elements, the volume of each mass element is $v_i = 4.447 \times 10^{-3}$ km^3.

For the second mass distribution, the same point mass model is used to compute the gravity and the density distribution is computed from the expansion of Legendre polynomials and associated functions determined by the pyramid model. The acceleration is thus given by,

$$\mathbf{A}_i = G\rho_i\, v_i \frac{\mathbf{r}_i - \mathbf{r}'}{|\mathbf{r}_i - \mathbf{r}'|^3}$$

$$G\rho_i = \sum_{n=0}^{\infty}\sum_{m=0}^{n} P_n^m(\sin\phi)[A_{nm}\ \cos m\lambda + B_{nm}\ \sin m\lambda]$$

$$\phi = \arctan \frac{r_{i\,z}}{\sqrt{r_{i\,x}^2 + r_{i\,y}^2}}$$

$$\lambda = \arctan \frac{r_{i\,y}}{r_{i\,x}}$$

The harmonic coefficients, A_{nm} and B_{nm}, are given in Table 2.2 through degree and order eight. Observe that the A_{00} term is G times the average density. When multiplied by the volume of the parallelepiped, the GM of the parallelepiped is obtained, 4.3909×10^{-4} km^3/s^2, which is the same as for Eros. The assumed density of the parallelepiped was scaled to give this result. For any position vector (r'), the computed accelerations from the two mass distributions are equal.

2.4.8 Comparison of Density Distributions for Eros-Shaped Gravity Models

The density of the asteroid Eros is assumed to be uniform as a comparison of shape-derived gravity harmonic coefficients with tracking data gravity harmonic coefficients indicates. They are essentially the same. If a layered mass distribution

Table 2.2 Parallelepiped density harmonic coefficients

($A_{00}, A_{10}, A_{11}, B_{11}, A_{20}, A_{21}, B_{21} A_{22}, B_{22}, A_{30}, A_{31}, B_{31} A_{32}, B_{32}, A_{33}, B_{33}$ . . . etc.)		
0.658265973900E-07	0.660720233674E-08	-0.779867085511E-10
0.212388242741E-22	-0.151579376268E-08	-0.450982091412E-12
0.555639014698E-23	-0.411934242871E-10	0.129182618080E-21
-0.364695372383E-08	-0.203512661705E-10	0.135078016923E-21
0.321466291566E-10	0.312979239778E-23	-0.116230925885E-11
0.157561296958E-22	-0.110879480257E-08	0.129465197091E-10
0.472975961526E-23	-0.294043480993E-11	-0.196546663309E-22
0.235272811137E-12	0.145599936330E-24	-0.496066386625E-12
-0.902867414825E-24	0.187235254050E-08	0.147796807825E-09
-0.748099227192E-22	-0.423378561105E-10	0.419822755398E-23
0.193361024420E-11	-0.470897875576E-23	-0.516303300732E-12
-0.698072254907E-25	0.745249490765E-13	-0.225316550913E-24
-0.504789625229E-09	0.172303293625E-10	0.297868751580E-23
0.195809393852E-11	-0.251065455280E-24	0.844796754753E-13
0.382935536118E-24	0.112830176929E-12	-0.256321277641E-24
0.137770067101E-14	-0.335817455339E-26	-0.377210797641E-14
-0.465770042459E-25	-0.268693316937E-09	0.786244282014E-10
0.190764832302E-22	0.438489664560E-10	0.201032828152E-23
-0.472226576246E-12	0.259427484565E-23	0.145826475100E-12
-0.309818128674E-25	-0.137045977460E-14	-0.606074934602E-26
-0.574898109647E-14	-0.778573660526E-27	0.118687218488E-14
-0.284595338760E-26	0.197674010708E-09	0.843894925391E-11
0.198244020084E-22	-0.101091521006E-11	-0.535960543951E-23
0.396992597132E-13	0.284556241654E-24	0.427821465123E-13
0.124788700925E-24	0.126538426055E-15	-0.173945340793E-26
0.345446641488E-15	0.334583981692E-26	0.750967584162E-17
-0.112491457849E-27	0.457906376166E-16	-0.776957252054E-28

is imposed on Eros where the density uniformly varies from 10% below the mean at the south pole to 10% above the mean at the north pole, the resultant surface density is shown in Fig. 2.8. The contour plot makes it difficult to discern the geometry. A cross section parallel to the x-z plane or y-z plane would show horizontal lines and a cross section parallel to the x-y plane would show constant density. The density harmonic coefficients for the pyramid model are computed as described above for the parallelepiped.

Figure 2.9 shows the density distribution of the modified Eros for the pyramid gravity model. Recall that for the pyramid model, the density is uniform from the surface to the center of Eros. The apparent mass concentrations do not exist. They are artifacts of the gravity modeling. Figure 2.8 is the assumed actual mass distribution. For extreme mass distributions, the modeled density may be negative. This is not a problem. If we constructed an Eros with a mass distribution indicated by the pyramid model, the external gravity field would be the same as for the real Eros. If the density is negative, we simply fill those places with some anti matter. The pyramid model makes no assumption about the actual mass distribution.

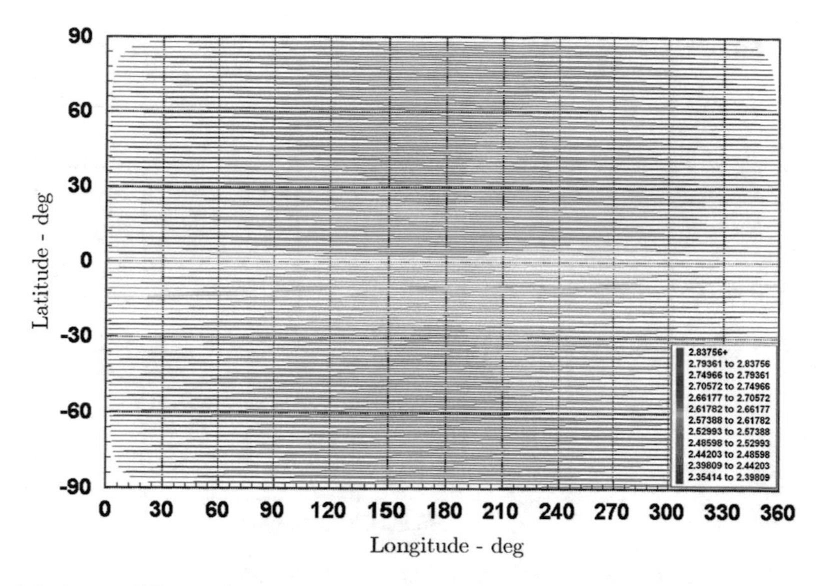

Fig. 2.8 Assumed Eros surface density

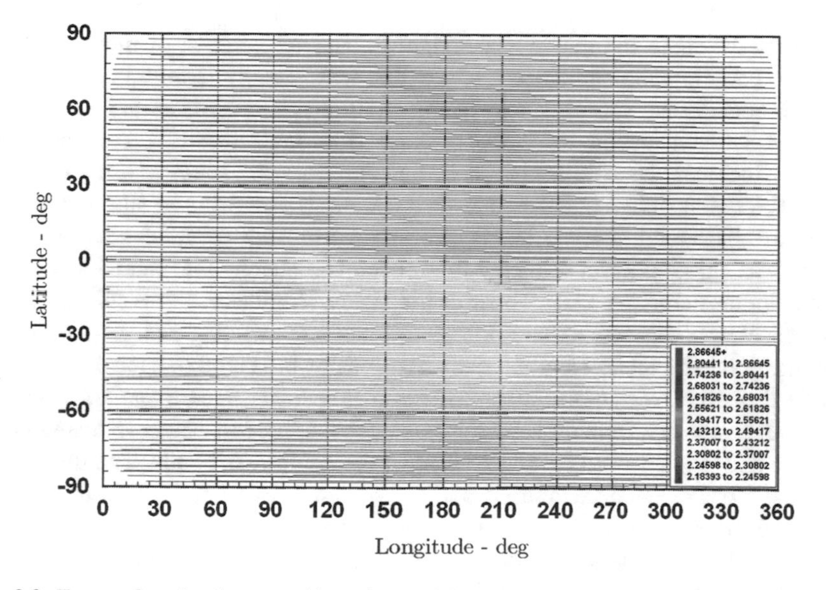

Fig. 2.9 Eros surface density-pyramid gravity model

2.4.9 Comparison of Gravity Model Accelerations

The output from the various gravity models of real interest is the acceleration of a point mass above the surface of the body. Contours of constant acceleration or potential are often plotted as a function of latitude and longitude. Since the difference in the models are small, the contours of constant acceleration would

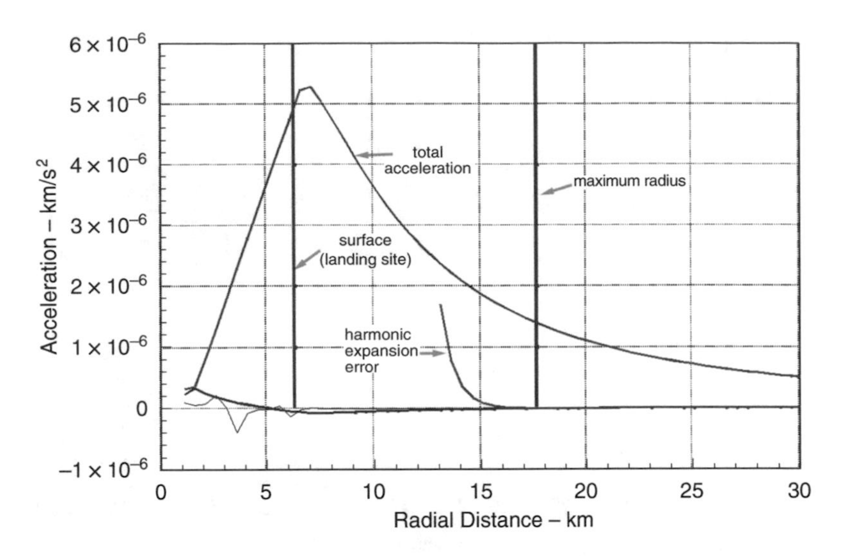

Fig. 2.10 Eros acceleration magnitude as function of radial distance

be nearly the same and thus not meaningful. The acceleration of most interest is at the NEAR landing site since a spacecraft was actually there and a great deal of high-quality data was obtained. In Fig. 2.10, the magnitude of the acceleration is displayed as a function of radial distance from the center of Eros. The model of Eros includes the 20% variation in density from the south pole to the north pole. The radial line segment passes through the NEAR landing site and extends to 30 km. The total acceleration is computed from the pyramid gravity model, which was selected as an arbitrary reference. The landing site radius and sphere of maximum radius are also shown in the figure. An interesting result is that the peak acceleration occurs about a kilometer above the landing site. At the landing site, the gravitational acceleration from the ends of Eros tend to cancel each other. Inside Eros, the acceleration magnitude is nearly linear. It is linear for a sphere. Inside Eros, the acceleration is really the acceleration of an object in a small cave or mine and is not related to the pressure of the compressed body. The differences between other model accelerations and the pyramid model are shown in Fig. 2.10 and the differences are small and can be easily explained.

Figure 2.11 shows the bottom part of Fig. 2.10 with the scale expanded by a factor of ten. Plotted is the difference between the pyramid model and other models as labeled. The curve labeled harmonic expansion represents the error in the harmonic expansion model obtained on the NEAR mission. The error increases dramatically as the spacecraft moves inside the sphere of maximum radius. During the NEAR mission, the spacecraft entered this region only once on landing. For orbit determination, the harmonic expansion error is significant out to 30 km, which is not obvious from the figure. Orbit determination is sensitive to acceleration errors down to 10^{-12} km/s^2. The curve labeled constant density was obtained from Werner's polyhedral model. This difference may be completely attributed to the assumption

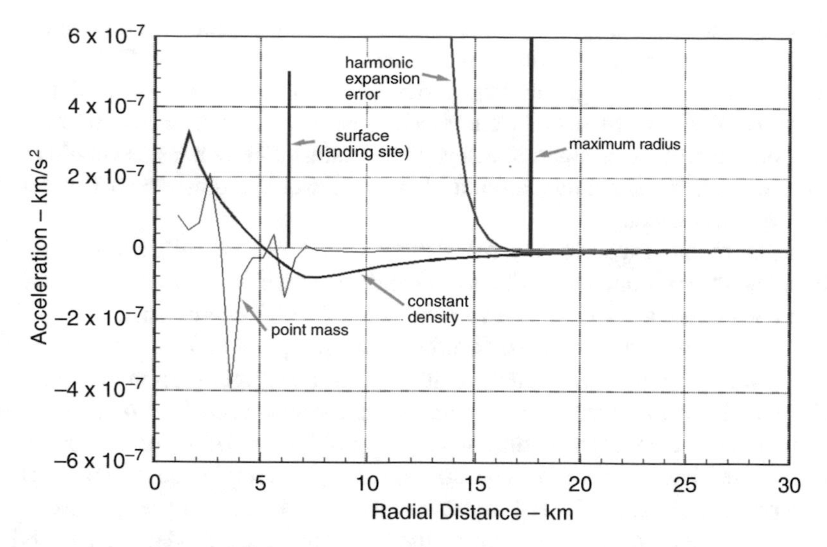

Fig. 2.11 Eros acceleration magnitude model errors

of constant density and vanishes if the density of Eros is assumed to be uniform. The point mass curve matches the pyramid model down to a kilometer above the surface. Here, the resolution of the pyramid model or the number of mass elements in the point mass model introduces error. If the resolution of the pyramid model and point mass model is increased, this difference vanishes. Inside Eros, the problem is magnified. A random error associated with the mass element that contains the point where the acceleration is being determined is introduced, depending on how far the evaluation point is from the center of the mass element. If we are unlucky and we hit the exact center, the error can go to infinity. This error source may be ameliorated by assuming a spherical mass distribution for this mass element.

2.4.10 Comparison of Gravity Model Accuracy and Computational Speed

The harmonic expansion model used for NEAR navigation operations about the asteroid Eros was sufficiently accurate for orbital operations down to 25 km radial distance from Eros. The time to make an orbit determination run, map the solution to 10 days in the future, and compute maneuvers could be about 20 min if time critical. Since the round trip light time was about 40 min and several days were allocated for navigation operations, computer run time was not a major consideration. For time critical missions, ground-based navigation operations cannot be performed faster than the round trip light time so 20 min for an orbit determination run is a good target for program design. During NEAR operations, a 12° and order harmonic expansion was used. This involved solving for 169 harmonic coefficients every 10 days and provided about 20° resolution. The pyramid model and Werner's polyhedral model were used for computing the landing trajectory. The shape model had 7700 triangular plates which provided about 5° resolution. This was more than adequate for propagating trajectories but would not have sufficed for orbit determination if the spacecraft were tracked inside the sphere of maximum radius. There was no orbit determination below 25 km radius. Future missions could require orbit determination at altitudes around 5 km which would require about 1° resolution for the gravity model.

The harmonic expansion model clearly would not be satisfactory for close orbital operations around a small body. One degree resolution would require a 200° and order harmonic expansion involving about 40,000 coefficients. A 40,000 by 40,000 matrix would have to be inverted. The computer run time would also be prohibitive. The point mass model would require about eight million mass elements and eight million acceleration computations for each total acceleration to achieve 1° resolution. The computer run time would be prohibitive. However, the point mass model is useful as a research tool because it is simple to implement and is exact both inside and outside the body. Polyhedral models are exact for any polyhedron shape that assumes constant density such as the triangular plate model used for NEAR operations. However, for variable density and 1° resolution, about 40,000 polyhedra would be required. A seamless transition in resolution would require changing the number of polyhedra and would be difficult to implement in computer code.

The pyramid model appears to be the best option for satisfying all accuracy and speed requirements. The resolution of the surface integration increases as the square of the number of function evaluations while the resolution of the point mass model increases as the cube. The number of computations required for one area patch is probably considerably less than required for one polyhedra. The polyhedra model computes an exact acceleration for every polyhedron. The base and edges require considerable computation and thus limits the polyhedral model to about 10,000 polyhedra. The pyramid model uses an approximation for each area patch and is exact only in the limit as the number of area patches approaches infinity. The pyramid model can be thought of as a porcupine quill model. A quill pierces the center of each area patch and has a cross section area proportional to the square of the radius. The mass per unit length of the quill increases as the square of the distance from the center. The mean value theorem gives a good approximation of the integrand over the integration interval but is exact only in the limit. Recall that π is also not exact on a computer and is only known to several hundred decimal places. Several hundred is a long way from infinity. Therefore, any mathematical equation becomes an approximation on a computer.

Another consideration related to computational speed is the ability to vary the resolution or accuracy of the gravity model as a function of the accuracy needed. For example, numerical integrators used for trajectory propagation vary the step size depending on the acceleration. Near a gravitating body where the trajectory curvature is great, the integration step size is reduced to minutes or hours. Far from a gravitating body, where the spacecraft moves in nearly a straight line, the step size may be several days. The pyramid model permits a simple adjustment of accuracy and computational speed by changing the size of the surface integration area patches. A similar adjustment of speed and accuracy for the polyhedral model could be achieved by increasing the number of triangles that comprise the shape model. A simple way to do this is to add additional vertices at the centroid of each triangle and thus replace each triangle with three smaller triangles.

A further consideration related to complexity and therefore computational speed is the need to compute variational partial derivatives for orbit determination. The partial derivatives of acceleration with respect to density harmonic coefficients and spacecraft position are needed. Consider that the exact solution for the potential of a parallelepiped requires about one page of Fortran code and there probably is not a more efficient language than Fortran for this application. The acceleration requires an additional differentiation of this potential function with respect to each coordinate axis and results in several pages of Fortran code. The variational partial derivatives would require many more pages of Fortran code. The variational partial derivatives for the harmonic expansion model have been derived and are available in computer code, either Fortran or C. This code could be modified to provide variational partial derivatives for the pyramid model. This job is much easier because the pyramid model does not require a potential function to be determined.

2.4.11 Gravitational Variational Equations

For orbit determination, the partial derivatives of acceleration with respect to the dynamic parameters are required. Recall that the translational variational equations are obtained by integrating

$$\frac{\partial \mathbf{A}}{\partial q} = \frac{\partial \mathbf{A}}{\partial \mathbf{r}}\frac{\partial \mathbf{r}}{\partial q} + \frac{\partial \mathbf{A}}{\partial \mathbf{v}}\frac{\partial \mathbf{v}}{\partial q} + \frac{\partial \mathbf{A}}{\partial q}|_{\mathbf{r},\mathbf{v}\text{ constant}}$$

Consider the following subset of constant dynamic parameters that pertain to specific columns of the above matrix.

$$q = (\mathbf{r}_0, \mathbf{v}_0, C_{n,m}, S_{n,m})$$

The gravitational variational equations are

$$\frac{\partial \mathbf{A}}{\partial \mathbf{r}} = \frac{\partial \mathbf{A}}{\partial \mathbf{A_b}}\frac{\partial \mathbf{A_b}(x,y,z)}{\partial \mathbf{A}_b(r,\lambda,\phi)}\frac{\partial \mathbf{A}_b(r,\lambda,\phi)}{\partial \mathbf{r}_b(r,\lambda,\phi)}\frac{\partial \mathbf{r}_b(r,\lambda,\phi)}{\partial \mathbf{r}_b(x,y,z)}\frac{\partial \mathbf{r}_b}{\partial \mathbf{r}}$$

$$\frac{\partial \mathbf{A}}{\partial \mathbf{v}} = 0$$

$$\frac{\partial \mathbf{A}}{\partial (C_{nm}, S_{nm})} = \frac{\partial \mathbf{A}}{\partial \mathbf{A_b}}\frac{\partial \mathbf{A_b}(x,y,z)}{\partial \mathbf{A}_b(r,\lambda,\phi)}\frac{\partial \mathbf{A}_b(r,\lambda,\phi)}{\partial (C_{nm}, S_{nm})}$$

The $\frac{\partial \mathbf{r}_b}{\partial \mathbf{r}}$ is simply the orthogonal transformation matrix from inertial EME J2000 coordinates to body fixed coordinates from the Rotational Equations of

Motion described in Chap. 1. The $\frac{\partial \mathbf{A}}{\partial \mathbf{A_b}}$ is simply the inverse of this transformation and transforms acceleration from body fixed to inertial coordinates. Similarly, the matrices $\frac{\partial \mathbf{r}_b(r,\lambda,\phi)}{\partial \mathbf{r}_b(x,y,z)}$ and $\frac{\partial \mathbf{A_b}(x,y,z)}{\partial \mathbf{A}_b(r,\lambda,\phi)}$ transform from Cartesian to spherical coordinates and back from spherical coordinates to Cartesian coordinates.

The $\frac{\partial \mathbf{A}_b(r,\lambda,\phi)}{\partial \mathbf{r}_b(r,\lambda,\phi)}$ is dependent on the particular gravity model whose partial derivatives are being computed. For the harmonic expansion model, we have

$$\frac{\partial \mathbf{A}_b(r, \lambda, \phi)}{\partial \mathbf{r}_b(r, \lambda, \phi)} = \frac{\partial \nabla U}{\partial \mathbf{r}_b(r, \lambda, \phi)} = \begin{bmatrix} \frac{\partial \nabla U}{\partial r} \\ \\ \frac{1}{r\cos(\phi)} \frac{\nabla U}{\partial \lambda} \\ \\ \frac{1}{r} \frac{\nabla U}{\partial \phi} \end{bmatrix}$$

2.5 Shape Model

The shape of an asteroid or comet nucleus may be determined by optical observation of landmarks and laser altimetry measurements. Optical observations must be stereoscopic which is achieved by imaging landmarks from different points in the spacecraft orbit around the body. Laser altimetry determines the distance from the spacecraft to the surface by measuring the round trip light time. From the attitude control system pointing angles, a vector may be determined that goes from the spacecraft to a point on the surface of the body. The vector from the center of the body to a point on the surface may be determined by simply adding the spacecraft position vector to the laser altimetry vector.

2.5.1 Triangular Plate Model

The vectors from the center of mass of the body to the surface are called vertices. A shape model may be determined by fitting a surface to these vertices. A simple way to do this is to connect a mosaic of polygons to the vertices. This process is called tessellation. A convenient way to do this is with rectangles. The problem with rectangles, or any polygon with more than three sides, is that the vertices are not in the same plane and the volume, moments of inertia, and gravity model parameters computed from the shape model are not exact. For this reason, a triangular plate model is used for navigation. Figure 2.12 shows a triangular plate model in the shape of the asteroid Eros.

This triangular plate model has 3872 vertices and 7740 plates. The vertices are numbered and written to a file. The plates are also written to the same file and are defined by three integer vertex numbers. An interesting relationship, determined by Euler and probably many mathematicians from antiquity, relates the number of

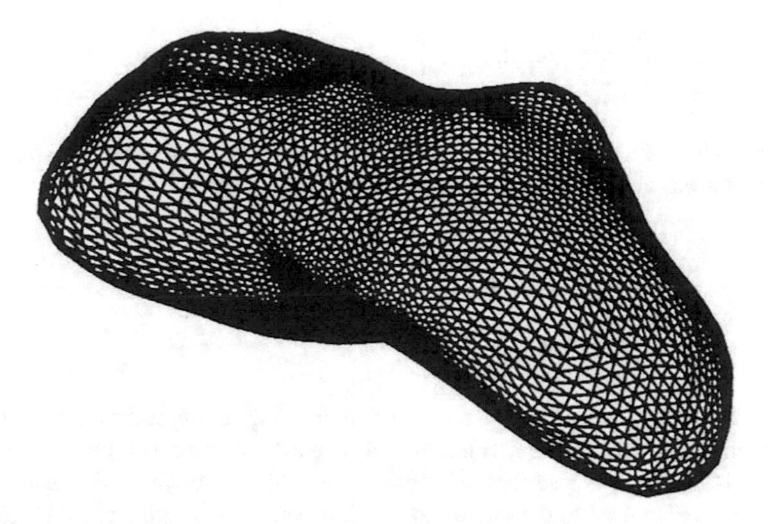

Fig. 2.12 Eros shape model

plates (p) and edges (e) to the number of vertices (v).

$$p = 2v - 4 \ (v > 3)$$
$$e = 3v - 6 \ (v > 3)$$

A triangular shape model can be designed to accurately represent any shape provided that the triangles are small enough. Consider a cube or a parallelepiped. A triangular shape model can be obtained by passing a diagonal through each face. Since there are six faces, the triangular shape model has 12 plates. The number of vertices is eight which is the same as for the cube. The number of edges is 18, 12 for the cube plus six for the diagonals that were added. Thus, the formulas work for a cube or parallelepiped. The resolution of the triangular plate model is inversely proportional to the number of plates. One degree resolution would require 41,253 plates. Therefore, the resolution of the shape model shown in Fig. 2.12 is about 41,253/p or 5.33°.

2.5.2 Harmonic Expansion Shape Model

The triangular plate model is useful for creating images and precision analysis of inertial properties but is not amenable to use for orbit determination. There are too many parameters for an orbit determination filter to assimilate. A more convenient surface can be defined by an expansion of Legendre polynomials and associated functions.

$$R = \sum_{n=0}^{\infty} \sum_{m=0}^{n} P_n^m(\sin\phi)[A_{nm}\ \cos m\lambda + B_{nm}\ \sin m\lambda] \tag{2.22}$$

R is the radius of a point on the surface at longitude λ and latitude ϕ. The harmonic coefficients are obtained by integrating over the unit sphere.

$$A_{nm} = \iint_S P_n^m(\sin\phi)\ \cos(m\lambda)\ R\ d\lambda\ d\phi$$

$$B_{nm} = \iint_S P_n^m(\sin\phi)\ \sin(m\lambda)\ R\ d\lambda\ d\phi$$

The integration is performed on a computer using an algorithm developed for determining the out-gassing acceleration of a spacecraft near a comet. The surface integral is not over the surface of a body, but rather over the unit sphere. This algorithm is also used to determine gravity harmonic coefficients of an irregularly shaped body and other applications that require surface integration. The idea behind this surface integration algorithm is to first cover the surface of the unit sphere with area patches whose sides are great circle lines of constant longitude or small circle lines of constant latitude. An exact tiling of the unit sphere is given by a soccer ball which is a dodecahedron projected onto a sphere. It is not possible to tile a unit sphere with small spherical rectangles or other polygons with more sides than the dodecahedron. We only need the area patches to have nearly the same area and sum to exactly 4π. In the limit as the area patches approach zero, it is not necessary for the area patches to have the exact size and shape. The procedure used here is to divide the northern and southern hemispheres into bands defined by small circles of constant latitude which define latitude bands. The width of the latitude bands determines the resolution. The latitude bands are divided into equal area patches as close to being square as possible. When we get to the poles, the small circle around the pole is divided into four equally shaped spherical triangles. For 1° resolution, there will be 41,253 area patches and they will sum to exactly 4π. In the limit as the resolution goes to zero and the number of patches goes to infinity, we get an exact surface integral. The reader may be troubled by summing a bunch of area patches of different sizes and shapes and getting a result that is exact. The fundamental theorem of calculus does not require that the width of all integration intervals be the same. The only requirement is that the width of all the intervals approaches zero as the number of intervals approaches infinity.

A spherical harmonic shape model for the asteroid Eros was obtained by integrating over the surface of the triangular plate model shown in Fig. 2.12. The degree and order of the expansion was 34. The resultant surface is shown in Fig. 2.13. The low degree and order coefficients were determined with high precision. Since orbit determination is most sensitive to coefficients below degree and order six, this model provided valuable *a prior* gravity harmonic coefficients during the NEAR mission. The higher degree and order coefficients tend to average out and do not contribute much to orbit determination accuracy.

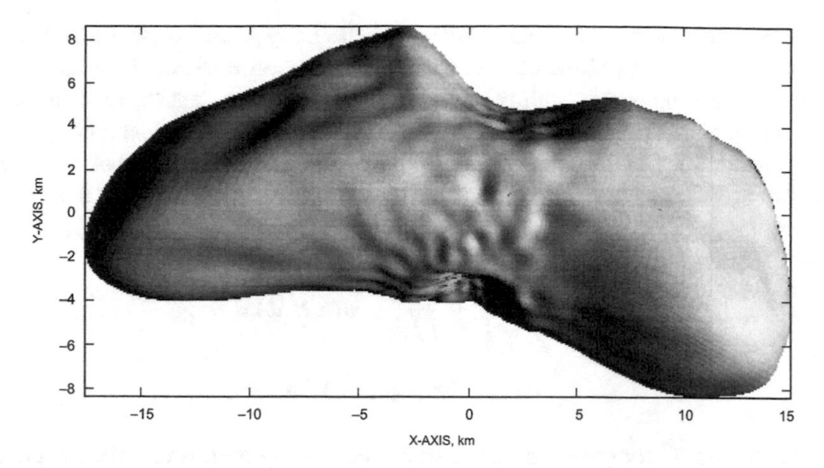

Fig. 2.13 Eros shape model (34×34 harmonic expansion)

2.5.3 Gravity Harmonic Expansion from Shape Model

The gravity field and inertial properties that are needed for analysis may be determined from the shape model of an irregularly shaped body. The volume, moments of inertia, and gravity harmonic coefficients are obtained by integrating over the volume of the body assuming constant density. It will be convenient to first determine the volume and center of mass of the body and then shift coordinates to the center of mass before determining the inertia tensor and gravity harmonics by another integration. Since the center of mass and inertia tensor place constraints on the gravity harmonic coefficients, these relationships may be used as a check on the numerical integration.

The mass is first determined by integration over the volume of the body and is given by

$$M = \iiint_V \rho(r, \lambda, \phi)\, dV \tag{2.23}$$

where ρ is the density and the surface R is defined by the function

$$R = f(\lambda, \phi)$$

The volume integral may be performed numerically on a computer by dividing the body into a finite number of concentric shells and then by partitioning the outer surface of each shell into a finite number of area patches. The volume within a given shell that is under a given area patch comprises an individual volume element. The volume integral is obtained by simply summing the value of the function evaluated at the center of each volume element over all the volume elements.

In order to obtain a reasonable accuracy, the body must be partitioned into a very large number of volume elements. For 1° resolution, each shell contains about 40,000 volume elements. With the assumption of uniform density, the number of function evaluations can be greatly reduced by extracting a spherical core from the center of the body and performing the volume integral from the outer surface of this core of radius r_c to the surface of the body (R). Thus, we have for the mass of the body

$$M = \frac{4}{3}\pi \bar{\rho} r_c^3 + \iiint_{r_c}^{R} \rho(r, \lambda, \phi)\, dV$$

$$dV = r^2 \cos\phi \; dr d\lambda d\phi$$

The amount of computation may also be reduced by performing the integration with respect to r analytically and then integrating over the unit sphere. Thus, we also have for the mass of the body

$$M = \bar{\rho} \iint_S \frac{R^3}{3}\, d\Omega \tag{2.24}$$

$$d\Omega = \cos\phi \; d\lambda \; d\phi$$

The center of mass is defined by the following three integrals.

$$\bar{x} = \frac{1}{M} \iiint_V x \, \rho(r, \lambda, \phi)\, dV$$

$$\bar{y} = \frac{1}{M} \iiint_V y \, \rho(r, \lambda, \phi)\, dV$$

$$\bar{z} = \frac{1}{M} \iiint_V z \, \rho(r, \lambda, \phi)\, dV \tag{2.25}$$

where

$$x = r \cos\lambda \cos\phi$$

$$y = r \sin\lambda \cos\phi$$

$$z = r \sin\phi$$

For the case of constant density, we may perform the integration with respect to r analytically as described above for the mass and we have

$$\bar{x} = \frac{\bar{\rho}}{M} \iint_S \cos\lambda \cos\phi \, \frac{R^4}{4} \, d\Omega$$

$$\bar{y} = \frac{\bar{\rho}}{M} \iint_S \sin\lambda \cos\phi \, \frac{R^4}{4} \, d\Omega$$

$$\bar{z} = \frac{\bar{\rho}}{M} \iint_S \sin\phi \, \frac{R^4}{4} \, d\Omega$$

The inertia tensor and gravity harmonic coefficients may be defined with respect to any origin. However, the dynamics of the translational and rotational motion of the body are best described with the origin at the center of mass. The true center of mass is the assumed center of the planetocentric coordinate system as determined by observation of spacecraft motion and the body translational and rotational dynamics. Thus, the center of mass determined directly from the figure of the body assuming constant density provides some insight as to the internal mass distribution. This interpretation of the data may be facilitated by transforming to the center of mass determined from the figure. The primed coordinates are defined with respect to the center of figure and we have for the translation from planetocentric coordinates to figure-centered coordinates

$$x' = r \cos\lambda \cos\phi - \bar{x}$$

$$y' = r \sin\lambda \cos\phi - \bar{y}$$

$$z' = r \sin\phi - \bar{z}$$

$$r' = \sqrt{x'^2 + y'^2 + z'^2}$$

$$\phi' = \sin^{-1} \frac{z'}{r'}$$

$$\lambda' = \tan^{-1} \frac{y'}{x'}$$

The elements of the inertia tensor with respect to the center of figure are defined by the following integrals:

$$I_{xx} = \iiint_V (y'^2 + z'^2) \, \rho(r', \lambda', \phi') \, dV'$$

$$I_{yy} = \iiint_V (z'^2 + x'^2) \, \rho(r', \lambda', \phi') \, dV'$$

$$I_{zz} = \iiint_V (x'^2 + y'^2)\, \rho(r', \lambda', \phi')\, dV'$$

$$I_{xy} = -\iiint_V x'y'\, \rho(r', \lambda', \phi')\, dV'$$

$$I_{yz} = -\iiint_V y'z'\, \rho(r', \lambda', \phi')\, dV'$$

$$I_{xz} = -\iiint_V x'z'\, \rho(r', \lambda', \phi')\, dV'$$

Performing the integration with respect to r analytically, we obtain

$$I_{xx} = \bar{\rho} \iint_S (\sin\lambda'^2 \cos\phi'^2 + \sin\phi'^2)\, \frac{R'^5}{5}\, d\Omega'$$

$$I_{yy} = \bar{\rho} \iint_S (\cos\lambda'^2 \cos\phi'^2 + \sin\phi'^2)\, \frac{R'^5}{5}\, d\Omega'$$

$$I_{zz} = \bar{\rho} \iint_S \cos\phi'^2\, \frac{R'^5}{5}\, d\Omega'$$

$$I_{xy} = \bar{\rho} \iint_S \cos\phi'^2 \sin\lambda' \cos\lambda'\, \frac{R'^5}{5}\, d\Omega'$$

$$I_{yz} = \bar{\rho} \iint_S \sin\lambda' \sin\phi' \cos\phi'\, \frac{R'^5}{5}\, d\Omega'$$

$$I_{xz} = \bar{\rho} \iint_S \cos\lambda' \sin\phi' \cos\phi'\, \frac{R'^5}{5}\, d\Omega'$$

The unnormalized gravity harmonic coefficients are computed in a similar manner. The coefficient generating functions are given by the following volume integrals:

$$C_{n0} = \frac{1}{M} \iiint_{V'} \left(\frac{r'}{r_o}\right)^n P_n(\sin\phi')\, \rho(r', \lambda', \phi')\, dV'$$

$$C_{nm} = \frac{2}{M} \frac{(n-m)!}{(n+m)!} \iiint_{V'} \left(\frac{r'}{r_o}\right)^n P_n^m(\sin\phi')\, \cos m\lambda'\, \rho(r', \lambda', \phi')\, dV'$$

$$S_{nm} = \frac{2}{M} \frac{(n-m)!}{(n+m)!} \iiint_{V'} \left(\frac{r'}{r_o}\right)^n P_n^m(\sin\phi')\, \sin m\lambda'\, \rho(r', \lambda', \phi')\, dV'$$

These volume integrals may also be converted to integrations over the unit sphere for the case of constant density by performing the r integration analytically.

$$C_{n0} = \frac{\bar{\rho}}{M} \iint_{S'} \frac{1}{n+3} \left(\frac{R'^{n+3}}{r_o{}^n} \right) P_n(\sin\phi')\, d\Omega'$$

$$C_{nm} = \frac{2\bar{\rho}}{M} \frac{(n-m)!}{(n+m)!} \iint_{S'} \frac{1}{n+3} \left(\frac{R'^{n+3}}{r_o{}^n} \right) P_n^m(\sin\phi')\, \cos m\lambda'\, d\Omega'$$

$$S_{nm} = \frac{2\bar{\rho}}{M} \frac{(n-m)!}{(n+m)!} \iint_{S'} \frac{1}{n+3} \left(\frac{R'^{n+3}}{r_o{}^n} \right) P_n^m(\sin\phi')\, \sin m\lambda'\, d\Omega'$$

The second degree gravity harmonics and elements of the inertia tensor share the same integrals and are thus not independent. The differences in the inertia tensor elements may be determined as a function of the second degree gravity harmonics. Since there are five differences and six parameters, a third equation is needed. This third equation could be the trace of the inertia tensor which would enable determining the inertia tensor and gravity harmonics as functions of each other and thus save six parameters that need to be estimated. Twelve parameters that are well understood would be replaced by six parameters that are not so well understood. A better approach is to estimate all twelve parameters and place an *a prior* constraint on the relationship between the parameters. This is accomplished by processing five dummy measurements with zero measurement error. The equations of constraint are

$$I_{xx} - I_{yy} = -4Ma^2C_{22}$$

$$I_{yy} - I_{zz} = Ma^2(C_{20} + 2C_{22})$$

$$I_{zz} - I_{xx} = -Ma^2(C_{20} - 2C_{22})$$

$$I_{xy} = -2Ma^2S_{22}$$

$$I_{yz} = -Ma^2S_{21}$$

$$I_{xz} = -Ma^2C_{21}$$

2.6 Comet Atmosphere

The comet nucleus is a source of a stream of dust and gas molecules that accelerate the spacecraft away from the nucleus. Although the actual size and mass of comets are generally unknown, most short-period comets considered for missions have estimated radii between 1 and 5 km. The solar radiation input to the surface of a comet nucleus heats an outer mantle of dust and debris that is generally a few centimeters thick. Heat is conducted through the mantle and results in the sublimation of ice and other volatile compounds that are covered by the mantle. Gas produced by sublimation, which is comprised mainly of water vapor, percolates through the mantle and escapes to the vacuum of space. The expanding gas molecules attain a velocity of several hundred meters per second immediately on leaving the surface and flow radially outward from the comet. A spacecraft orbiting the comet nucleus will experience dynamic pressure with a resultant force that is nearly radial. Since the spacecraft velocity is much smaller than the gas molecule velocity, the normal dynamic pressure attributable to the spacecraft motion may be neglected. The aerodynamics of a spacecraft orbiting a comet nucleus are more analogous to a sail boat than to an airplane.

As the comet approaches the sun and the mantle is heated to higher temperatures, the percolation of gas through the mantle becomes so intense that large pieces of the mantle are sloughed off exposing the bare ice to direct solar radiation and the vacuum of space. The local activity becomes very intense and huge jets appear where the surface activity is perhaps an order of magnitude greater than the surrounding surface covered by the mantle. The model of comet nucleus outgassing must accommodate several discrete jets in addition to the normal background outgassing of the comet.

A spacecraft orbiting a comet nucleus will experience a force from the pressure of expanding gas and dust that varies widely depending on the position of the spacecraft relative to the comet, the distance of the comet from the Sun, and the activity of the local comet surface in response to solar energy input. For navigation, a model must be developed that takes all of these factors into account but is simple enough to be incorporated into existing navigation software.

A simplified two-part empirical model of accelerations on the spacecraft due to comet nucleus outgassing and dust emissions is defined. The first part describes the accelerations acting on the spacecraft from the comet outgassing that results from the solar radiation input. For this model, the spacecraft acceleration is assumed to be directed radially from the comet nucleus and varies with the cosine of the sun angle and inversely with the square of the distance. The second part describes the behavior of a gas vent or jet and the spacecraft acceleration is described in the same manner as for the outgassing, except that the acceleration is directed away from a specific region on the comet nucleus surface. Thus the outgassing model could be interpreted as the integration of many gas and dust jets over the entire surface of the comet.

2.6.1 Outgassing Model

The outgassing model assumes that the spacecraft acceleration is directed radially from the comet nucleus and varies with the cosine of the sun angle and inversely with the square of the distance. It is defined by three variables A_D, A_T, and A_N, which represent acceleration magnitudes acting on a spacecraft, each at the reference radius $\mathrm{r}_{ref} = 10\,\mathrm{km}$ from the center of the comet. A_D is the acceleration magnitude directly over the subsolar point (the position on the comet directly under the sun). A_T, the acceleration magnitude over the terminator, which is the edge of the sunlit side of the comet, was assumed to be $0.5\,\mathrm{A}_D$. A_N is the acceleration over the anti-subsolar point, the position on the night side of the comet directly opposite

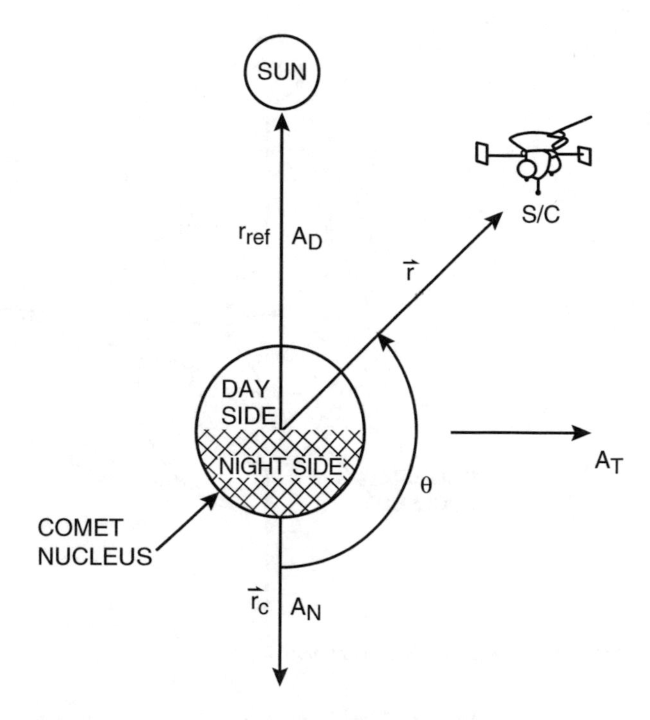

Fig. 2.14 Empirical outgassing model

to the subsolar point. A_N was assumed to be 0.1 A_D. Intermediate accelerations are defined as follows, using the angle θ measured from the anti-subsolar direction.

$$A = (A_T - A_D)\cos\theta + A_T \quad \text{for } 90 < \theta < 180$$

$$A = (A_N - A_T)\cos\theta + A_T \quad \text{for } 0 < \theta < 90$$

$$\cos\theta = \frac{\mathbf{r}\cdot\mathbf{r}_c}{r\, r_c}$$

The acceleration vector of the spacecraft due to outgassing is thus given by

$$\mathbf{A} = A\, r_{ref}^2\, \frac{\mathbf{r}}{r^3}.$$

Figure 2.14 displays the comet outgassing geometry.

2.6.2 *Jet Model*

The vent or gas jet model follows the steady gas pressure model with some minor revisions. Here, the acceleration of the jet $\mathbf{A_j}$ is described as a function of the angle ϕ and represents the acceleration from a region surrounding the body fixed coordinates of the jet. The activity of the jet is thus determined by its exposure to the sun attaining a maximum when the sun is directly over head and decreasing to a minimum on the dark side of the comet. The geometry is illustrated in Fig. 2.15.

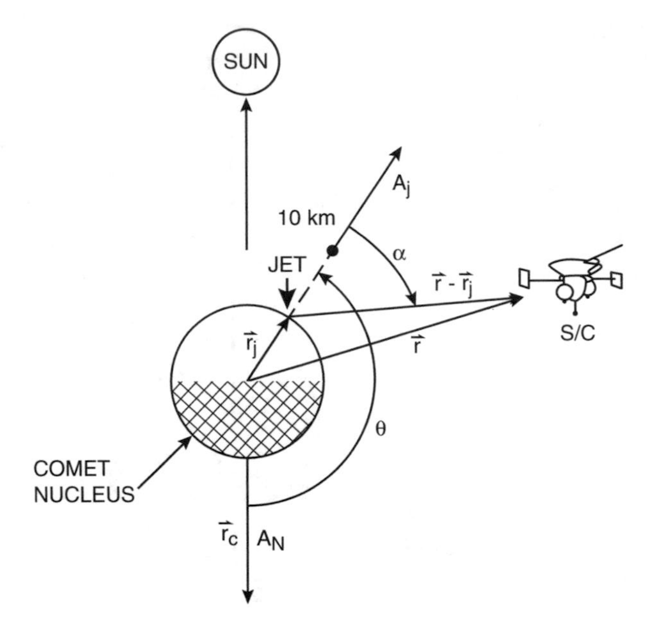

Fig. 2.15 Internal mass element model

The body fixed position of the jet is first converted to inertial coordinates:

$$\mathbf{r}_j = T^{\mathrm{T}}\,\mathbf{r}_{jb}$$

where T is the transformation matrix describing the attitude of the comet in inertial space. The angle ϕ describes the location of the jet and is given by

$$\cos\phi = \frac{\mathbf{r}_j \cdot \mathbf{r}_c}{r_j\, r_c}.$$

Intermediate acceleration magnitudes due to the jet are defined as follows, using the angle ϕ:

$$A_j = (A_{T_j} \;-\; A_{D_j})\cos\phi \;+\; A_{T_j} \quad \text{for } 90 < \phi < 180$$

$$A_j = (A_{N_j} \;-\; A_{T_j})\cos\phi \;+\; A_{T_j} \quad \text{for } 0 < \phi < 90$$

where the parameters A_{T_j}, A_{D_j}, and A_{N_j} are defined in a similar manner to the corresponding definitions of the parameters A_T, A_D, and A_N for the outgassing model. The position of the spacecraft relative to the gas jet is described by the angle α given by

$$\cos\alpha = \frac{(\mathbf{r} - \mathbf{r}_j)\cdot\mathbf{r}_j}{r_j|\mathbf{r} - \mathbf{r}_j|}$$

and the acceleration of the spacecraft is directed away from the jet and given by

$$\mathbf{A_j} = A_j\, r_{ref}^2\;\cos\alpha\;\frac{\mathbf{r} - \mathbf{r}_j}{|\mathbf{r} - \mathbf{r}_j|^3}.$$

2.6.3 Thermodynamic Model

The thermodynamic model of the nucleus of a comet relates the surface temperature and gas production rate to the solar energy input. The heat equation describes the flow of heat within the core and mantle where the heat transfer is dominated by conduction. Heat transfer into the nucleus can be modeled by partitioning the nucleus into a set of concentric shells surrounding a central core and then partitioning each shell into individual volume elements so that a heat equation can be written for each element. The thickness of these shells depends on the conductivity and heat capacity.

The governing equation of the heating rate per unit area of the dust that comprises the mantle is given by:

$$c_m \rho_m \Delta r_m \dot{T}_{surf} = (1 - A)I_n - (1 - A)\sigma_b T_{surf}^4 - \frac{2k_d(t_{surf} - Tsub)}{\Delta r_m} \tag{2.26}$$

where c_m, ρ_m, and Δr_m are the specific heat, density, and thickness of the mantle. A is the albedo, I_n is the incident solar intensity, σ_b is the Stefan-Boltzmann constant, and k_d is the constant of proportionality for heat conduction. T_{sub} is the temperature of the comet mantle below the top layer of ice. The first term on the right of Eq. (2.26) is the input solar energy, the second term is the heat radiation to space, and the third term is the conduction into the mantle. The heating rate, $\dot{T}_{surf}$, is given by

$$\dot{T}_{surf} = \frac{(1 - A)}{c_m \rho_m \Delta r_m} I_n - \frac{(1 - A)}{c_m \rho_m \Delta r_m} \sigma_b T_{surf}^4 - \frac{2k_d(t_{surf} - Tsub)}{c_m \rho_m \Delta r_m^2}$$

$$I_n = I_s \left(\frac{r_{earth}}{r_{comet}} \right)^2 \hat{\mathbf{r}}_{comet} \cdot \hat{\mathbf{r}}_n$$

where I_s is the solar energy flux at the Earth, r_{earth} is the distance from the Earth to the Sun, r_{comet} is the distance from the comet to the Sun, and $\hat{\mathbf{r}}_{comet}$ and $\hat{\mathbf{r}}_n$ are unit vectors from the comet to the sun and normal to the surface of the nucleus.

The heat equation must be integrated with respect to both space and time. In order to compute the variation of temperature as a function of space and time, we need to compute the Laplacian $\nabla^2 T$. The Laplacian is an operator that appears in many engineering applications and states that the volumetric accumulation of some physical quantity is equal to the flow across the boundary of an elementary volume element. This physical quantity is conserved in this process and the flow is proportional to the gradient of a scalar potential function. In our case, the physical quantity is heat and the scalar potential function is temperature. In Cartesian coordinates, the Laplacian is given by:

$$\nabla^2 T = \frac{\partial^2 T}{\partial x^2} + \frac{\partial^2 T}{\partial y^2} + \frac{\partial^2 T}{\partial z^2} = 0 \tag{2.27}$$

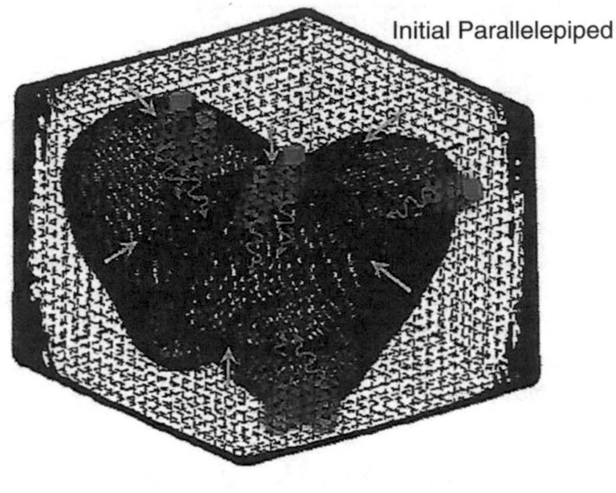

Fig. 2.16 Volume elements

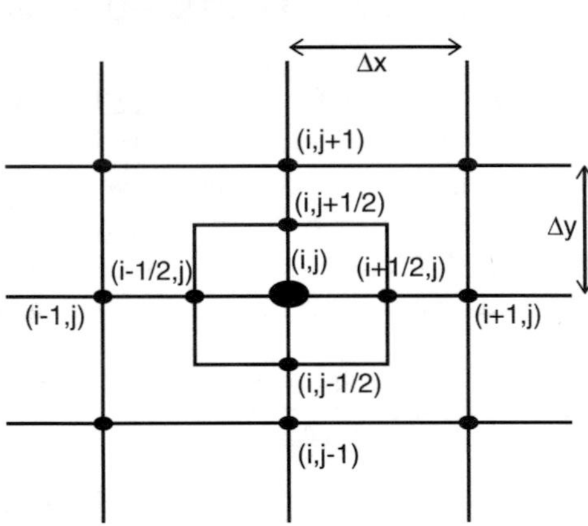

Fig. 2.17 Volume element cross section

A computer solution of this equation can be achieved by breaking the comet nucleus into many cubical volume elements and developing the required second partial derivatives by finite differences. When using this approach, the volume elements must be very small or varied in size to accommodate variations in the gradient of temperature. For a comet nucleus, ideally we could use small volume elements near the surface, where the thermal gradients are relatively large, and large volume elements near the center, where the thermal gradients are not so large. A simple algorithm for defining these volume elements is to enclose the comet in a parallelepiped as illustrated in Fig. 2.16.

The coordinates of the volume elements x, y, and z may be related to the indices i, j, and k. In computer terms, a triple-indexed do loop may be programmed to access each of the volume elements individually. Those volume elements outside of the comet surface are discarded. The required partial derivatives are computed by finite difference. Figure 2.17 shows a typical cross section of volume elements.
The thermal gradient with respect to the x coordinate is

$$\frac{\partial T}{\partial x(i - \frac{1}{2}, j, k)} = \frac{T_{i,j,k} - T_{i-1,j,k}}{\Delta x}$$

$$\frac{\partial T}{\partial x(i + \frac{1}{2}, j, k)} = \frac{T_{i+1,j,k} - Ti, j, k}{\Delta x}$$

and

$$\frac{\partial^2 T}{\partial x^2} = \frac{T_{i+1,j,k} - 2T_{i,j,k} + T_{i-1,j,k}}{\Delta x^2}$$

$$\frac{\partial^2 T}{\partial y^2} = \frac{T_{i,j+1,k} - 2T_{i,j,k} + T_{i,j-1,k}}{\Delta y^2}$$

$$\frac{\partial^2 T}{\partial z^2} = \frac{T_{i,j,k+1} - 2T_{i,j,k} + T_{i,j,k=1}}{\Delta z^2}$$

The heating rate of the comet nucleus is then given by

$$\dot{T} = \frac{k_d}{c_m \rho_m} \left[\frac{\partial^2 T}{\partial x^2} + \frac{\partial^2 T}{\partial y^2} + \frac{\partial^2 T}{\partial z^2} \right]$$

subject to the boundary condition that $T = T_{sub}$ on the surface of the nucleus.

For navigation, the temperature variation of the nucleus is of interest over a time interval of several weeks to several years. The temperature of the surface, which controls the amount of outgassing, varies periodically with respect to a baseline which evolves over hundreds of years. In order to establish this baseline, the thermodynamic model must be integrated over hundreds of years. This integration was performed for the asteroid Eros and the results are illustrated in Fig. 2.18.

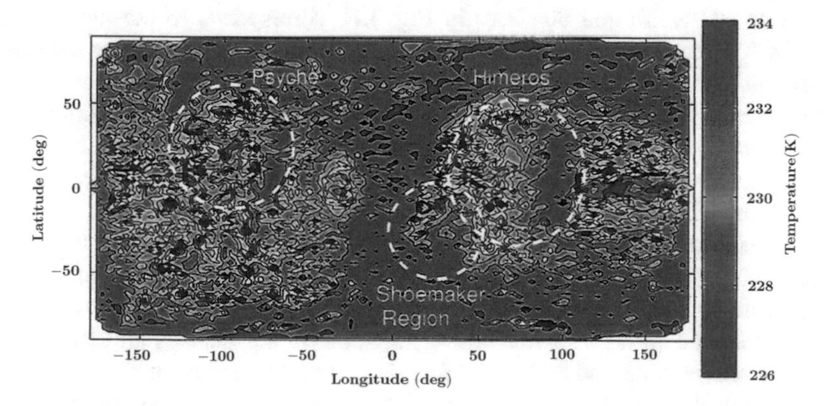

Fig. 2.18 Eros surface temperature distribution

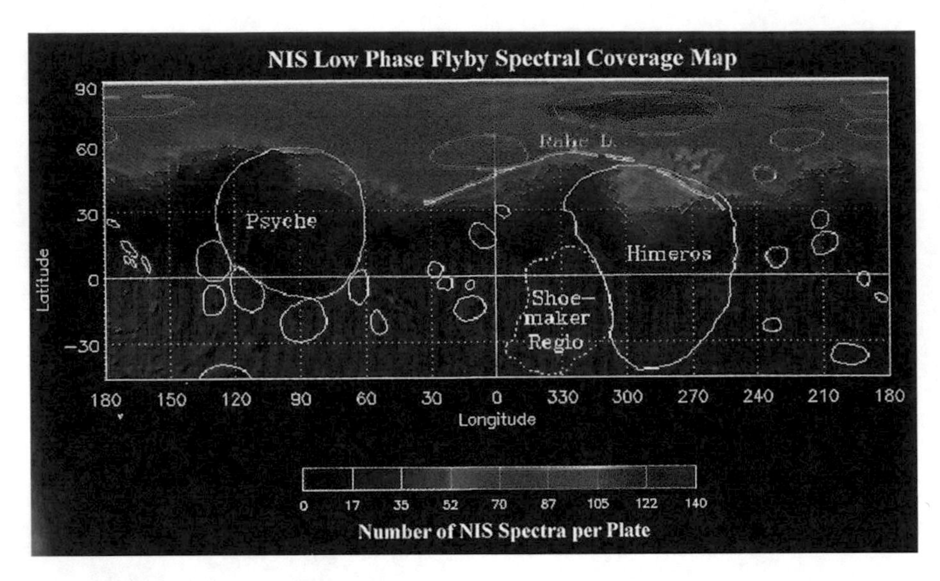

Fig. 2.19 Eros surface temperature distribution

For this integration, the temperature was integrated using the known shape, orbit, and attitude of Eros. The diurnal temperature variation was averaged over a single revolution of Eros as a function of the Sun's latitude. Given the direction of Eros's spin axis, the poles are warmer than the equatorial regions as shown in Fig. 2.18.

During the NEAR mission, Eros was extensively mapped in the infrared. Figure 2.19 shows a map of Eros infrared spectral parameters for the northern hemisphere. The southern hemisphere was dark during the time that the data was acquired. In Fig. 2.19, longitude is defined as positive West. Navigation coordinate systems are right handed and longitude is defined positive East. The warm region located at 80° North and 60° East in Fig. 2.18 corresponds to the high-spectral intensity located at 80° North and 300° West in Fig. 2.19. The cooler regions of Psyche and Himeros located at 30° North also correspond.

In order to compute acceleration of the spacecraft, the thermodynamic model can be interfaced with the empirical jet model. It is envisioned that in navigation operations a number of models would be available that vary in accuracy and number of parameters that need to be estimated. These would range from simple empirical models that would suffice during approach to a comet to high-precision physical models that may be needed for low-altitude orbits. Integration of the thermodynamic model with the jet model involves integration over the surface of the comet. Since the albedo varies as a function of location on the surface, an expansion of Legendre polynomials and associated functions may be employed. The coefficients of this expansion would be parameters to be determined by the orbit determination software.

2.7 Summary

Force models are accessed by the numerical integrator in order to compute the acceleration of the spacecraft required by the equations of motion. Since the equations of motion use the vector sum of all the force models, each force model may be packaged in a separate subroutine. For convenience, the force is divided by the mass of the spacecraft and acceleration vectors are computed. In addition, the partial derivatives of acceleration with respect to spacecraft position, velocity, and constant parameters are included in the subroutine. If there are a large number of force models, the number of constant parameters may be too many for the orbit determination filter to process. The NEAR mission had over 600 parameters. Therefore, the force models must be designed to provide only the number of parameters to accurately model the acceleration. The accuracy of the required force model will depend on the mission phase. For example, the gravity model requires one coefficient when the spacecraft is far from the central body but may require several hundred coefficients when in a close orbit.

The solar pressure model is needed for long-term orbit prediction during cruise. In orbit, the solar pressure is overwhelmed by gravity harmonic uncertainties and does not have much effect on short-term predictions. Atmospheric drag models are only needed when the spacecraft is in an atmosphere. Propulsion system models generally use the rocket equation and assume constant thrust. Another consideration in designing force models is computation of the partial derivatives. These partial derivatives are needed for integration of the variational equations and can impose a significant burden on computer time.

Exercises

2.1 A small unguided rocket weighs 26 pounds of which 4.86 pounds is fuel. The I_{sp} of the fuel is 120 s. Determine the range and maximum altitude if it is launched at an angle $\theta = 25°$ with respect to the local horizontal plane. Assume that the velocity is applied as an impulse and the trajectory is a parabola. The range (R) and maximum altitude (H) are given by

$$R = \frac{\Delta v_0^2 \sin(2\theta)}{g_0} \qquad H = \frac{\Delta v_0^2 \sin^2(\theta)}{2g_0}$$

2.2 The rocket in Exercise 2.1 is 2.75 in in diameter and the burn time was 5 s. Determine the thrust and drag force at burn out. The density of air is 2.508×10^{-3} slugs/ft^3 and the drag coefficient is 0.6. Navigation receives input from many sources that work in different systems of units. Engineers working in wind tunnels measuring drag like to work in English units. Making conversions is an important part of navigation operations.

2.3 Derive the equations of motion of the rocket in Exercise 2.1, determine the range and maximum altitude, and show that the trajectory is a parabola.

2.4 A spent rocket is found that weighs 30 pounds and is 6 ft long and 3 in in diameter. It is estimated that the fuel would weigh 5 pounds and probably have an I_{sp} of 200 s. Determine the maximum range for this rocket assuming a launch angle of 45°. This was a real problem.

2.5 The rocket in Exercise 2.1 was planned to be fired at a hillside that is 1042 ft downrange and 275 ft high. The target is at an elevation of 250 ft above the launch site. Determine the launch angle. Just before the launch, the launcher was moved 225 ft back so the range was now 1267 ft. Determine the new launch angle. Where does the rocket go if the correction to the launch angle is not made and the rocket is launched with the original launch angle for the 1042 ft range. This was also a real problem and the actual rocket went as predicted. Hint, the hill was only 275 ft high and there was ocean behind it but no fisherman.

2.6 A spacecraft is in a circular orbit about the sun with a radius of 1 AU (149×10^6 km). A solar sail is unfurled with an area of 800 m^2 and oriented facing the sun. The spacecraft weighs 500 kg and the solar sail is a perfect specular reflector. Determine the orbit of the spacecraft which is a conic section.

2.7 Determine the size of a solar sail in Exercise 2.6 that would enable the spacecraft to escape from the sun. Assume that the total mass remains at 500 kg. Neglecting general relativity, determine the size of a sail that would leave the solar system at one tenth the speed of light.

2.8 A spacecraft is maneuvered into a circular orbit about an asteroid. The plane of the spacecraft orbit faces the sun and the spacecraft is 10 km from the asteroid center. The asteroid is 1 AU from the sun and has a radius of 5 km and specific gravity of three. The universal gravitational constant is $6.674 \times 10^{-20}\,\mathrm{km^3\,kg^{-1}\,s^{-2}}$. The spacecraft is a black body with an area of 10 m^2 and mass of 500 kg. Determine the orbit of the spacecraft and the period of the orbit.

2.9 The integrand of the volume integral for the second degree gravity harmonic coefficients may be expressed as Cartesian components of the volume elements. Show that $I_{yy} - I_{zz} = Ma^2(C_{20} + 2C_{22})$.

Bibliography

Balmino, G., B. Moynot *J. Geophys. Res.* 87, 9735, 1982.

Bordi, J. J., P. G. Antreasian, J. K. Miller and B. G. Williams, "Altimeter Range Processing Analysis for Spacecraft Navigation about Small Bodies", AAS00-165, AAS/AIAA Space Flight Mechanics Meeting, Clearwater FL, January 23, 2000.

Bordi, J. J., J. K. Miller, B. G. Williams, R. S, Nerem and F. J. Pelletier, "The Impact of Altimeter Range Observations on NEAR Navigation", AIAA 2000–4423, AIAA/AAS Astrodynamics Specialist Conference, Denver, CO, August 14, 2000

Garmier, R.and J. P. Barriot, Ellipsoidal Harmonic Expansions of the Gravitational Potential: Theory and Application. Celes. Mech. and Dyn. Astron. (in press),2000.

Heiskanen, W.A. and H. Moritz 1967. Physical Geodesy. W.H. Freeman and Company, San Francisco, CA.

Kaula, W.M. 1966. Theory of Satellite Geodesy. Blaisdell, Waltham, MA.

Llanos, P. J., J. K. Miller and G. R. Hintz, "Comet Thermal Model for Navigation", AAS 13–259, 23rd Space Flight Mechanics Meeting, Kauai, Hawaii, February 10, 2013.

Miller, J. K., P. J. Llanos and G. R. Hintz, "A New Gravity Model for Navigation Close to Comets and Asteroids", AIAA 2014–4144, AIAA/AAS Astrodynamics Specialist Conference, San Diego, CA, 2014.

Miller, J. K. and G. R. Hintz, "A Comparison of Gravity Models used for Navigation Near Small Bodies", AAS17-557, 2017 AAS/AIAA Astrodynamics Specialist Conference, Stevenson, WA, 2017.

Miller, J. K., "Planetary and Stellar Aberration", EM 312-JKM-0311, Jet Propulsion Laboratory, June 16, 2003.

Weeks, C. J. and Miller, J. K., "A Gravity Model for Navigation Close to Asteroids and Comets", *The Journal of the Astronautical Sciences*, Vol 52, No 3, July-September 2004, pp 381–389.

Weeks, C. J., "The Effect of Comet Outgassing and Dust Emission on the Navigation of an Orbiting Spacecraft", AAS 93–624, AIAA/AAS Astrodynamics Specialist Conference, Victoria B.C., Canada, August 16, 1993.

Werner, R. A., "The Gravitational Potential of a Homogeneous Polyhedron or Don't Cut Corners", *Celestial Mechanics & Dynamical Astronomy*, Vol 59, 1994, pp 253–278.

Chapter 3
Trajectory Design

The problem of trajectory design requires the determination of spacecraft position and velocity as a function of time that satisfy design constraints. The constraints that must be satisfied are supplied to the trajectory designer as parameters that are generally functions of the Cartesian state. Thus, the main interest in developing solutions of the equations of motion for navigation is to enable computation of parameters that satisfy mission constraints and state vectors that may be used to initialize numerical integration for further refinement of the trajectory design. Analytic solutions of the equations of motion are of intrinsic interest because of their mathematical elegance. However, when applied to trajectory design, solutions are sought that enable the full Cartesian state to be determined with high precision and these solutions are numerical.

3.1 Restricted Two-Body Trajectories

The solution of the equations of motion for a point mass that is accelerated by a spherical central body was first obtained by Kepler. The trajectory is an ellipse, hyperbola, circle, or parabola depending on the initial conditions. Kepler's solution reveals that the trajectory shape is dependent on the energy and angular momentum of the spacecraft. Since the circle and parabola are limiting cases of an ellipse or hyperbola, only the solution for the ellipse or hyperbola is needed. Circular orbits are generally avoided because of singularities in determining the orbit and parabolic orbits are generally encountered only during the transition between elliptical and hyperbolic motion when the spacecraft is being accelerated by propulsive thrust or atmospheric drag. The solution of the equations of motion could be obtained by numerical integration from an initial state vector. Since there are six degrees of freedom associated with the initial conditions, six orbit parameters are required to describe the trajectory in addition to the central body gravity constant. Since there

J. Miller, *Planetary Spacecraft Navigation*, Space Technology Library 37,
https://doi.org/10.1007/978-3-319-78916-3_3

are many parameters that describe an ellipse or hyperbola and its orientation in space, a set of parameters is desired that will permit determination of the Cartesian state at various points along the trajectory and thus obtain the same result as could be obtained by numerical integration. Two of these parameters must be shape parameters, three are needed to orient the orbit in space, and an additional parameter is needed to specify the position of the point mass or body in the orbit at a particular time.

3.1.1 Elliptical Orbit

The equations of motion are first developed for an ellipse. The state vector (X), or state column matrix to be more precise, is comprised of the elements or components of the position and velocity vectors.

$$X = [x, y, z, \dot{x}, \dot{y}, \dot{z}]^T$$
$$\mathbf{r} = (x, y, z)$$
$$\mathbf{v} = (\dot{x}, \dot{y}, \dot{z})$$

The magnitudes of the position and velocity vectors are

$$r = \sqrt{x^2 + y^2 + z^2}$$
$$v = \sqrt{\dot{x}^2 + \dot{y}^2 + \dot{z}^2}$$

The angular momentum vector, which is also the pole of the orbit plane, is given by,

$$\mathbf{h} = \mathbf{r} \times \mathbf{v}$$

The magnitude of the angular momentum vector (h) is the angular momentum orbit parameter. The energy parameter is obtained by summing the kinetic and potential energy.

$$C_3 = v^2 - \frac{2GM}{r} \tag{3.1}$$

The actual energy is obtained by multiplying C_3 by one half of the mass of the body. Since the body is assumed to be a point mass, the body mass is assumed to be zero or small compared to the central body mass. Since the acceleration is the ratio of force to body mass, in the limit as the body mass approaches zero, it cancels from the acceleration. The force equation contains the product of the central body mass and body mass leaving GM instead of GMm for the gravity parameter. The factor of one half, that accounts for acceleration from rest to v, is omitted from the energy parameter. The spacecraft mass is also removed from the angular momentum parameter.

$$h = r^2 \dot{\eta} \tag{3.2}$$

where η is the true anomaly or polar angle that specifies the angular position of the body in the plane of the orbit. The angle η is measured counterclockwise from periapsis in the plane of the orbit. The velocity magnitude (v) may be computed from the radial and azimuthal components of velocity and is given by

$$v^2 = \dot{r}^2 + r^2 \dot{\eta}^2 \tag{3.3}$$

Substituting Eq. (3.3) into Eq. (3.1), the *vis viva* equation is obtained.

$$\dot{r}^2 + r^2 \dot{\eta}^2 = C_3 + \frac{2GM}{r} \tag{3.4}$$

The time parameter may be eliminated from the *vis viva* equation by substituting $\dot{\eta}$ from Eq. (3.2) and making use of

$$\dot{r} = \frac{dr}{d\eta} \dot{\eta}$$

yielding

$$\frac{h^2}{r^4} \left(\frac{dr}{d\eta} \right)^2 + \frac{h^2}{r^2} = C_3 + \frac{2GM}{r} \tag{3.5}$$

The *vis viva* equation may be put into an integrable form by substituting

$$\frac{dr}{d\eta} = -r^2 \frac{d(\frac{1}{r})}{d\eta}$$

and

$$d\eta = \frac{d\left(\frac{h}{r}\right)}{\sqrt{C_3 + \frac{2\,GM}{r} - \frac{h^2}{r^2}}} \tag{3.6}$$

Completing the square in the denominator,

$$d\eta = \frac{d\left(\frac{h}{r}\right)}{\sqrt{\left(C_3 + \frac{GM^2}{h^2}\right) - \left(\frac{h}{r} - \frac{GM}{h}\right)^2}}$$

A change of variable to ϕ defined by

$$\phi = \frac{h}{r} - \frac{GM}{h}$$

$$d\phi = d\left(\frac{h}{r}\right)$$

gives

$$d\eta = \frac{d\phi}{\sqrt{\left(C_3 + \frac{GM^2}{h^2}\right) - \phi^2}}$$

The solution is the *vis viva* integral.

$$\cos\eta = \frac{\phi}{\sqrt{\frac{GM^2}{h^2} + C_3}}$$

Replacing the dummy variable ϕ and solving for r, the equation of an ellipse in polar coordinates is obtained.

$$r = \frac{\frac{h^2}{GM}}{1 + \sqrt{1 + \frac{h^2 C_3}{GM^2}} \cos\eta} \tag{3.7}$$

The equation for an ellipse in polar coordinates is given by,

$$r = \frac{p}{1 + e\cos\eta} \tag{3.8}$$

Comparing Eq. (3.7) and Eq. (3.8), the parameter of orbit (p) and eccentricity (e) may be written from inspection.

$$p = \frac{h^2}{GM} \tag{3.9}$$

$$e = \sqrt{1 + \frac{h^2 C_3}{GM^2}} \tag{3.10}$$

From the geometry shown in Fig. 3.1, the following geometric parameters may be computed.

$$e = \sqrt{1 - \frac{b^2}{a^2}}$$

$$r_p = \frac{p}{1+e}$$

$$r_a = \frac{p}{1-e}$$

$$c = ae$$

$$p = a(1-e^2)$$

These geometric parameters have names that describe the geometry of an ellipse. These names can be found in the geometry literature and can be derived by inspection of Fig. 3.1. Since the same names are applied to different geometrical parameters for a hyperbola that share the same equations for a two-body orbit, it is left as an exercise for the reader to sort out all the possible orbit elements. Introducing the results obtained by integration of the *vis viva* integral, some additional parameters that are of interest may be determined.

$$\begin{aligned} C_3 &= -\frac{GM}{a} \\ \cos\eta &= \frac{p-r}{re} \\ \dot{r} &= e\sqrt{\frac{GM}{p}}\sin\eta \end{aligned} \tag{3.11}$$

The radial component of velocity may be obtained from the dot product of the position and velocity vectors yielding an equation for $\sin\eta$.

$$\sin\eta = \frac{\mathbf{r}\cdot\mathbf{v}}{re}\sqrt{\frac{p}{GM}} \tag{3.12}$$

The true anomaly (η) is obtained from a four-quadrant arctangent evaluation of $\sin\eta$ and $\cos\eta$. The quadrants are selected from the signs of the sine and cosine functions.

$$\eta = \tan^{-1}(\sin\eta, \cos\eta) \tag{3.13}$$

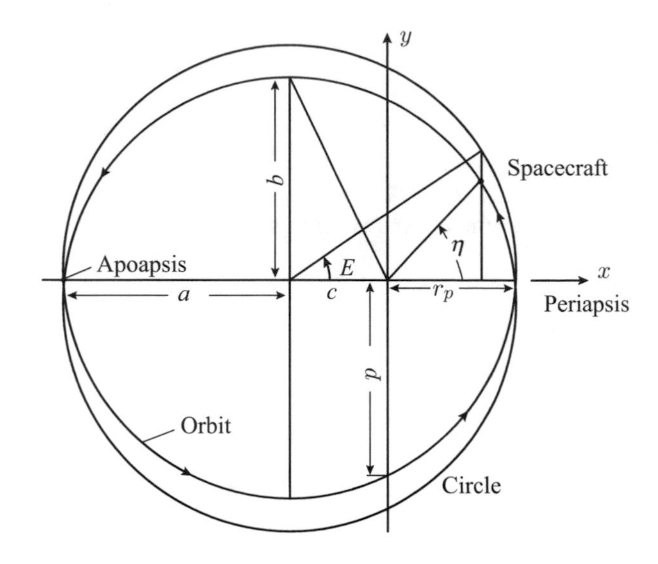

Fig. 3.1 Elliptical orbit geometry

The two shape parameters, p and e, are selected for inclusion in the set of six parameters that are used to describe the orbit. The third parameter that is needed to define the orbit solution is the time elapsed from the last periapsis passage to the epoch of the point in the orbit defined by the Cartesian state vector. The equation for true anomaly as a function of time is not integrable with simple functions. A change of variable to the angle E, referred to as the eccentric anomaly, yields an equation that can be integrated and the resulting equation is called Kepler's equation. From the geometry shown in Fig. 3.1, the following equation relates r and $\cos\eta$ to $\cos E$.

$$r\cos\eta = a\cos E - ae \tag{3.14}$$

Substituting Eq. (3.1.1) into (3.14), another equation for r as a function of E is obtained.

$$r = a - ae\cos E \tag{3.15}$$

Squaring Eqs. (3.14) and (3.15) and subtracting gives an equation that relates r and $\sin\eta$ to $\sin E$,

$$r\sin\eta = b\sin E$$

An equation that may be integrated for time as a function of E may be obtained by eliminating r from the *vis viva* integral. The *vis viva* integral may be put into a form that involves only r, $\dot{r}$, and constant parameters.

$$r^2\dot{r}^2 = -\frac{r^2 GM}{a} + 2GM\,r - GM\,a(1-e^2) \tag{3.16}$$

Differentiating Eq. (3.15) with respect to time gives

$$\dot{r} = ae \sin E \; \dot{E} \tag{3.17}$$

Substituting Eqs. (3.17) and (3.15) into Eq. (3.16) gives, after many cancelations,

$$\frac{a^3}{GM}(1 - e \cos E)^2 \; \dot{E}^2 = 1 \tag{3.18}$$

and

$$dt = \sqrt{\frac{a^3}{GM}}(1 - e \cos E) \, dE \tag{3.19}$$

The integral from periapsis, $E = 0$, to the point in the orbit of interest is Kepler's equation.

$$t - t_p = \sqrt{\frac{a^3}{GM}}(E - e \sin E) \tag{3.20}$$

The period of the orbit may be obtained by integrating over one complete revolution and

$$P = 2\pi \sqrt{\frac{a^3}{GM}} \tag{3.21}$$

The parameter t_p is the epoch of periapsis passage and is selected as the third parameter to characterize the orbit solution. The final set of parameters are the longitude of the ascending node (Ω), inclination (i), and argument of periapsis (ω). The geometry is shown in Fig. 3.2. The longitude of the ascending node and inclination may be determined from the components of the pole or angular momentum vector.

$$\Omega = \tan^{-1}\left(\frac{h_x}{-h_y}\right) \tag{3.22}$$

$$i = \tan^{-1}\left(\sqrt{\frac{h_x^2 + h_y^2}{h_z}}\right) \tag{3.23}$$

The argument of periapsis is obtained by first computing the argument of latitude (ω_n), the angle from the nodal crossing to the orbit point, and subtracting the true anomaly. A coordinate system is defined with $\hat{\mathbf{x}}_1$ in the direction of the ascending node and $\hat{\mathbf{z}}_1$ in the direction of the orbit pole.

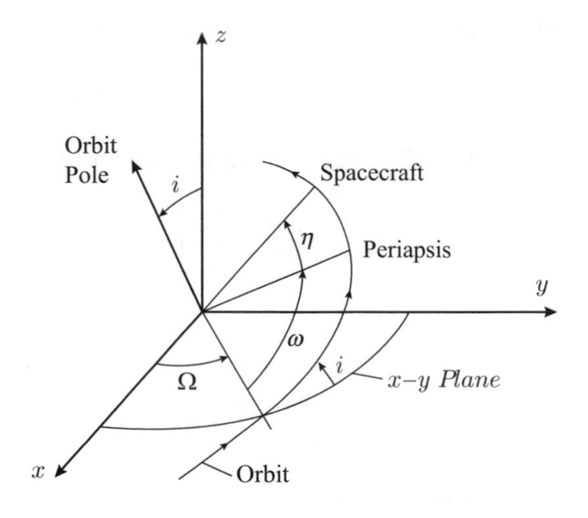

Fig. 3.2 Orbit orientation in space

$$\hat{\mathbf{x}}_1 = \frac{\hat{\mathbf{z}} \times \mathbf{h}}{h}$$

$$\hat{\mathbf{y}}_1 = \frac{\mathbf{h} \times \hat{\mathbf{x}}_1}{h}$$

The angle ω_n is defined by the components of the position vector ($\mathbf{r}$) and

$$\omega_n = \tan^{-1}\left(\frac{\sin\omega_n}{\cos\omega_n}\right) = \omega + \eta$$

where

$$\sin\omega_n = \hat{\mathbf{y}}_1 \cdot \hat{\mathbf{r}}$$

$$\cos\omega_n = \hat{\mathbf{x}}_1 \cdot \hat{\mathbf{r}}$$

and

$$\omega = \omega_n - \eta \tag{3.24}$$

The orbit element set that has been derived to describe two-body motion is

$$O_c = [p, e, t_p, \Omega, i, \omega] \tag{3.25}$$

and these elements may be obtained as a function of the state at some point in the orbit. The orbit element set (O_c) will be referred to here as classical orbit elements. In the literature, the semimajor axis (a) is often specified in place of the parameter of orbit (p).

$$O_c(t) = f_c(X, GM, t) \tag{3.26}$$
$$X = (\mathbf{r}, \mathbf{v})$$

The inverse function may be obtained by solving the above equations for the state as a function of the orbit elements. Starting with Kepler's equation, the eccentric anomaly (E) may be obtained from the time of periapsis (t_p) and the time (t). A closed form solution for Kepler's equation in terms of elementary functions cannot be obtained. Therefore, it is necessary to iterate using Newton's method to obtain E as a function of t and t_p. The true anomaly (η) and radius (r) are then computed from the eccentric anomaly and orbit elements p and e. The position and velocity is determined in the orbit plane coordinate system with $\mathbf{x}_2$ in the direction of periapsis and $\mathbf{z}_2$ in the direction of the orbit pole (see Fig. 3.2). The position of the body in the plane of the orbit coordinate system is

$$\mathbf{r}_2 = (r\cos\eta, r\sin\eta, 0)$$

and the velocity may be obtained by differentiating $\mathbf{r}_2$.

$$\mathbf{v}_2 = (-\sqrt{\frac{GM}{p}}\sin\eta,\ \sqrt{\frac{GM}{p}}(e+\cos\eta),\ 0)$$

The position and velocity in the reference coordinate system is obtained by rotating through the angles defined in Fig. 3.2.

$$\mathbf{r} = \mathrm{R}_c^T\,\mathbf{r}_2$$
$$\mathbf{v} = \mathrm{R}_c^T\,\mathbf{v}_2$$

$$\mathrm{R_c} = \begin{bmatrix} \cos\omega & \sin\omega & 0 \\ -\sin\omega & \cos\omega & 0 \\ 0 & 0 & 1 \end{bmatrix} \begin{bmatrix} 1 & 0 & 0 \\ 0 & \cos i & \sin i \\ 0 & -\sin i & \cos i \end{bmatrix} \begin{bmatrix} \cos\Omega & \sin\Omega & 0 \\ -\sin\Omega & \cos\Omega & 0 \\ 0 & 0 & 1 \end{bmatrix}$$

The inverse function permits transformation of the classical orbit element set into the state (X) at t.

$$X(t) = f_c^{-1}(O_c, GM, t)$$

The function f_c and its inverse permit a one to one mapping from state vector to classical orbit elements and back to state vector. These functions may be coded into subroutines on a computer and used to propagate a spacecraft trajectory from some time t_1 to a later time t_2. For example, the orbit elements may be computed from the state vector at t_1

$$O_c(t_1) = f_c(X(t_1), GM, t_1)$$

The time is advanced to t_2 and $X(t_2)$ is computed from the inverse function.

$$X(t_2) = f_c^{-1}(O_c, GM, t_2)$$

Trajectory propagation is thus accomplished with two calls to subroutines which can be accomplished with three lines of Fortran code or several lines of C code.

3.1.2 Hyperbolic Orbit

The transformation of a state vector to classical orbit elements for an ellipse may be modified to transform to a hyperbola. The eccentric anomaly is replaced by the hyperbolic eccentric anomaly (F) defined by

$$\sinh F = \frac{r \sin \eta}{b}$$

$$\cosh F = \frac{(a + r)}{ae}$$

and

$$F = \ln(\sinh F + \cosh F)$$

Kepler's equation for the hyperbola becomes

$$t - t_p = \sqrt{\frac{a^3}{GM}}(e \sinh F - F) \tag{3.27}$$

For the inverse transformation from classical orbit elements to state vector, it is necessary to solve Kepler's equation for the hyperbola by iteration.

The classical orbit elements (O_c) are not convenient for describing a hyperbolic trajectory during flyby of a planet or other celestial body. A different set of six elements are defined for this purpose. The geometry is illustrated in Fig. 3.3 in the plane of the orbit. The parameters are essentially the same as for an ellipse. The classical shape parameters p and e are replaced by the hyperbolic impact parameter (b), which is also the semiminor axis of the hyperbola, and the hyperbolic excess velocity (V_∞).

$$b = \sqrt{ap}$$

$$V_\infty = \sqrt{C_3}$$

The orientation of the hyperbola in space is defined with respect to the approach asymptote (**S**) and pole of the orbit plane. In the plane of the orbit, the limiting true

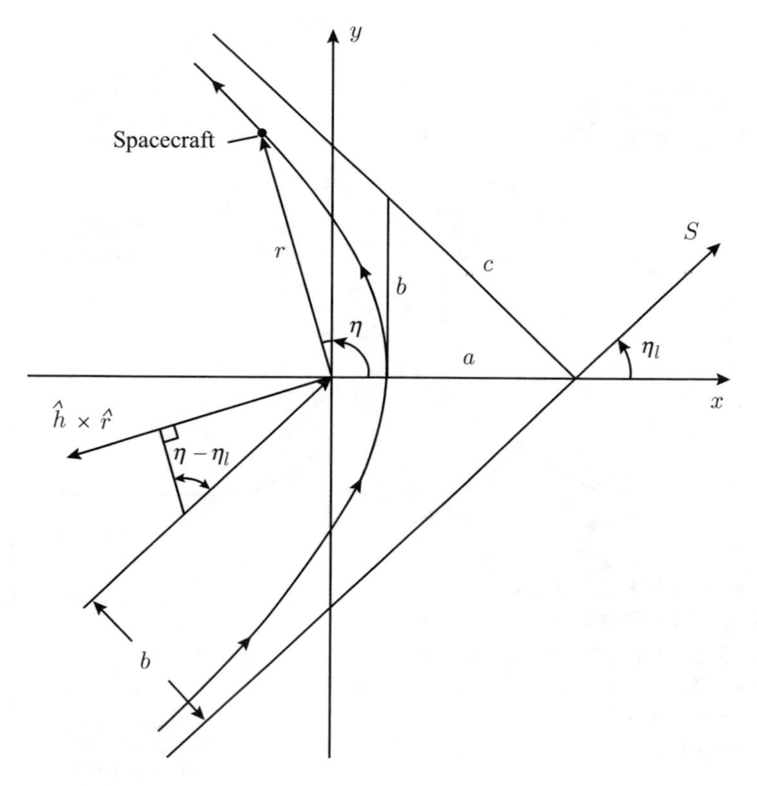

Fig. 3.3 Hyperbolic orbit geometry

anomaly as the body goes to infinity is

$$\eta_l = \cos^{-1}\left(\frac{1}{e}\right)$$

The direction of the approach asymptote, see Fig. 3.3, is given by the following vector sum,

$$\hat{\mathbf{S}} = (\hat{\mathbf{h}} \times \hat{\mathbf{r}}) \sin(\eta - \eta_l) - \hat{\mathbf{r}} \cos(\eta - \eta_l)$$

The **T** coordinate axis is perpendicular to **S** and in the $x - y$ plane of the reference coordinate system. The unit vector in the direction of **T** is given by,

$$\hat{\mathbf{T}} = \frac{\hat{\mathbf{S}} \times \hat{\mathbf{z}}}{|\hat{\mathbf{S}} \times \hat{\mathbf{z}}|}$$

and the unit vector in the direction of the **R** coordinate axis, that completes the right-hand system, is given by,

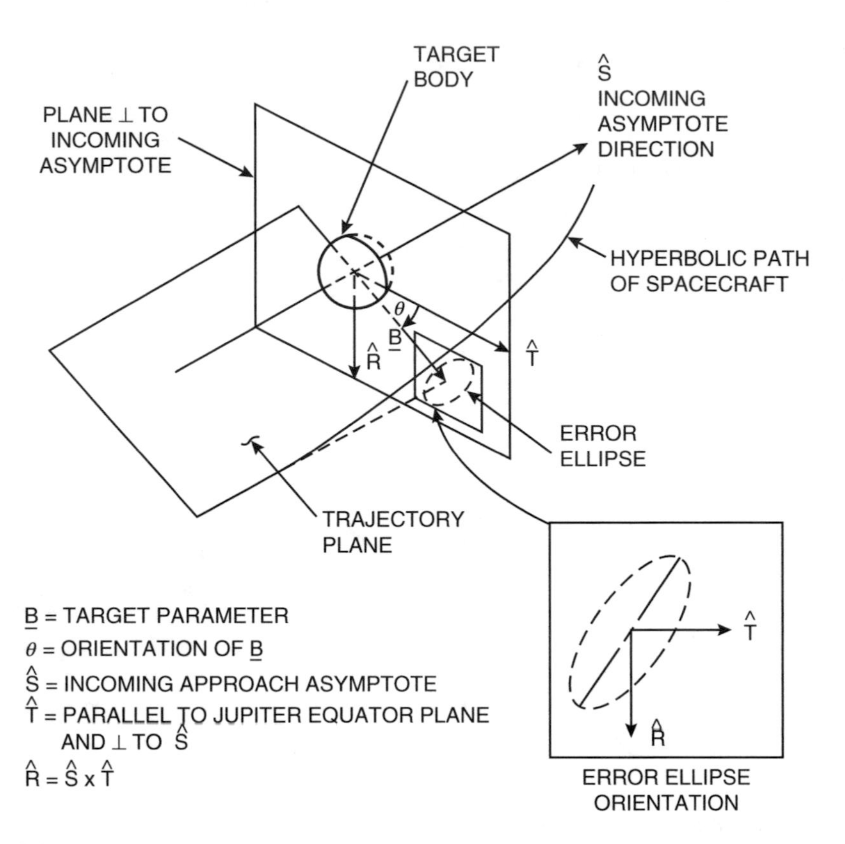

Fig. 3.4 B-plane definition

$$\hat{\mathbf{R}} = \hat{\mathbf{S}} \times \hat{\mathbf{T}}$$

The orientation of the hyperbola in space is shown in Fig. 3.4. The B-plane is defined perpendicular to the approach asymptote and passes through the center of the central body. The orientation of the plane of the orbit is defined by the angle θ between the **T** coordinate axis and the **B** vector that is in the plane of the orbit. The angle θ is determined from a four-quadrant arctangent function where the signs of the numerator and denominator are used to place the angle in the proper quadrant.

$$\theta = \tan^{-1}\left(\frac{\hat{\mathbf{T}} \cdot \hat{\mathbf{h}}}{-\hat{\mathbf{R}} \cdot \hat{\mathbf{h}}}\right)$$

The direction of the approach asymptote is defined by the right ascension (α_∞) and declination (δ_∞). The four-quadrant arctangent is used to place α_∞ in the proper quadrant.

$$\alpha_\infty = \tan^{-1}\left(\frac{\hat{S}_y}{\hat{S}_x}\right)$$

$$\delta_\infty = \sin^{-1}\hat{S}_z$$

A modified orbit element set has been derived to describe two-body hyperbolic motion.

$$O_h = [b, \theta, t_p, V_\infty, \alpha_\infty, \delta_\infty]$$

O_h may also be obtained as a function of the state at some point in the orbit as was done for the classical orbit element set.

$$O_h(t) = f_h(X, GM, t)$$

The inverse function for the hyperbola is obtained by solving the above equations for the state as a function of the orbit elements as done for the ellipse. Kepler's equation is solved by iteration to obtain F as a function of t and t_p. The true anomaly (η) and radius (r) are then computed from F and the orbit elements p and e. The elements p and e may be computed from b and V_∞. The position and velocity in the plane of the orbit are computed as for the classical elements. The final transformation from orbit plane coordinates to the reference coordinate system is given by,

$$\mathbf{r} = \mathrm{R}_h^T\,\mathbf{r}_2$$

$$\mathbf{v} = \mathrm{R}_h^T\,\mathbf{v}_2$$

and

$$\mathrm{R_h} = \begin{bmatrix} \sin\eta_l & 0 & \cos\eta_l \\ -\cos\eta_l & 0 & \sin\eta_l \\ 0 & 1 & 0 \end{bmatrix} \begin{bmatrix} \cos\theta & \sin\theta & 0 \\ -\sin\theta & \cos\theta & 0 \\ 0 & 0 & 1 \end{bmatrix}$$
$$\begin{bmatrix} 1 & 0 & 0 \\ 0 & \sin\delta_\infty & -\cos\delta_\infty \\ 0 & \cos\delta_\infty & \sin\delta_\infty \end{bmatrix} \begin{bmatrix} \sin\alpha_\infty & -\cos\alpha_\infty & 0 \\ \cos\alpha_\infty & \sin\alpha_\infty & 0 \\ 0 & 0 & 1 \end{bmatrix}$$

From the inverse function, the state may be computed as a function of the hyperbolic orbit elements and GM.

$$X(t) = f_h^{-1}(O_h, GM, t)$$

For interplanetary trajectory design, the outgoing hyperbolic asymptote is often needed. A modified set of hyperbolic orbit elements may be defined that has the outgoing departure asymptote direction as part of the set of orbit elements replacing the incoming approach asymptote direction. For the conversion of a state vector to the modified hyperbolic elements, the following procedure may be used. First, change the direction of the velocity vector and compute the incoming hyperbolic elements.

$$O_{hi} = f_h(X_i, GM, t)$$

$$X_i = [x, y, z, -\dot{x}, -\dot{y}, -\dot{z}]$$

$$O_{hi} = [b_i, \theta_i, t_{p_i}, V_{\infty i}, \alpha_{\infty i}, \delta_{\infty i}]$$

The outgoing elements may then be computed from the incoming elements.

$$O_{ho} = [b_o, \theta_o, t_{p_o}, V_{\infty o}, \alpha_{\infty o}, \delta_{\infty o}]$$

where the direction of the outgoing asymptote is

$$\alpha_{\infty o} = 180 - \alpha_{\infty i}$$

$$\delta_{\infty o} = -\delta_{\infty i}$$

The remaining outgoing parameters are the same as the incoming parameters. The inverse function may be obtained by simply reversing the procedure defined above. In summary, the hyperbolic elements with respect to the outgoing asymptote as a function of the state and the inverse function are given by,

$$O_{ho}(t) = f_{ho}(X, GM, t)$$

$$X(t) = f_{ho}^{-1}(O_{ho}, GM.t)$$

3.1.3 Injection Flight Plane Hyperbolic Trajectory

The design of the Earth departure hyperbola must be interfaced with the launch vehicle ascent trajectory design. The interface point is generally defined shortly after final stage burnout and the spacecraft position and velocity at this point are referred to as injection conditions. The injection conditions are described by a convenient set of parameters called injection flight plane coordinates. The departure hyperbolic orbit elements are also used to design Earth departure trajectories. However, hyperbolic orbit elements do not relate very well to the launch vehicle ascent trajectory and are generally not used for describing the trajectory when the position of the spacecraft near the Earth is of interest. The injection flight plane coordinates are shown schematically in Fig. 3.5. The departure hyperbola is shown propagated back from the injection point defined by the vector **r** to the Earth's

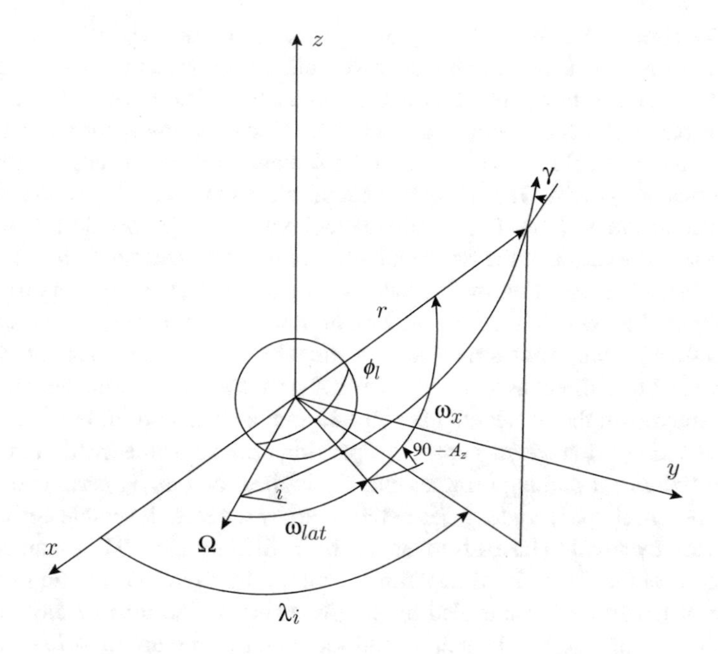

Fig. 3.5 Orbit injection geometry

equator. This two-body conic trajectory does not pass through the launch site but flies over the latitude of the launch site (ϕ_l) as shown in the figure. At the launch site latitude overfly point, the azimuth angle (A_z) is defined by measuring clockwise from north in the local tangent plane. In Fig. 3.5, A_z is shown with respect to the local East vector. The hyperbolic trajectory continues from launch latitude overfly to the injection point. The central angle (ω_x) from launch site latitude overfly to injection is the third injection flight plane coordinate. The remaining parameters are the velocity magnitude (V_i), the flight path angle (γ_i), and the inertial longitude (λ_i). The flight path angle is the angle between the velocity vector and local horizontal plane. The inertial longitude is measured from the vernal equinox to the projection of the injection vector on the Earth's equator. This unusual set of coordinates was devised by General Dynamics to interface Atlas launch vehicle trajectories with injected payloads and has continued in use to the present time.

The rationale behind the selection of injection flight plane parameters is related to launch vehicle constraints. The individual parameters are a function of both position and velocity at injection. In the design of interplanetary trajectories, it is necessary to separate position from velocity. The spacecraft velocity is sought that results in a trajectory between two positions determined by the location of the planets. Thus, the injection flight plane parameters must be separable into parameters that are related to position and are fixed and parameters that are related to velocity and are permitted to vary. For design of interplanetary trajectories, the fixed parameters are r, γ_i and either ω_x or A_z and the variable parameters are V, λ_i and either ω_x

or A_z. The choice of fixing either ω_x or A_z depends on the type of launch vehicle ascent trajectory that is used. The launch vehicle ascent trajectory may be direct or into a parking orbit. For a direct ascent, the launch vehicle rises vertically until it clears the gantry and is then tipped a fixed angle to start a gravity turn. For a gravity turn, the launch vehicle is accelerated in the direction of the velocity vector and is slowly turned by gravity. The initial horizontal direction is roughly in the direction of the launch azimuth (A_z). The detailed launch vehicle trajectory design must take into account the rotation of the Earth and the actual Earth fixed azimuth differs some from the inertial azimuth. During the launch vehicle ascent, the trajectory traverses a central angle of ω_x which is a characteristic of the launch vehicle and corresponds to about 3000 km downrange from the launch site which is assumed to be the Kennedy space center. For a direct ascent, ω_x is fixed. For some launch vehicles, the ability exists to shut down the rocket engine at the instant a circular orbit is achieved. The spacecraft and upper stage may coast in a parking orbit for some fraction of an orbit and restarted. For a parking orbit ascent, A_z is fixed and ω_x is permitted to vary. The injection flight path angle (γ_i) is determined by the launch vehicle performance and is generally small to take advantage of the Earth's rotation. The parameter λ_i is used to control the direction of departure from the Earth in the ecliptic plane. The direction of departure is controlled by simply selecting the time of day to launch or the launch window. The launch azimuth and coast time are used to control the velocity component out of the Ecliptic plane.

The injection flight plane coordinates may be computed as a function of the Cartesian state at injection. The classical elements O_c and hyperbolic elements O_h and their inverses may be used to propagate the spacecraft along the ascent hyperbola and determine state vectors at launch site overfly and injection. From these state vectors, the injection flight plane parameters may be computed.

$$O_i(r_i) = f_i(X, GM) \tag{3.28}$$

where

$$O_i = [\phi_l, r_i, \omega_x, A_z, V, \gamma_i, \lambda_i]$$

Also, the state at injection may be computed from the inverse relationships as a function of the injection flight plane parameters.

$$X(r_i) = f_i^{-1}(O_i, GM) \tag{3.29}$$

3.1.4 Lambert's Problem

An important problem relating to the determination of orbits and design of interplanetary trajectories was defined by Lambert and Euler. Given the flight time between two position vectors, Lambert's problem is to determine the orbit that transfers from the first position vector to the second in the given flight time. Lambert's problem is fundamental to interplanetary trajectory design since the position vectors of interest are generally the ephemerides of two planets and the problem is to design a trajectory that will go from one planet to another in a specified interval of time. Problems of this type are referred to as two-point boundary value problems. In the current age of computers, Lambert's problem may be easily solved by targeting the second position vector. The velocity at the first position vector is varied iteratively using Newton's method until the propagated trajectory intersects the second position vector at the time specified. The partial derivatives required by the three-parameter search can be computed by finite difference and the trajectory propagation performed by solution of the two-body equations of motion as described in Sects. 3.1.1 and 3.1.2.

An analytic solution of Lambert's problem was provided by Lagrange, who was proud of this accomplishment as well he should be, and showed his solution to Lambert about a year before he died. Since that time, a considerable amount of research has been expended identifying singularities and developing efficient algorithms for digital computers. Even in the modern era of high-speed computers, where analytic methods have generally given way to much simpler numerical methods, analytic solutions of Lambert's problem are often preferred for designing large numbers of trajectories that are required for surveying possible missions to the planets and other celestial bodies.

Figure 3.6 shows the geometry of Lambert's problem. A trajectory is sought that transfers a spacecraft from position $\mathbf{r}_1$ at time t_1 to position $\mathbf{r}_2$ at time t_2. The transfer time is given by solution of Kepler's equation at the two end points.

$$t_2 - t_1 = \sqrt{\frac{a^3}{GM}}\,[(E_2 - E_1) - e(\sin E_2 - \sin E_1)]$$

Making use of the identity

$$\cos\left(\frac{E_1 + E_2}{2}\right)\sin\left(\frac{E_2 - E_1}{2}\right) = \frac{1}{2}(\sin E_2 - \sin E_1)$$

$$t_2 - t_1 = \sqrt{\frac{a^3}{GM}}\left[(E_2 - E_1) - 2e\cos\left(\frac{E_1 + E_2}{2}\right)\sin\left(\frac{E_2 - E_1}{2}\right)\right]$$

Lagrange defined two angles, α and β, for his solution and these are related to the eccentric anomaly at the end points by

$$\frac{\alpha - \beta}{2} = \frac{E_2 - E_1}{2} \quad \text{and} \quad \cos\left(\frac{\alpha + \beta}{2}\right) = e\cos\left(\frac{E_2 + E_1}{2}\right)$$

Kepler's equation becomes

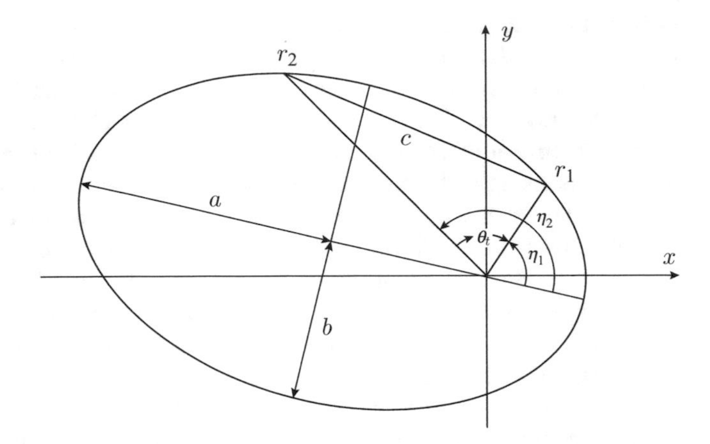

Fig. 3.6 Orbit transfer geometry

$$t_2 - t_1 = \sqrt{\frac{a^3}{GM}}\left[\alpha - \beta - 2\cos\left(\frac{\alpha+\beta}{2}\right)\sin\left(\frac{\alpha-\beta}{2}\right)\right]$$

$$t_2 - t_1 = \sqrt{\frac{a^3}{GM}}\left[(\alpha - \sin\alpha) - (\beta - \sin\beta)\right] \tag{3.30}$$

The solution involves developing equations for α and β as a function of known parameters and a, the semimajor axis, and iterating on a using Newton's method until the desired transfer time $(t_2 - t_1)$ is achieved. From the geometry shown in Fig. 3.6, the law of cosines may be applied to the triangle and

$$c^2 = r_1^2 + r_2^2 - 2r_1 r_2 \cos\theta_t$$

$$c^2 = (r_1 + r_2)^2 - 4r_1 r_2 \cos^2\left(\frac{\theta_t}{2}\right)$$

The chord may also be obtained by simply differencing the position vectors and computing the magnitude.

$$c = |\mathbf{r_2} - \mathbf{r_1}| \tag{3.31}$$

It can be shown from the geometry that

$$r_1 + r_2 + c = 2a(1 - \cos\alpha)$$

$$r_1 + r_2 - c = 2a(1 - \cos\beta)$$

and

$$\sin^2\left(\frac{\alpha}{2}\right) = \frac{r_1 + r_2 + c}{4a} \tag{3.32}$$

$$\sin^2\left(\frac{\beta}{2}\right) = \frac{r_1 + r_2 - c}{4a} \tag{3.33}$$

Before Eq. (3.30) can be solved iteratively for the semimajor axis (a) of the transfer orbit, the quadrant of the angles α and β must be determined. The pole of the orbit is computed by taking the cross product of the two position vectors. In order to resolve the transfer angle, it is assumed that the orbit is direct and the spacecraft goes from $\mathbf{r_1}$ to $\mathbf{r_2}$. If a retrograde orbit is desired, the solution is obtained for the direct orbit in the opposite direction and the resulting inclination and node are adjusted to give the desired retrograde orbit. With this assumption, the pole vector (P) is opposite to the cross product of the position vectors. If this cross product is in the southern hemisphere, the transfer angle (θ_t) is assumed to be greater than 180°. For a direct orbit, the pole vector is in the northern hemisphere. The pole vector and transfer angle for the cross product of the position vectors in the northern hemisphere is given by

$$\mathbf{P} = \mathbf{r_1} \times \mathbf{r_2} \tag{3.34}$$

$$\theta_t = \cos^{-1}\left(\frac{\mathbf{r}_1 \cdot \mathbf{r}_2}{r_1 r_2}\right)$$

and for the cross product of the position vectors in the southern hemisphere,

$$\mathbf{P} = -\mathbf{r_1} \times \mathbf{r_2} \tag{3.35}$$

$$\theta_t = 360 - \cos^{-1}\left(\frac{\mathbf{r}_1 \cdot \mathbf{r}_2}{r_1 r_2}\right)$$

The quadrant of the angle β is assigned following the convention determined by Battin.

$$0 \le \beta \le \pi \quad \text{for} \quad \theta_t \le \pi$$

$$-\pi \le \beta \le 0 \quad \text{for} \quad \theta_t \ge \pi$$

The inclination and longitude of the ascending node are computed in the usual manner from the pole vector which is in the same direction as the angular momentum vector.

$$\Omega = \tan^{-1}\left(\frac{P_x}{-P_y}\right) \tag{3.36}$$

$$i = \tan^{-1}\left(\sqrt{\frac{P_x^2 + P_y^2)}{P_z}}\right) \tag{3.37}$$

Equation (3.30) is solved iteratively for a. Some experimentation may be required to determine the quadrant of α and select either the hyperbolic or elliptical version of Kepler's equation. Once a solution is found, the energy may be determined from a. This effectively proves Lambert's theorem but the solution is in terms of parameters that are not convenient. The angular momentum, or parameter of orbit (p), requires some further solution of the orbit. The following identities may be determined from the geometry. One could draw an ellipse and spot the positions r_1 and r_2 on the ellipse and verify these identities by direct measurement. For example, one could measure a and b and compute e from its definition.

$$p\left(\frac{r_1 + r_2}{r_1 r_2}\right) = 2 + 2e\cos\left(\frac{\eta_2 + \eta_1}{2}\right)\cos\left(\frac{\eta_2 - \eta_1}{2}\right) \tag{3.38}$$

$$\sqrt{r_1 r_2}\cos\left(\frac{\theta_t}{2}\right) = 2a\sin\frac{\alpha}{2}\sin\frac{\beta}{2} \tag{3.39}$$

$$\sqrt{r_1 r_2}\cos\left(\frac{\eta_2 + \eta_1}{2}\right) = \frac{a}{e}\cos\left(\frac{\alpha + \beta}{2}\right) - ae\cos\left(\frac{\alpha - \beta}{2}\right) \tag{3.40}$$

Multiplying Eq. (3.40) by e, adding Eq. (3.39), and replacing $a(1 - e^2)$ by p gives

$$e\sqrt{r_1 r_2}\cos\left(\frac{\eta_2 + \eta_1}{2}\right) = p\cos\left(\frac{\alpha - \beta}{2}\right) - \sqrt{r_1 r_2}\cos\left(\frac{\theta_t}{2}\right) \tag{3.41}$$

Substituting Eq. (3.41) into Eq. (3.38) and solving for p,

$$p = \frac{2r_1 r_2 \sin^2\left(\frac{\theta_t}{2}\right)}{r_1 + r_2 - 2\sqrt{r_1 r_2}\cos\left(\frac{\theta_t}{2}\right)\cos\left(\frac{\alpha - \beta}{2}\right)} \tag{3.42}$$

and the eccentricity is given by

$$e = \sqrt{1 - \frac{p}{a}}$$

From the geometry, a compact formula for p can be derived.

$$p = \frac{4ar_1 r_2}{c^2}\sin^2\left(\frac{\alpha + \beta}{2}\right)\sin^2\left(\frac{\theta_t}{2}\right) \tag{3.43}$$

The true anomaly may then be computed using the four-quadrant arctangent

$$\eta_1 = \tan^{-1}(\sin\eta_1, \cos\eta_1) \tag{3.44}$$

where

$$\cos\eta_1 = \frac{p - r_1}{r_1 e}$$
$$\cos\eta_2 = \frac{p - r_2}{r_2 e}$$
$$\sin\eta_1 = \frac{cos\eta_1 \cos\theta_t - \cos\eta_2}{\sin\theta_t}$$

The argument of periapsis ω is the angle from the ascending node ($\mathbf{\Omega}$) to periapsis. The true anomaly (η_1) is the angle from periapsis to the position vector ($\mathbf{r_1}$). We need the angle from the ascending node to $\mathbf{r_1}$. The ascending node vector is given by the cross product of the z axis with the pole vector.

$$\mathbf{\Omega} = \hat{\mathbf{z}} \times \mathbf{P}$$

the angle from the ascending node to $\mathbf{r_1}$ is

$$\omega_n = \tan^{-1}(\sin\omega_n, cos\omega_n)$$
$$\sin\omega_n = -P_x P_z r_{1x} + P_y P_z r_{1y} + (P_x^2 + P_y^2) r_{1z}$$
$$\cos\omega_n = -r_{1x} P_y + r_{1y} P_x$$

The argument of periapsis is then

$$\omega = \omega_n - \eta_1 \tag{3.45}$$

The final parameter needed to describe the transfer orbit is the time of periapsis passage.

$$t_p = t_1 - \sqrt{\frac{a^3}{GM}}(E_1 - e\sin E_1) \tag{3.46}$$

The solution of Lambert's problem yields a classical set of orbit elements (O_l) defined as a function of the time and position vectors of two points relative to a central body with gravitational parameter GM.

$$O_l(t_1) = f_l(t_1, \mathbf{r}_1, t_2, \mathbf{r}_2, GM) \tag{3.47}$$
$$O_l = [p, e, t_p, \Omega, i, \omega] \tag{3.48}$$

The solution of Lambert's theorem was coded into a subroutine by the author in 1966. This subroutine, along with subroutines previously described for transforming orbit elements into state vectors and state vectors into orbit elements, was coded in Fortran 2. At that time, the aerospace industry operated with closed shops for computer programming so the work was done under the table. These subroutines have been incorporated into many programs for analysis and flight operations and are in their original form except for some minor corrections. The archaic "if" statements are still in the code. An important principle of computer programming is not to tamper with a program that works.

An alternative to the above mathematical solution is to solve Lambert's problem by targeting. Lambert's problem reduces to finding the velocity at t_1 that results in the desired position at t_2 when the trajectory is propagated from t_1 to t_2. The complete state $(\mathbf{r}_1, \mathbf{v}_1)$ can then be converted to the required orbit elements (O_l) using the subroutine described earlier. The targeting procedure involves first making an initial guess for v_1. The velocity vector in a circular orbit passing through $\mathbf{r_1}$ provides an initial guess, but zero velocity also works. The state at t_1 is converted to the orbit elements $O_c(t_1)$ as described in Sect. 3.1.1. The trajectory is propagated to t_2 by computing the inverse at t_2.

$$X(t_2) = O_c^{-1}(t_2)$$

The position at t_2 is compared with the desired position. If they do not agree, a Newton Raphson iteration is performed. The state transition matrix is obtained as described in Sect. 1.4 only conic propagation of the trajectory is used. The upper-right 3×3 partition of the state transition matrix is inverted and multiplied by the position miss at t_2 to obtain a correction to the velocity at t_1. After several iterations, the miss goes to zero. When compared with the mathematical solution, the targeted solution agreed within ten decimal places. The advantage of the targeted solution is the avoidance of complexity associated with resolving issues associated with transfer angles, energy, and direction that plagues the mathematical solution. A solution to Lambert's problem may be obtained by using the orbit element transformation subroutines exclusively.

3.2 Interplanetary Transfer

The problem of interplanetary trajectory design is initially concerned with finding a trajectory that will transfer a spacecraft from one planet to another where the calendar date is specified at the beginning and end. This problem is a two-point boundary value problem where the two points are the positions of the first planet at the start time and the second planet at the end time.

3.2.1 *Hohmann Transfer*

The existence of two-body transfer orbits solves only part of the interplanetary trajectory design problem. The transfer trajectory is generally initiated by a large rocket motor burn at the first planet, the Earth, and terminated by a large rocket motor burn at the target planet that inserts the spacecraft into orbit. For a planetary flyby, the second motor burn is omitted. For preliminary trajectory design, the two rocket motor burns are generally computed as impulsive burns and a trajectory is desired that minimizes the magnitudes of the velocity changes at the end points to achieve transfer. The velocity change is from the orbital velocity of the planets to the velocity of the spacecraft and is generally specified in terms of the energy parameter (C_3) associated with the hyperbolic departure and approach conic orbits. This interplanetary orbit transfer problem is referred to as the optimum two-impulse transfer problem and a solution was first obtained by the German rocket engineer Walter Hohmann in 1925.

The geometry of the Hohmann transfer orbit is shown in Fig. 3.7 for a transfer orbit from Earth to Mars. The Hohmann transfer orbit is tangential to the Earth orbit at launch and tangential to Mars orbit at encounter. The transfer angle is 180°. The optimality appears obvious from Fig. 3.7 since energy is added in the direction of the Earth's orbital velocity and subtracted in the direction of Mars orbital velocity. The proof of optimality is a bit mathematically tedious and is best demonstrated numerically by obtaining the Hohmann transfer orbit as a solution to the problem of constrained trajectory optimization which is described in a later chapter. The opportunity for a Hohmann transfer from Earth to Mars occurs when the planets have the alignment shown in Fig. 3.7. This alignment occurs at a frequency determined by the difference of Earth and Mars angular orbital rates or about every 1.64 years. The launch opportunity occurs on the day when Mars is at a particular angle with respect to the Earth. The spacecraft must be launched at the time that the transfer orbit will encounter Mars at the encounter time. Thus, Mars must lag the encounter point by an angle of about 136° at the time of launch. The lag angle may be computed from the periods of the spacecraft and Mars orbits.

A problem with the Hohmann transfer orbit is that the solution is at a singular point associated with elliptic transfer orbits. A 180° orbit transfer trajectory must be in the plane of the Earth, Mars, and Sun. If Mars is slightly out of the ecliptic plane, as it generally is except at the nodal crossings, the transfer orbit must also be out of the ecliptic plane. For Mars, near the 180° transfer point, the transfer orbit plane may be far out of the ecliptic plane requiring an enormous expenditure of rocket fuel to make the plane change. For some cases, the Hohmann transfer trajectory is over the Sun's ecliptic pole. One simple remedy is to introduce a small midcourse plane change maneuver and another remedy is to simply avoid the singularity. For most interplanetary trajectory designs, the latter remedy is used and the launch and encounter dates are biased a few days. Fortunately, since the optimum solution is obtained for launch and encounter dates where the derivative of the cost function is zero with respect to these times, substantial deviations from optimality may be made with a small performance penalty.

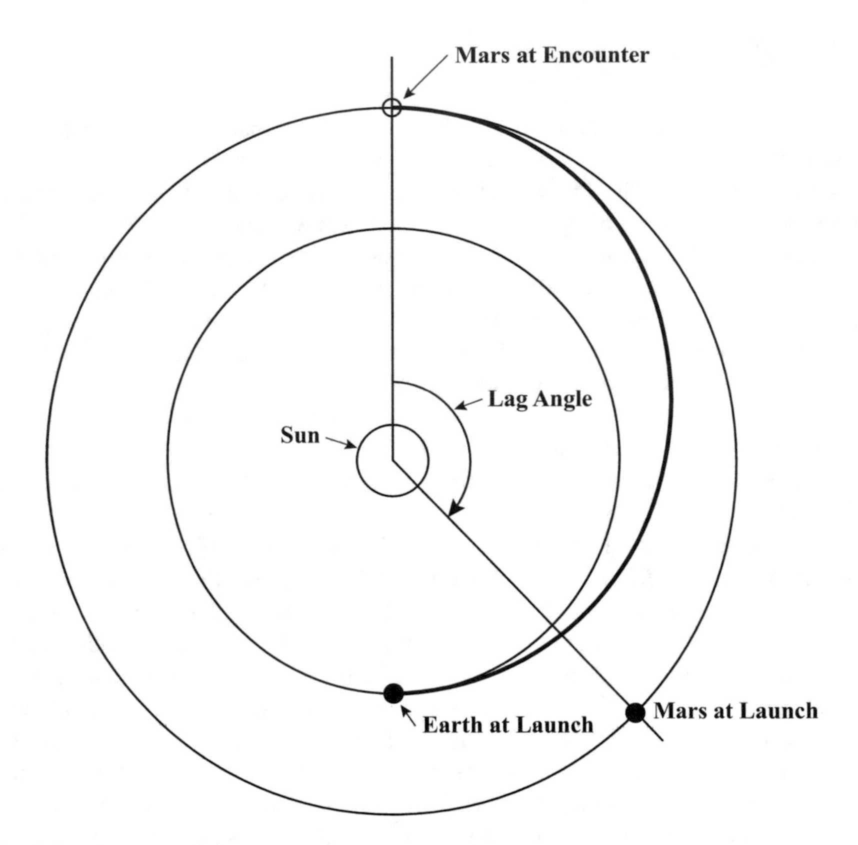

Fig. 3.7 Earth–Mars Hohmann transfer

Once the launch date and encounter date are determined, a preliminary trajectory design may be implemented using the tools described in Sects. 3.1.1 through 3.1.4. The design process involves interfacing the interplanetary transfer trajectory with departure and approach hyperbolic trajectories that satisfy design constraints at Earth and the target planet. The Earth injection hyperbola must be achievable by the launch vehicle and the encounter hyperbola must result in the desired geometry at the target. A simple method to design an interplanetary trajectory is by patched conics. The principal behind patched conics is that the motion of the spacecraft may be computed by including the acceleration of only the dominant central body and ignoring the tidal accelerations associated with all other bodies. The error introduced by ignoring tidal acceleration is generally small enough that the resulting trajectory may be used to evaluate other mission constraints and to provide an initial guess for precision targeting programs that are employed for the final design.

Figure 3.8 illustrates schematically the patched conic design process for an Earth to Mars trajectory. In order to initialize the process, the Earth position vector at the Earth injection time (t_i) and the Mars position vector at the Mars encounter time are obtained from a planetary ephemeris file. The conic Earth ephemeris and Mars ephemeris orbit elements are computed as described is Sect. 3.1.

$$O_{ce}(t_i) = f_c(X_e(t_i), GM_s, t_i)$$
$$O_{cm}(t_m) = f_c(X_m(t_m), GM_s, t_m)$$

From Lambert's orbit transfer solution, a two-body sun-centered conic is computed from the center of Earth to the center of Mars.

$$O_l = f_l(t_e, \mathbf{r}_e, t_m \mathbf{r}_m, GM_s)$$

The position vectors $\mathbf{r}_e$ and $\mathbf{r}_m$ are obtained from the inverse conic element transformations O_{ce}^{-1} and O_{cm}^{-1}. Two patch points are defined at t_1 and t_2 about 2.5 days from Earth injection and 2 days from Mars encounter. At these times, the spacecraft is about 2 million km from the planets where the tidal acceleration of the Sun is about equal to the central body acceleration of the planets. The tidal acceleration of the Sun is simply the difference between the Sun's acceleration of the spacecraft and the Sun's acceleration of the planet. The state vector of the spacecraft and planets are computed at the patch points.

$$X(t1) = f_c^{-1}(O_l, GM_s, t_1)$$
$$X_e(t1) = f_c^{-1}(O_{ce}, GM_s, t_1)$$
$$X(t2) = f_c^{-1}(O_l, GM_s, t_2)$$
$$X_m(t2) = f_c^{-1}(O_{cm}, GM_s, t_2)$$

The hyperbolic orbit elements of the spacecraft are computed with respect to Earth and Mars at the patch points from the planet relative state vectors. The incoming asymptote is computed for Mars and the outgoing asymptote for Earth.

$$O_{hi} = f_{hi}(X(t2) - X_m(t2), GM_m, t_2)$$
$$O_{ho} = f_{ho}(X(t1) - X_e(t1), GM_e, t_1)$$

The encounter conditions at Mars (b_i, θ_i) are set equal to the target encounter conditions (b_i^*, θ_i^*) and the outgoing encounter conditions at Earth (b_o, θ_o) are set equal to the injection conditions (b_o^*, θ_o^*). For the initial iterations, the injection conditions at Earth are $b_o^* = 20{,}000$ km and $\theta_o^* = 180°$. Planet relative state vectors are then computed at the patch points and another Lambert transfer orbit is computed connecting the new patch points.

$$X_1(t_1) = f_{ho}(O_{ho}, GM_e, t_1)$$
$$X_2(t_2) = f_{hi}(O_{hi}, GM_m, t_2)$$
$$O_l = f_l(t_1, \mathbf{r}(t_1) + \mathbf{r}_e(t_1), t_2, \mathbf{r}(t_2) + \mathbf{r}_m(t_2), GM_s)$$

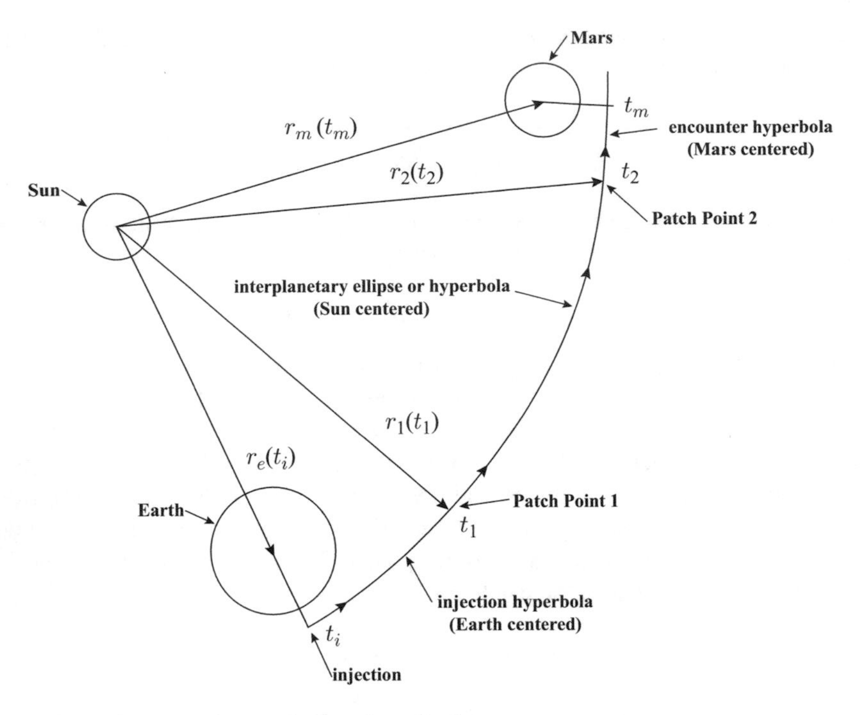

Fig. 3.8 Earth–Mars trajectory design schematic diagram

New Sun-centered state vectors are computed and the above process is repeated several times until the solution converges to the target encounter conditions. The Earth departure hyperbola target conditions are then modified to bring the injection position into alignment with the launch vehicle ascent trajectory as specified by injection flight plane coordinates. The true anomaly of the injection point is computed from the defined injection radius (r_i). From the true anomaly and the departure hyperbolic orbit elements, the eccentric anomaly is computed and the time of injection is determined from Kepler's equation. The injection state vector is then computed from,

$$X_i = f_{ho}(O_{ho}, GM_e, t_i)$$

The injection flight plane coordinates are then computed from the injection state vector (Eq. 3.28).

$$O_i = f_i(X, GM)$$

The injection flight plane coordinates are then replaced with the target values. For example, the injection flight path angle may be set equal to the target value ($\gamma_i = \gamma_i^*$) and the central angle from the launch site to injection is set equal to its target value ($\omega_x = \omega_x^*$) to define O_i^*. The injection state vector and new outgoing hyperbolic encounter target parameters are then computed.

Table 3.1 Earth–Mars trajectory design iterations

Earth injection flight plane						Mars encounter hyperbola				
i	ω_x [a] (deg)	γ_i [a] (deg)	V_i (km/s)	A_z (deg)	λ_i (deg)	b [a] (km)	θ [a] (deg)	V_∞ (km/s)	α_∞ (deg)	δ_∞ (deg)
1	32.68	28.74	10.94	127.97	113.54	2219.34	43.72	3.00	253.96	−14.64
2	21.10	23.26	10.99	133.79	119.00	28887.02	257.29	3.44	254.24	−12.19
3	19.53	22.44	10.99	134.43	119.37	10633.94	253.18	3.45	254.17	−11.72
4	19.32	22.33	10.99	134.50	119.40	6965.54	235.98	3.45	254.18	−11.66
5	19.29	22.31	10.99	134.51	119.41	6489.04	233.95	3.45	254.18	−11.66
6	19.29	22.31	10.99	134.51	119.41	6429.07	233.65	3.45	254.18	−11.65
7	19.29	22.31	10.99	134.52	119.41	6421.31	233.62	3.45	254.18	−11.65
8	19.29	22.31	10.99	134.52	119.41	6420.32	233.61	3.45	254.18	−11.65
9	19.29	22.31	10.99	134.52	119.41	6420.20	233.61	3.45	254.18	−11.65
10	15.92	18.00	10.99	134.75	112.99	6420.18	233.61	3.45	254.18	−11.65
11	16.01	17.98	10.99	134.77	112.78	6479.39	233.95	3.46	254.18	−11.65
12	16.06	18.00	10.99	134.77	112.77	6428.60	233.65	3.46	254.18	−11.65
13	16.06	18.00	10.99	134.77	112.77	6421.32	233.62	3.46	254.18	−11.65
14	16.06	18.00	10.99	134.77	112.77	6420.34	233.61	3.46	254.18	−11.65

[a] Target parameters

$$X_{i2} = f_i^{-1}(O_i^*, GM_e)$$
$$O_{ho}^* = f_{ho}(X_{i2}, GM_e, t_i)$$

The outgoing hyperbola target parameters are reset to correspond to the desired injection flight plane coordinates. After several iterations, the solution converges to the target injection coordinates at Earth and the encounter parameters at the target planet.

An example of the design of an interplanetary trajectory is the Mars Odyssey spacecraft launched on April 7, 2001 and arrived at Mars on October 24, 2001. The flight time of 199.4 days is considerably less than the nominal Hohmann transfer time of about 258 days. The transfer angle of 140.2° is also considerably less than the 180° Hohmann transfer angle. The odyssey spacecraft was launched about 3 months later than the Hohmann optimum transfer time. Launches late in the launch period, surrounding the optimum transfer time of Jan 11, 2001, require additional launch energy to achieve the required injection conditions. However, the cost savings associated with shortening the flight time and consequently mission operations time may well offset the additional launch energy cost provided that the launch vehicle has the additional capability. Table 3.1 shows the result of implementing the interplanetary trajectory design procedure described above. Tabulated are the injection conditions at Earth and encounter conditions at Mars following each iteration cycle. The converged solution agrees reasonably well with the trajectory that was actually designed. The velocities at injection and at encounter agree within about 50 m/s.

3.3 Three-Body Trajectory

Gravity assist trajectories are an important class of trajectories that have been used by Voyager, Galileo, Cassini, and other missions to tour the solar system. The accessibility of a target planet, particularly those beyond the orbit of Jupiter, depends on finding a transfer orbit with energy relative to the Earth, within the capability of the launch vehicle. The use of gravity assist to increase the transfer orbit energy has opened up the exploration of planets that otherwise would not be accessible with current launch vehicle capability.

Most interplanetary and planetary orbiter mission trajectories, since the beginning of the space age, have used Keplerian two-body motion in their design. Missions requiring three-body transfers are generally limited to those involving the satellites of the major planets, for example, missions to Lagrange points in the Earth–Moon system. Even though gravity assist trajectories can be designed by repeated application of two-body theory, they are included in the three-body classification because the gravity assist requires a simultaneous exchange of energy among three bodies. The three-body theory employed for the design of gravity assist trajectories involves the use either of vectors defining the approach and departure hyperbolic asymptotes with respect to the gravity assist planet or of Tisserand's criterion which pertains to the interplanetary Keplerian orbits connecting the launch, gravity assist, and target planets. It will be shown that while both design techniques follow from the Jacobi integral, they yield significantly different results, since they represent different approximations of the true equations of motion.

3.3.1 Jacobi Integral

An important integral describing constraints on energy transfer for the restricted three-body problem was discovered by Carl Gustav Jacob Jacobi in the Nineteenth century. A point mass moving in the vicinity of two massive bodies in circular orbits about their barycenter will conserve a certain function of the state and gravitational parameters of the massive bodies referred to as Jacobi's integral. The constant of integration is called Jacobi's constant. The equations of motion for a spacecraft near two massive bodies are given by,

$$\begin{aligned}
\ddot{x} &= GM_1\frac{x_1 - x}{r_1^3} + GM_2\frac{x_2 - x}{r_2^3} \\
\ddot{y} &= GM_1\frac{y_1 - y}{r_1^3} + GM_2\frac{y_2 - y}{r_2^3} \\
\ddot{z} &= GM_1\frac{z_1 - z}{r_1^3} + GM_2\frac{z_2 - z}{r_2^3}
\end{aligned} \tag{3.49}$$

The two massive bodies rotate around the barycenter and the rotation rate is simply 2π divided by the period of the orbit,

$$\omega = \sqrt{\frac{GM_1 + GM_2}{\rho^3}} \tag{3.50}$$

where ρ is the distance separating the two massive bodies. The geometry is illustrated in Fig. 3.9. The primed coordinate system (x', y', z') represents a rotating coordinate system in which the two massive bodies lie on the x' axis, with

$$\begin{aligned} x &= x' \cos \omega t - y' \sin \omega t \\ y &= x' \sin \omega t + y' \cos \omega t \\ z &= z' \end{aligned} \tag{3.51}$$

After differentiating Eq. (3.51) twice and substituting into Eq. (3.49), the following result is obtained. The sine and cosine terms are elimination by first rotating Eq. (3.51) its alignment with the rotating frame.

$$\begin{aligned} \ddot{x}' - 2\omega\dot{y}' - \omega^2 x' &= -GM_1 \frac{x_1' - x'}{r_1^3} - GM_2 \frac{x_2' - x'}{r_2^3} \\ \ddot{y}' + 2\omega\dot{x}' - \omega^2 y' &= -\left(\frac{GM_1}{r_1^3} + \frac{GM_2}{r_2^3}\right) y' \\ \ddot{z}' &= -\left(\frac{GM_1}{r_1^3} + \frac{GM_2}{r_2^3}\right) z' \end{aligned} \tag{3.52}$$

Equation (3.52) may be put into a form that can be integrated by defining the function

$$U = \frac{1}{2}\omega^2 (x'^2 + y'^2) + \frac{GM_1}{r_1} + \frac{GM_2}{r_2} \tag{3.53}$$

and substituting into Eq. (3.52).

$$\begin{aligned} \dot{x}'\ddot{x}' - 2\omega\dot{x}'\dot{y}' &= \dot{x}' \frac{\partial U}{\partial x'} \\ \dot{y}'\ddot{y}' + 2\omega\dot{x}'\dot{y}' &= \dot{y}' \frac{\partial U}{\partial y'} \\ \dot{z}'\ddot{z}' &= \dot{z}' \frac{\partial U}{\partial z'} \end{aligned} \tag{3.54}$$

Adding Eqs. (3.54),

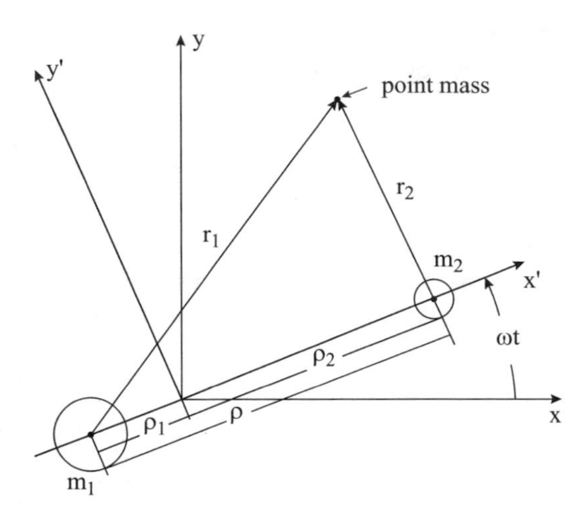

Fig. 3.9 Restricted two-body geometry

$$\dot{x}'\ddot{x}' + \dot{y}'\ddot{y}' + \dot{z}'\ddot{z}' = \dot{x}'\frac{\partial U}{\partial x'} + \dot{y}'\frac{\partial U}{\partial y'} + \dot{z}'\frac{\partial U}{\partial z'} = \frac{dU}{dt} \tag{3.55}$$

The integral of Eq. (3.55), called the Jacobi integral, is

$$\dot{x}'^2 + \dot{y}'^2 + \dot{z}'^2 = 2U - C$$

or

$$\dot{x}'^2 + \dot{y}'^2 + \dot{z}'^2 = \omega^2 x'^2 + \omega^2 y'^2 + 2\frac{GM_1}{r_1} + 2\frac{GM_2}{r_2} - C \tag{3.56}$$

where C is the constant of integration.

3.3.2 *Tisserand's Criterion*

Francois Felix Tisserand was a nineteenth century astronomer who discovered a unique application of Jacobi's integral to identify comets. In the restricted three-body problem, a certain function of the orbit elements before and after a planetary encounter is conserved. If this function is computed for two comet observations on different orbits and the results are the same, one may conclude that the observations are of the same comet and the comet has encountered a planet between the observations. This may be confirmed by propagating the orbits forward or backward in time to see if they encountered a planet.

In the application of Tisserand's criterion to gravity assist trajectory design, the procedure is reversed. Transfer trajectories from the launch planet to the intermediate planet and from the intermediate planet to the target planet are computed using

Lambert's theorem. These trajectories are matched based on Tisserand's criterion to identify viable launch and encounter opportunities. Tisserand's criterion follows directly from Jacobi's integral. The Jacobi integral is transformed back to inertial coordinates (the unprimed coordinates in Fig. 3.9).

$$\dot{x}^2 + \dot{y}^2 + \dot{z}^2 = 2\omega(x\dot{y} - y\dot{x}) + 2\frac{GM_1}{r_1} + 2\frac{GM_2}{r_2} - C \tag{3.57}$$

For GM_1 much greater than GM_2, the z component of the angular momentum vector is given by,

$$x\dot{y} - y\dot{x} = h_z = h\cos i \tag{3.58}$$

$$h = \sqrt{GM_1\, a(1-e^2)}$$

and from the *vis viva* integral the energy is given by,

$$\dot{x}^2 + \dot{y}^2 + \dot{z}^2 = GM_1\left(\frac{2}{r_1} - \frac{1}{a}\right) \tag{3.59}$$

Substituting Eqs. (3.58) and (3.59) into Eq. (3.57) gives

$$GM_1\left(\frac{2}{r_1} - \frac{1}{a}\right) - 2\omega\sqrt{GM_1\, a(1-e^2)}\cos i = 2\frac{GM_1}{r_1} + 2\frac{GM_2}{r_2} - C$$

Substituting Eq. (3.50) for ω and for small GM_2 compared with GM_1,

$$C \approx \frac{GM_1}{a} + 2GM_1\sqrt{\frac{a(1-e^2)}{\rho^3}}\cos i \tag{3.60}$$

In the literature, Tisserand's criterion is often developed in dimensionless coordinates and the Jacobi constant modified to remove constant parameters. If a is divided by ρ to define $\bar{a}$ and Eq. (3.60) is multiplied through by ρ and divided by GM_1, Tisserand's criterion in dimensionless coordinates becomes

$$\frac{C\rho}{GM_1} \approx \frac{1}{\bar{a}} + 2\sqrt{\bar{a}(1-e^2)}\cos i$$

If the first observation of a spacecraft or comet has orbit elements a_1, e_1, and i_1 and the second observation after a planetary encounter has orbit elements a_2, e_2, and i_2, then

$$\frac{1}{a_1} + 2\sqrt{\frac{a_1(1-e_1^2)}{\rho^3}}\cos i_1 \approx$$

$$\frac{1}{a_2} + 2\sqrt{\frac{a_2(1-e_2^2)}{\rho^3}}\cos i_2$$

3.3.3 Gravity Assist Vector Diagram

Figure 3.10 shows the encounter geometry in the vicinity of the intermediate planet that supplies the gravity assist energy boost to the spacecraft. The incoming velocity of the spacecraft ($\mathbf{V}_1$) is subtracted from the planet velocity ($\mathbf{V}_p$) to obtain the planet relative approach velocity ($\mathbf{v}_i$) as shown in the upper vector diagram in Fig. 3.10. The lower vector diagram shows the same relationship for the outgoing velocity vectors. If the incoming and outgoing velocities are computed far from the planet yet close enough to the planet that the heliocentric energy may be assumed constant, the velocities $\mathbf{v}_i$ and $\mathbf{v}_o$ are approximately the $\mathbf{v}_\infty$ vectors associated with the two-body hyperbola about the planet. In the limit of two-body motion assumed for patched conic trajectories, $\mathbf{v}_i$ and $\mathbf{v}_o$ are equal in magnitude. Since the planet velocity is also assumed to be constant during the relatively short-time interval of the planet encounter, the outgoing vector diagram may be superimposed on the incoming vector diagram as shown in Fig. 3.10. The outgoing heliocentric spacecraft velocity magnitude is greater than the incoming velocity magnitude and the spacecraft has acquired additional orbit energy relative to the Sun. The energy acquired by the spacecraft comes from the Sun and the planet.

Consider the triangle formed by the spacecraft and planet heliocentric velocity vectors and the incoming velocity vector. From the law of cosines,

$$v_i^2 = V_p^2 + V_1^2 - 2V_p V_1 \cos A \tag{3.61}$$

The orbit of the planet about the Sun may be approximated by a circle with velocity magnitude given by,

$$V_p = \sqrt{\frac{GM_s}{\rho}}$$

The heliocentric orbit of the spacecraft may be regarded as a two-body conic. The velocity magnitude is given by

$$V_1 = \sqrt{\frac{2GM_s}{\rho} - \frac{GM_s}{a_1}}$$

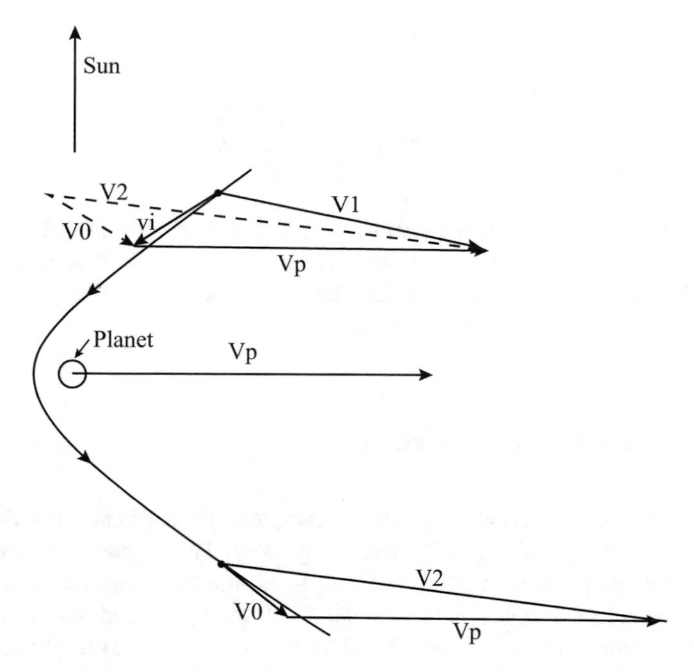

Fig. 3.10 Gravity assist vector diagram

In the plane of the orbit, the angle A is simply the flight path angle (γ). For the general case, the angle A is a function of γ and the inclination of the spacecraft orbit plane with respect to the planet orbit plane i_1 and

$$\cos A = \cos\gamma \cos i_1$$

$$\cos\gamma = \frac{\sqrt{GM_s a_1(1-e_1^2)}}{V_1 r_1}$$

Making these substitutions into Eq. (3.61) gives

$$v_i^2 = \frac{2GM_s}{\rho} - \frac{GM_s}{a_1} + \frac{GM_s}{\rho} -$$

$$2V_1\sqrt{\frac{GM_s}{\rho}}\sqrt{\frac{GM_s a_1(1-e_1^2)}{\rho^2 V_1^2}} \cos i_1$$

The energy of the spacecraft relative to the planet, the potential energy of the spacecraft relative to the Sun and the velocity of the planet relative to the sun may be regarded as constant. Collecting these "constant" terms on the left side gives

$$C \approx \frac{3GM_s}{\rho} - v_i^2$$

$$\approx \frac{GM_s}{a_1} + 2GM_s \sqrt{\frac{a_1(1-e_1^2)}{\rho^3}} \cos i_1 \tag{3.62}$$

Equation (3.62) provides an interesting insight into the geometrical meaning of Jacobi's constant in the limit where one of the gravitating bodies is much more massive than the other. The terms in the Jacobi integral are related to the velocity vector diagram of the participating bodies.

3.3.4 Cassini Trajectory Design

The Cassini mission to Saturn provides an example of the application of Tisserand's criterion to the design of a gravity assist trajectory. The segments of the Cassini trajectory that are of interest are from Earth to Jupiter and from Jupiter to Saturn. The first step is to determine the encounter times at Jupiter and Saturn. An initial guess of the encounter times of Jupiter and Saturn is made based on the approximate flight times associated with a Hohmann transfer. Point-to-point conic solutions for the trajectory segments from Earth to Jupiter and from Jupiter to Saturn are computed using the solution of Lambert's theorem. A point to point conic solution assumes zero mass for the planets and only the gravity of the sun is included. The solution of Lambert's theorem gives the two-body conic connecting two position vectors where the flight time is known. The two position vectors are obtained from the planetary ephemerides and the conic trajectory is computed from planet center to planet center as shown in Fig. 3.11.

The next step is to compute the velocity vectors relative to Jupiter, one for the incoming trajectory segment ($\mathbf{v}_i$) and one for the outgoing trajectory ($\mathbf{v}_o$). If the Jacobi constants for the trajectory segments do not match, then the following procedure can be used to find potentially viable encounter time solutions. The encounter time of Jupiter is fixed and the encounter times of Earth and Saturn are permitted to vary over a suitable range of times. For each pair of Earth–Jupiter encounter times and Jupiter–Saturn encounter times, a Lambert solution is computed and the Jacobi constant is computed from the orbit elements. The Jacobi constants for the two interplanetary trajectory legs are matched and the results are cross plotted in Fig. 3.12. Several approximations may be used for computing the Jacobi constant. Results for Tisserand's criterion and the Jupiter energy criterion are shown in Fig. 3.12 as dashed lines. The Jupiter energy criterion (Eq. 3.62) is equivalent to matching the incoming and outgoing velocity magnitudes relative to Jupiter. A criterion is used that matches the average of Tisserand's criterion and the Jupiter energy criterion and is shown in Fig. 3.12 as the solid line. The equation for this criterion, after simplification to remove constant parameters, is given by,

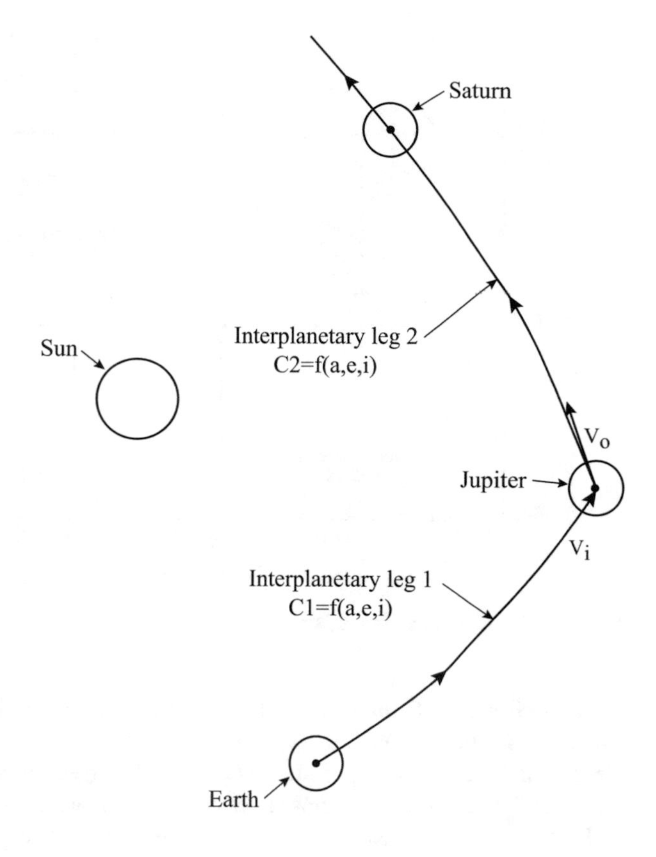

Fig. 3.11 Earth–Jupiter–Saturn encounter

$$\frac{1}{a_1} + 2\sqrt{\frac{a_1(1-e_1^2)}{\rho^3}}\cos i_1 - \frac{v_i^2}{GM_s}$$
$$\approx \frac{1}{a_2} + 2\sqrt{\frac{a_2(1-e_2^2)}{\rho^3}}\cos i_2 - \frac{v_o^2}{GM_s} \tag{3.63}$$

For a given pair of Earth launch and Saturn encounter times indicated in Fig. 3.12, the approach and departure velocity vectors at Jupiter are obtained and the hyperbolic conic relative to Jupiter is computed. A preliminary assessment of the viability of the Jupiter centered hyperbola is performed. A trajectory that intersected the surface of Jupiter, for example, or hits one of Jupiter's satellites would not be viable. Next, the encounter conditions at Earth and Saturn are examined for viability. If the energy at Earth or Saturn is unacceptable, the trajectory is not viable. If a viable trajectory is not found for all the launch date encounter date pairs indicated by Fig. 3.12, the above procedure is repeated for another Jupiter encounter time.

Once a viable set of encounter times has been determined, a patched conic trajectory is designed that connects Earth, Jupiter, and Saturn. The procedure involves computing the approach and departure velocity vectors at the patch points

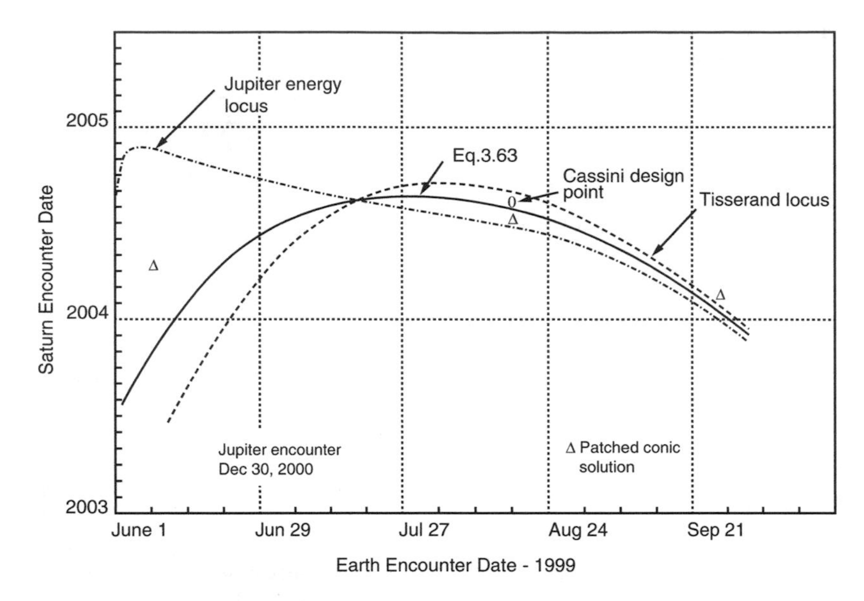

Fig. 3.12 Earth–Jupiter–Saturn loci

shown in Fig. 3.13 from the point-to-point conic solution. The two-body hyperbolic trajectory is then computed with respect to each of the participating planets. For Earth and Saturn, the departure and approach target plane positions are given. A new set of patch point positions relative to the Sun are computed. The states relative to Earth, Jupiter, and Saturn are added to the respective planetary ephemerides at the appropriate times. The patch point times are selected such that the spacecraft position is near the sphere-of-influence of the planets. The planetary ephemerides may be computed from two-body orbit elements with respect to the Sun. This procedure is repeated several times for the new patch points until a ballistic trajectory is obtained from Earth to Saturn. It will be necessary to allow the Saturn encounter time to vary a small amount from the point-to-point solution. The results, shown in Fig. 3.12 for three launch dates, compare favorably with the point-to-point solutions. Also, the Cassini design point, obtained by numerical integration, is shown in Fig. 3.12 for comparison.

The patched conic solution is used as a starting point for targeting an integrated trajectory. A comparison of the Cassini integrated trajectory and the patched conic solution is shown in Fig. 3.14. State vectors are computed from the patched conic trajectory and differenced with state vectors obtained from the integrated Cassini ephemeris. The magnitude of the position difference is plotted as a function of time and the heliocentric range of the spacecraft is also plotted for comparison. The maximum error is less than one percent of the heliocentric range. Since the period of the Saturn orbit is 29 years, an error of several months in the predicted encounter time at Saturn from the point-to-point conic solutions should be expected. This error in computing the encounter times is exacerbated by accelerations from the

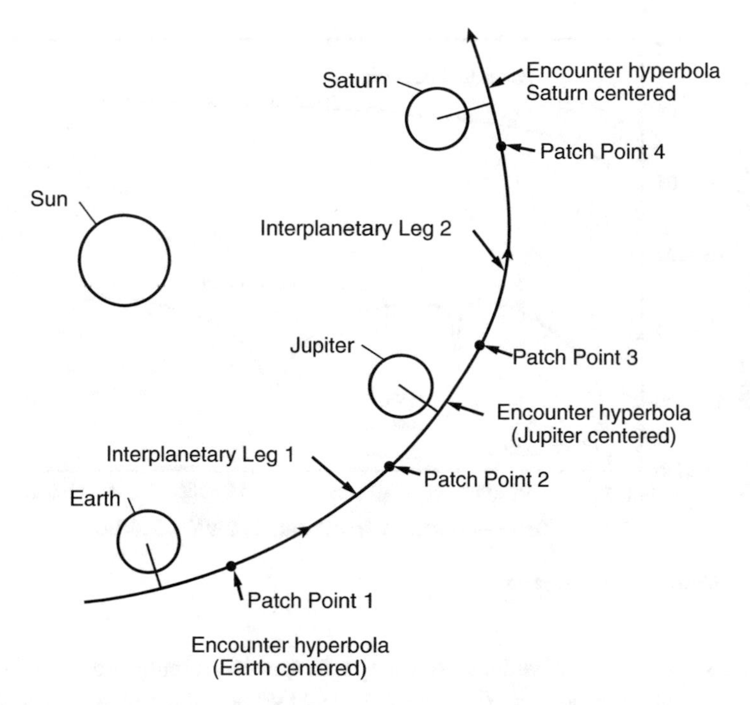

Fig. 3.13 Earth–Jupiter–Saturn encounter

third body that has been ignored for the two-body computations. However, a design error of only one percent enables a fairly accurate assessment of mission design constraints from the conic solution.

3.4 Four-Body Trajectory

An early investigation of flight to the Moon by V. A. Egorov in 1958 identified several problems relating to the design and navigation of translunar trajectories. These included hitting the Moon, circumnavigation of the Moon with a return to Earth at a flat entry angle, using the Moon's gravity for assist in reaching the planets, and the possibility of the Moon capturing a projectile launched from the Earth. Based on consideration of the three-body problem and its associated Jacoby integral, solutions can be demonstrated for these problems with the exception of the Moon capturing a projectile launched from Earth. For the problem of lunar capture, Egorov concluded that the Moon could not possibly capture a projectile launched from the Earth on the first circuit of the trajectory no matter what initial conditions are specified. This conclusion was based on analysis of the three-body problem and did not consider the Sun's gravity.

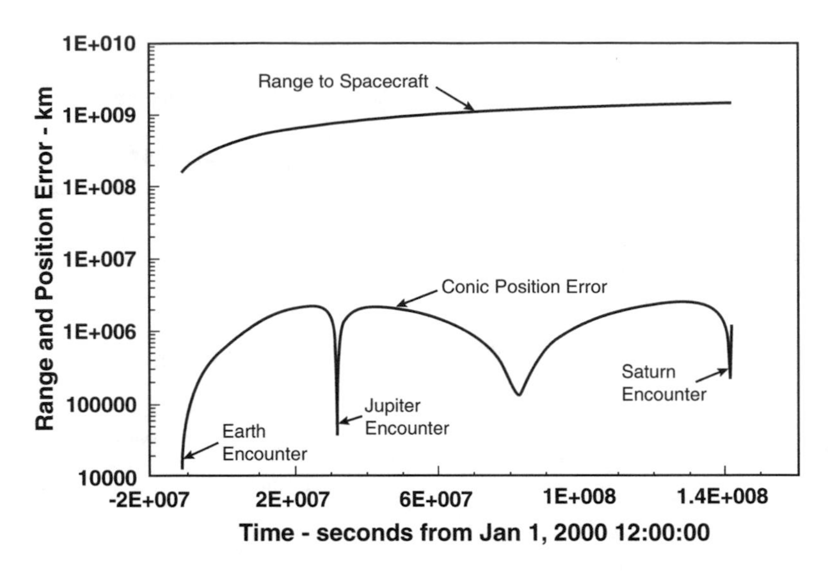

Fig. 3.14 Cassini conic design error

The first example of a ballistic trajectory of a spacecraft launched from the Earth into orbit about the Moon was discovered in 1990 while analyzing a plan to salvage the Muses A (Hiten) spacecraft in a highly eccentric orbit about the Earth. The key to the discovery was the utilization of the Sun's gravity to affect the transfer to a lunar capture orbit. The result was a numerical solution to the restricted four-body problem of the Earth, Moon, Sun, and a point mass spacecraft. Lunar transfer trajectories that require analysis that goes beyond that provided by three-body theory and the Jacobi integral are referred to as four-body trajectories. Examples are Hiten, Lunar A, and the Genesis return trajectory from the vicinity of the Moon to Earth. These trajectory designs cannot be fully explained or analyzed using three-body theory and the Jacobi integral. As is the case for the three-body problem, a complete analytic solution of the four-body problem has not been obtained. Furthermore, an integral relationship similar to the Jacobi integral has not been found for the four-body problem and the prospects for finding such an integral are dim. Current theories, such as Weak Stability Theory, are explanatory and not predictive and thus cannot be used for design of trajectories that require a simultaneous four-body solution without some intervention by the trajectory designer.

In the absence of a predictive four-body theory, the trajectory designer may use the existing solution of the two-body problem and the Jacobi integral to piece together trajectory segments and achieve the desired result. Indeed, most lunar transfer trajectory designs are obtained by patching together conic orbits where the Earth's gravity dominates to conic orbits where the Moon's gravity dominates. By extension, the trajectory segment dominated by the Earth, Moon, and spacecraft Jacobi integral may be pieced together with the trajectory segment dominated by the

Sun, Earth, and spacecraft Jacobi integral to obtain continuous ballistic trajectories that connect Earth departure or arrival with capture orbits about the Moon and the nearby Lagrange points.

3.4.1 Moon Capture of Projectile Launched from Earth

A spacecraft in a lunar capture orbit will approach the Moon in a nearly circular orbit about the Earth that is just inside the Moon's orbit or just outside the Moon's orbit. As the spacecraft approaches the Moon, the Moon's gravity provides the necessary acceleration to slow down or speed up the spacecraft depending on whether the approach orbit is inside or outside the Moon's orbit. If the spacecraft has just the right approach velocity, it is drawn into orbit about the Moon. The orbital mechanics of capture orbits are well documented in the literature. A spacecraft is placed in an orbit that is loosely bound to the Moon and whose semimajor axis is just inside the Moon's sphere of influence. The lunar periapsis is directed toward the Earth and apoapsis is therefore directed away from the Earth. The orbit is integrated for several revolutions about the Moon and if it remains captured the apoapsis altitude is raised slightly. A convenient orbit parameter for raising apoapsis is the eccentricity which will tend to keep the energy of the orbit about the Moon constant. After several trys, the spacecraft will escape from the Moon and enter into an orbit about the Earth. Since the equations of motion are reversible, a capture trajectory can be obtained by repeating the above procedure only integrating the equations of motion backward.

The resulting capture orbits are generally nearly circular about the Earth and either inside or outside the Moon's orbit. For a critical value of the starting eccentricity of the orbit about the Moon, the spacecraft will just escape the Earth–Moon system and go into orbit about the Sun. Raising the eccentricity slightly will result in an eccentric orbit with a periapsis radius relative to the Earth that is inside the Moon's orbit. The results of generating several capture orbits are shown in Fig. 3.15. For a range of starting eccentricities from 0.94151 to 0.943, most of the capture orbits either escape from the Earth–Moon system or fall into an uninteresting eccentric orbit about the Earth with periapsis radius less than that of the Moon's orbit. This behavior of capture orbits including the reduction in periapsis radius with respect to the Earth has been observed and is common knowledge.

A remarkable result was obtained during study of the Hiten trajectory during Memorial day weekend of 1990. If the starting eccentricity of the Moon's orbit was adjusted to 0.94171, the spacecraft falls into a highly eccentric orbit that returns to Earth as shown in Fig. 3.15. This was a surprising result, but Ed Belbruno was not surprised because of his work with weak stability theory. Previous studies of the possibility of the Moon capturing a projectile launched from the Earth indicated that this result was very improbable. Analysis by Fesenkov based on the Jacobi integral concluded that this result was impossible. Egorov introduced a term not considered by Fesenkov that opened the possibility of capture after more than one circuit. He acknowledged that the Sun may provide a perturbation that could enable capture.

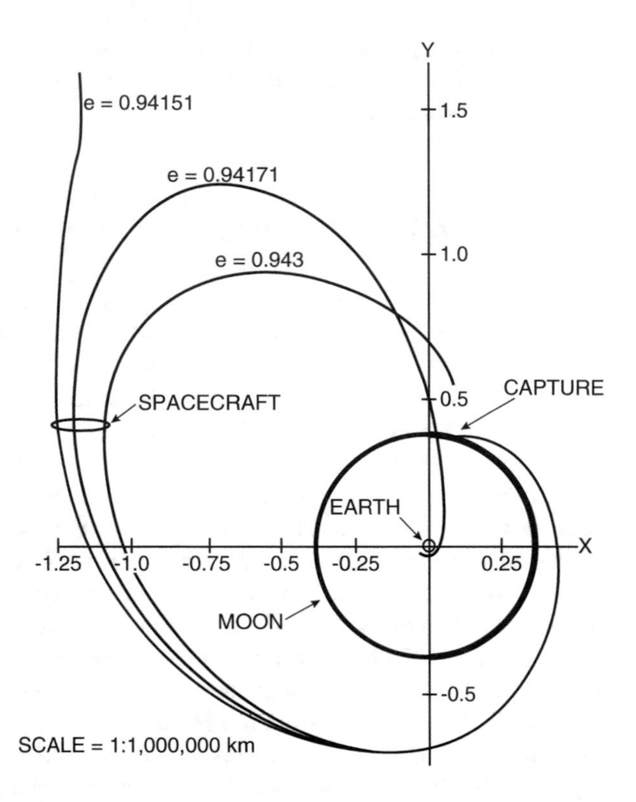

Fig. 3.15 Examples of lunar capture orbits

The discovery of a capture orbit was not made by systematically perturbing eccentricity, as suggested by Fig. 3.15, until the result was observed. The approach used in finding this orbit was from a different direction. In attempting to design a trajectory for the Hiten spacecraft to get to the Moon, a bielliptic transfer was attempted. The idea was to design a capture orbit that escapes from the Earth–Moon system and intersects a direct trajectory from Earth orbit. The capture trajectory was integrated backward and the trajectory from Earth orbit was integrated forward. At the intersection, a maneuver was performed to join the two trajectory segments. It was soon discovered that an escape trajectory would not work. The velocity correction required at the intersection was too big. It was also observed that the minimum velocity at the intersection point near the boundary of escape to orbit about the sun was about 250 m/s. While fine-tuning the eccentricity of the capture orbit, it was observed that the velocity change began to drop. It became apparent that the minimum velocity change was zero and the result was an orbit similar to the orbit shown in Fig. 3.15.

Attempts to extend the result shown in Fig. 3.15 to other initial orbit conditions revealed a strong dependence on the location of the Sun relative to the Earth and Moon. Clearly, the tidal acceleration of the Sun was the vehicle for transforming a nearly circular orbit coincident with the Moons orbit into a highly eccentric orbit that intersects the Earth. The affect of the Sun on the transfer trajectory can be seen

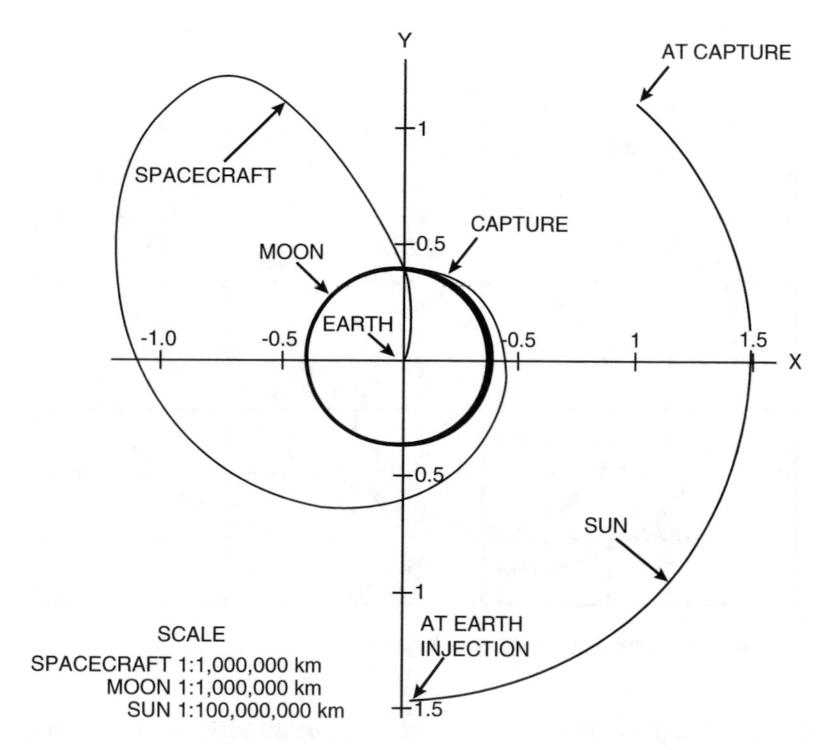

Fig. 3.16 Earth–Moon ballistic transfer

from inspection of Fig. 3.16. Shown is an Earth to Moon ballistic transfer trajectory with the orbit of the Sun in Earth-centered inertial coordinates superimposed. The Sun's orbit has been reduced by a scale factor of 100. As the backward integrated trajectory spirals outward from the Moon, the Sun is on the opposite side of the Earth from the spacecraft. The spacecraft is in the second quadrant near apoapsis while the Sun is in the fourth quadrant. The net effect of the solar tide is to reduce the angular momentum sufficiently to lower periapsis radius to the radius of the Earth. Reversing the direction of integration gives the desired lunar capture trajectory.

The lunar transfer trajectory from the Earth's surface to capture by the Moon may be modified slightly to enable transfer from a variety of Earth orbits to lunar capture. Also, Fig. 3.16 suggests that capture orbits may be designed to escape from the Earth to the Sun–Earth Lagrange points. With a little imagination, these capture orbits may be pieced together with the Earth transfer trajectory to design orbits that go from near Earth orbit to the Lagrange points briefly capturing the Moon along the way. An example of a lunar capture transfer trajectory with modified initial conditions near Earth orbit is shown in Fig. 3.17. The spacecraft is launched into an elliptic staging orbit about the Earth with apoapsis radius that reaches the orbit of the Moon. The spacecraft remains in the staging orbit until the Sun is in the right position for a lunar capture orbit. The spacecraft is timed to arrive at the Moon for a

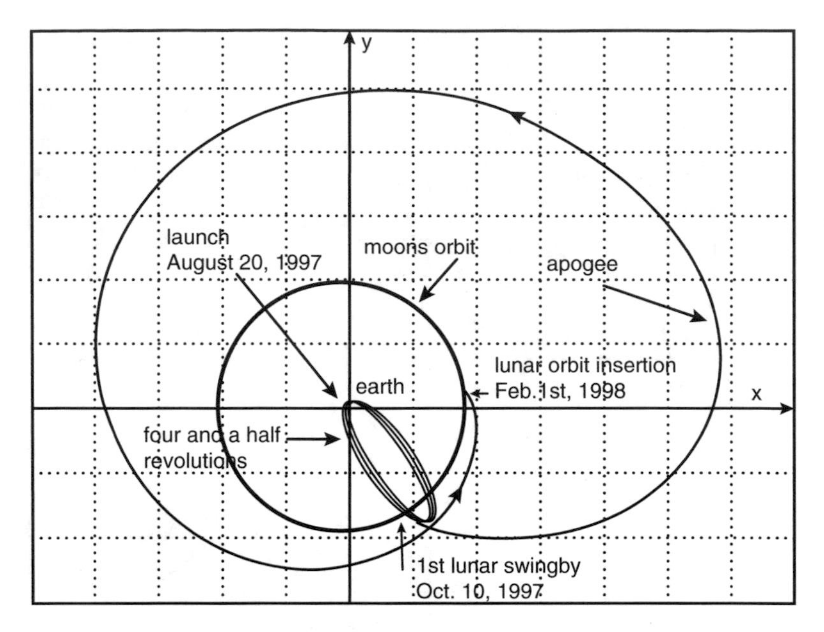

Fig. 3.17 Proposed Lunar A Mission transfer trajectory

gravity assist that places the spacecraft on the capture orbit. Figure 3.17 displays the characteristic kidney shape often associated with the 3-month variety of the capture orbit.

3.4.2 *Angular Momentum and Energy Management*

An important tool for design of lunar capture orbits is the management of angular momentum. Raising of the periapsis radius of the Earth-centered orbit from near the surface of the Earth to the radius of the Moon's orbit requires the addition of angular momentum to the orbit. This requires placing the spacecraft in a region of space where the angular momentum rate of increase from the solar tide can raise the angular momentum to that required for capture. The energy and angular momentum management is accomplished by starting from a lunar capture orbit with the correct angular momentum and energy (approximately the same as the Moon) and integrating backward to a region of space where the angular momentum is reduced to a small enough value to intersect the Earth's surface. The spacecraft then falls back to the Earth and the energy required is supplied by the launch vehicle when the direction of integration is reversed.

The angular momentum of the orbit relative to the Earth is given by

$$h = \sqrt{p\,GM_e} \tag{3.64}$$

where p is the parameter of orbit and GM_e is the Earth's gravitational constant. For a spacecraft launched from the Earth that nearly escapes the Earth–Sun system, the orbit is nearly parabolic and p is approximately twice the radius of the Earth (12,000 km). At lunar capture, p is approximately the radius of the Moon's orbit (384,000 km). Equation (3.64) requires raising the angular momentum (h) from 69,000 km^2/s to 390,000 km^2/s, a net increase of 321,000 km^2/s.

The angular momentum orbit parameter (h) is the magnitude of the angular momentum vector given by,

$$\mathbf{h} = \mathbf{r} \times \mathbf{v}$$

Consider an Earth-centered rotating coordinate system with the x axis pointing at the Sun and the z axis in the direction of the orbit angular momentum vector. The geometry is shown in Fig. 3.18. Neglecting the rotation about the Sun and the tidal acceleration of the Moon, the rate of change of angular momentum is given by

$$\dot{\mathbf{h}} = \dot{\mathbf{r}} \times \mathbf{v} + \mathbf{r} \times \dot{\mathbf{v}} \tag{3.65}$$

where

$$\mathbf{r} = (x, y, 0)$$
$$\dot{\mathbf{r}} = \mathbf{v} = (\dot{x}, \dot{y}, \dot{z})$$
$$\dot{\mathbf{v}} = (a_x, 0, 0)$$

Carrying out the indicated substitutions, the angular momentum rate is approximately

$$\dot{h} = -y\, a_x \tag{3.66}$$

The tidal acceleration (a_x) is approximately in the x direction since the Sun is far from the Earth at the scale shown in Fig. 3.18. The tidal acceleration of the Sun is simply the difference between the acceleration of the spacecraft and the acceleration of the Earth caused by the Sun's gravity and

$$a_x = \frac{GM_s}{(r_s + x)^2} - \frac{GM_s}{r_s^2}$$

which may be approximated by

$$a_x = a_x(x = 0) + \frac{da_x}{dr_s}\,\delta x$$
$$a_x = \frac{2GM_s}{r_s^3}\, x$$

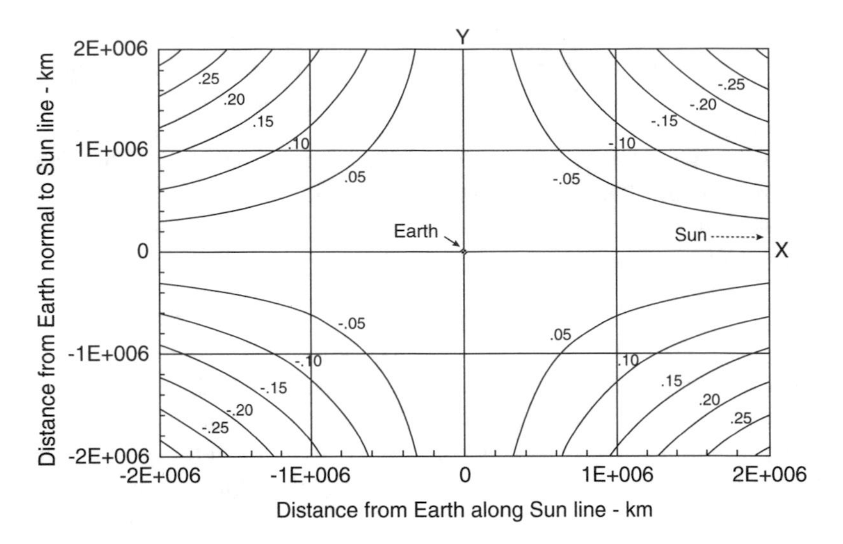

Fig. 3.18 Sun–Earth angular momentum transfer contours

Substituting into the equation for angular momentum rate (Eq. 3.66) yields

$$\dot{h} = -\frac{2GM_s}{r_s^3}\, x\, y \tag{3.67}$$

Equation (3.67) is the equation for a hyperbola as a function of x and y. A family of hyperbolas are plotted in Fig. 3.18 for various values of the angular momentum rate in the units of km^2/s^2. As an example of the application of the angular momentum contours, consider a spacecraft launched from Earth into the second or fourth quadrant of Fig. 3.18 where the angular momentum rate attributable to the solar tide is positive. At coordinate $x = 1,400,000\,\text{km}$ and $y = -750,000\,\text{km}$, the angular momentum rate of increase is $0.084\,\text{km}^2/\text{s}^2$. In order to raise the periapsis radius from the Earth surface to the radius of the Moon's orbit, an increase in angular momentum of $321{,}000\,\text{km}^2/\text{s}$ is required. Thus, the spacecraft would need to dwell near the indicated coordinates for 3,821,000 s or about 44 days. The actual time required to achieve the required angular momentum increase can be obtained by performing a line integral along the actual flight path and include the tidal acceleration of the Moon. For an actual trajectory integration, the average value of the angular momentum rate would be about half the value used in this example and the Moon's tidal acceleration contribution would be small. The total flight time is therefore approximately 90 days.

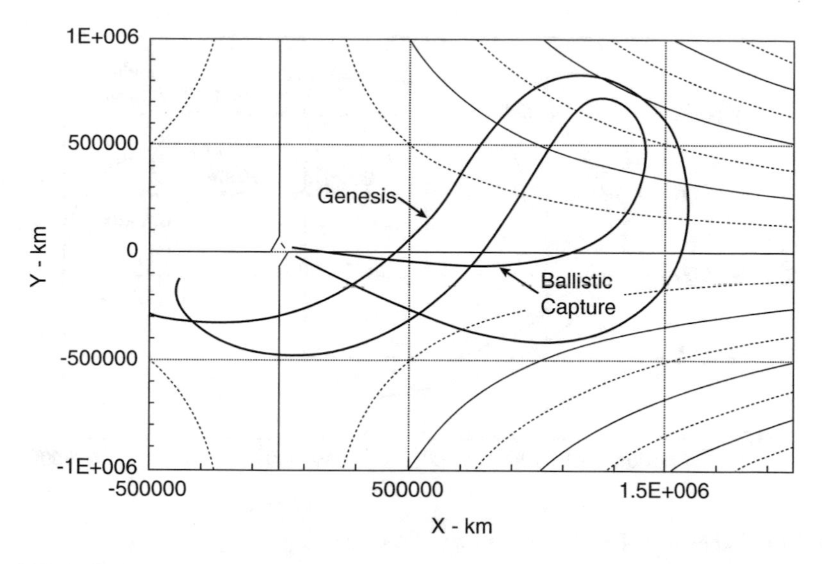

Fig. 3.19 Ballistic capture and angular momentum contours

3.4.3 Genesis Earth Return Trajectory

The Genesis return trajectory starts from a Lagrange point and flies by the orbit of the Moon on a trajectory that is nearly captured and proceeds on a transfer orbit to the Earth. The portion of the orbit from near the Moon's orbit to the Earth is an example of a four-body transfer. The Genesis return trajectory is plotted in Fig. 3.19 along with the ballistic capture trajectory. The coordinate frame is the same as shown in Fig. 3.18 with the Earth at the center and the Sun in the $+x$ direction. Both trajectories go from the vicinity of the Moon's orbit to the Earth. In the rotating coordinate system, both trajectories execute a slow loop in the first quadrant where the maximum rate of angular momentum removal is about $0.1\,\text{km}^2/\text{s}^2$ as indicated by the hyperbolic contours shown in Fig. 3.18. The Genesis trajectory experiences a higher rate of angular momentum removal in the first quadrant which is partially restored in the fourth quadrant where the sign changes to positive. The total angular momentum removal is about the same for both trajectories which is characteristic of the four-body transfer.

Since the trajectories shown in Fig. 3.19 are initiated at different times, the position of the Moon in its orbit relative to the Genesis trajectory is not clear. The ballistic capture orbit originates at the Moon. The Genesis trajectory in the vicinity of the Moon's orbit comes under significant influence of the Moon's gravity. The boundary between domination by Earth–Sun gravity and domination by Earth–Moon gravity is a region of space that has been referred to as the Weak Stability Boundary.

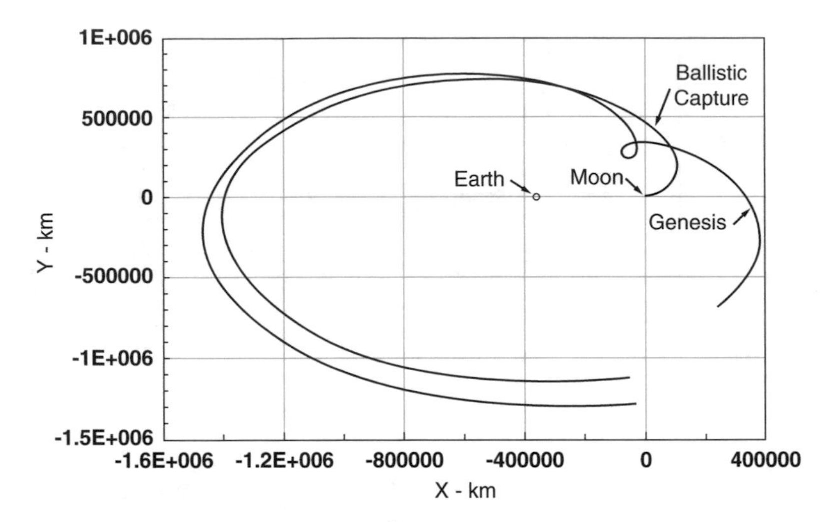

Fig. 3.20 Comparison of genesis orbit with ballistic capture

In order to gain some insight into the behavior of the trajectory dynamics near the Moon's orbit, a coordinate transformation is performed to a rotating coordinate system centered at the Moon with the $+x$ axis in the direction of the Earth–Moon vector and the z axis in the direction of the angular momentum vector. The Genesis return trajectory and the ballistic capture orbit[4] are plotted in this coordinate system as shown in Fig. 3.20. The departure from the vicinity of the Moon of the two trajectories are essentially the same. The motion near the Moon requires further investigation. The ballistic capture orbit enters into a close capture orbit of the Moon and the Genesis trajectory comes within 300,000 km of the Moon and executes a strange loop.

3.4.4 *Jacobi Integral and Capture*

When the spacecraft comes close to the Moon, the tidal perturbation from the Sun is small compared to the perturbations from the Moon and Earth. In this region of space, the trajectory may be analyzed using restricted three-body theory. In the rotating primed coordinates, a certain integral relating to the energy of the point mass, referred to as the Jacobi Integral, is constant. The Jacobi integral in the rotating coordinate frame is given by

$$\dot{x}'^2 + \dot{y}'^2 + \dot{z}'^2 = \omega^2 x'^2 + \omega^2 y'^2 + 2\frac{GM_1}{r_1} + 2\frac{GM_2}{r_2} - C \tag{3.68}$$

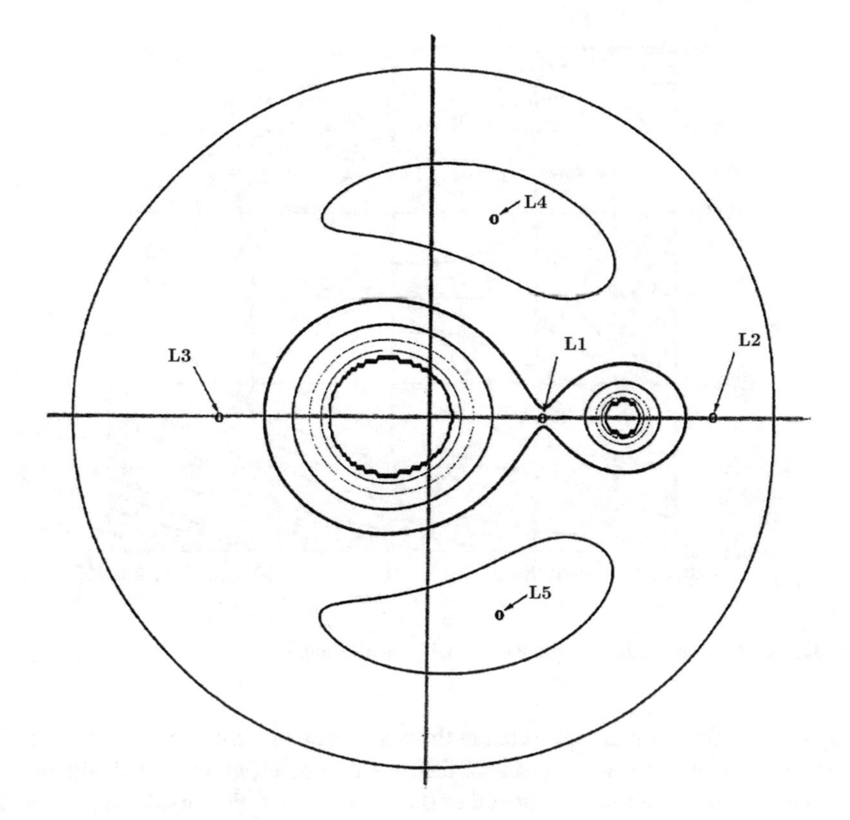

Fig. 3.21 Jacobi zero velocity contours

Consider a point mass or spacecraft moving with zero velocity relative to the massive bodies. In inertial space, the rotation of both massive bodies about each other is ω. For zero velocity relative to the massive bodies, the Jacoby integral reduces to

$$\omega^2 x'^2 + \omega^2 y'^2 + 2\frac{GM_1}{r_1} + 2\frac{GM_2}{r_2} = C \tag{3.69}$$

A spacecraft moving with velocity or kinetic energy that is small compared to the gravitational potential energy will tend to move in a direction that keeps C constant. Thus, contours of constant C will describe the motion in the rotating coordinate frame. Contours of constant C, referred to as Jacobi zero velocity contours or Hill's surfaces, may be plotted in rotating coordinates as shown in Fig. 3.21. The familiar zero velocity contours are for two massive bodies that are of the same order of magnitude in mass. The five stable Lagrange points are labeled as L1 through L5. A spacecraft placed at one of the stable Lagrange points will stay there unless perturbed by some external force. The zero velocity contours suggest other stable

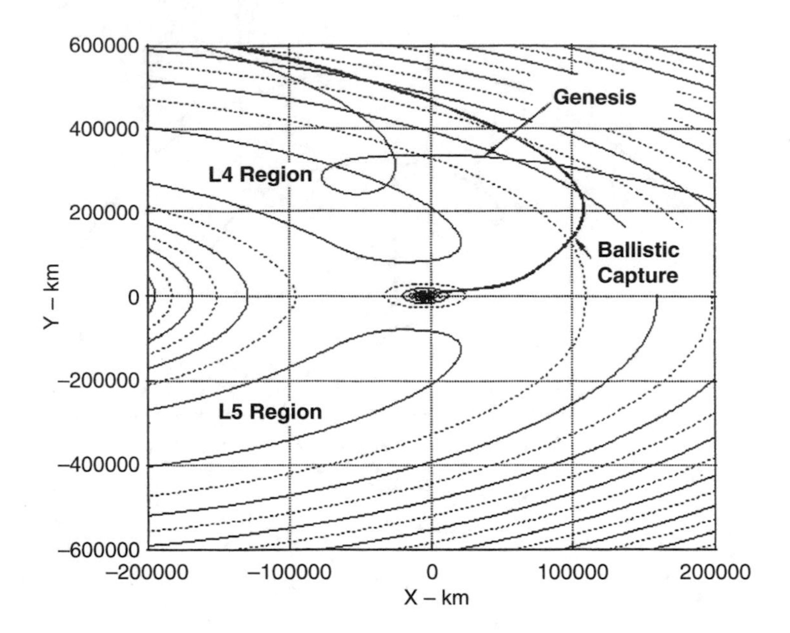

Fig. 3.22 Genesis and ballistic capture orbits with Jacobi contours

trajectories such as circular orbits about the massive bodies, a circular orbit about the center of mass and outside the orbits of the massive bodies, and circumnavigation of one massive body and return to the other on a free return trajectory. It also appeared to Fesenkov and Egorov that a direct trajectory from one body to a close orbit about the other would not be possible because of the stricture near L1.

The zero velocity contours for the Earth–Moon system are highly distorted from the contours shown in Fig. 3.21. Since the Earth is about 80 times more massive than the moon, the teardrop regions around L4 and L5 encircle the Earth and are joined through L3. A spacecraft in orbit near L4 or L5 can migrate back and forth between L4 and L5 through L3 without encountering the Moon. A trajectory of an asteroid in the Earth–Sun system has been recently discovered that exhibits this motion.

The actual zero velocity contours for the Earth–Moon system in the vicinity of the Moon are shown in Fig. 3.22. Also, plotted are the Genesis return trajectory and a ballistic capture orbit. As both orbits approach zero velocity relative to the Earth–Moon rotating frame, they fall onto the same Jacobi contour indicating that the orbits have essentially the same Jacobi constant or energy. The bifurcation that separates the two trajectories on departure from the Moon's orbit is a property of chaotic trajectories.

3.5 NEAR Orbit Phase Trajectory Design

Trajectory design of the orbit phase of the Near Earth Asteroid Rendezvous (NEAR) mission involves procedures that depart significantly from those used for previous missions. On previous missions, the trajectory design involved finding a flight path that satisfied a rigid set of spacecraft and mission design constraints. A precise spacecraft trajectory was designed well in advance of arrival at the target body. For NEAR, the uncertainty in the dynamic environment did not permit a precise spacecraft trajectory to be defined in advance of arrival at Eros. The principal cause of this uncertainty is limited knowledge of the gravity field and rotational state of Eros. As a result, the concept for NEAR trajectory design was to define a number of rules for satisfying spacecraft and mission constraints and to apply these rules to various assumptions for the model of Eros.

3.5.1 Spacecraft and Mission Constraints

The spacecraft constraints that apply to Eros trajectory design include limits on fuel consumption, solar panel illumination, and momentum wheel management. Other constraints define the flexibility and speed with which mission operations may be conducted. Probably the most important spacecraft constraint is to perform the prime mission within the allocated propellant budget. The propellant consumption constraint translates into about a 50–100 m/s delta velocity change during the orbit phase of the primary mission. This was a fairly generous allocation and was not difficult to satisfy.

The most difficult spacecraft constraint to satisfy relates to solar panel illumination. Since the science instruments are fixed with respect to the spacecraft body, it is necessary to tum the spacecraft to point these instruments at Eros. In order to satisfy spacecraft power requirements, the solar panels cannot be turned more than about 30° off the Sun-line. If the angle between the line to nadir and the plane perpendicular to the Sun-line is greater than 30°, the nadir point cannot be imaged without turning the spacecraft more than 30°. A coordinate frame is defined with the z axis pointing away from the Sun and the equatorial or x-y plane perpendicular to the Sun-line and is referred to as the Sun Plane-Of-Sky (POS) coordinate frame. In the Sun POS coordinate frame, orbits with inclination less than 30° direct or greater than 150° retrograde will not violate the solar panel constraint permitting imaging of nadir from any point in the orbit.

Another important constraint relates to the time to conduct mission operations. In order to conduct the mission smoothly without resorting to round the clock operations, the minimum time between spacecraft propulsive maneuvers is limited to 1 week. The real limitation is the time to compute accurate orbit determination solutions in support of propulsive maneuvers that are required to keep the spacecraft on course. The differential velocity change resulting from maneuver execution errors corrupts the orbit solution. A rapid redetermination of the orbit places a large amount of pressure on the Mission Operations team to deliver accurate data and process this data into reliable solutions for the spacecraft orbit. By allowing a minimum of 1 week between maneuvers, this pressure is considerably reduced. A further benefit is that the amount of data available for the orbit solution is increased and the data quality is increased. The more maneuvers that are performed, the more the orbit is corrupted and the more the quality of science is compromised. In addition, more risk to the mission is incurred because of poor trajectory control.

A trajectory design constraint related to orbit stability is that all low-inclination orbits retrograde with respect to the asteroid equator. Retrograde orbits are more stable because the faster relative motion of the spacecraft with respect to the asteroid tends to average out the effects of gravity harmonics. For this reason, synchronous direct orbits are particularly unstable since the spacecraft lingers over the same point on the asteroid's surface and may exchange enough energy to escape from or collide with the asteroid. In low orbit, even retrograde synchronous orbits may be unstable.

Science constraints on the trajectory design take the form of desires to obtain some particular orbital geometry and are generally not easily quantified. The requirement of the gamma ray spectrometer to obtain low-altitude orbits drove the trajectory design to achieve these orbits in a timely manner. The plan to stage the trajectory through a series of successively smaller circular orbits seems to satisfy most science and navigation requirements and makes the trajectory relatively simple to design. The general plan is to spend a specified amount of time in a series of circular orbits of predetermined radius. This keeps the mission on schedule and enables a general imaging or mapping plan to apply for any Eros gravity field that may be encountered. Transfer orbits between the circular orbits may also provide a unique opportunity for science observations from a perspective different from the circular orbits. However, the need to get to desired circular orbits may also make the transfer orbits unattractive for science. In any event, the transfer orbits need to be designed to achieve circular orbits and only limited science constraints can be accommodated in these orbits.

3.5.2 Targeting Strategy

The general approach to the targeting strategy is to develop a broad set of objectives and compute a series of propulsive maneuvers that will steer the spacecraft in the direction of satisfying these objectives. This differs substantially from the traditional approach of defining a number of constraints and searching for the trajectory that globally maximizes some performance index. The NEAR approach is to compute a maneuver that satisfies a local set of constraints and then propagate the trajectory into the future. At the appropriate time, a minimum of 1 week in the future, the constraints are reevaluated and another maneuver is computed. This strategy is repeated until all the science objectives are achieved.

The spacecraft is first placed in an orbit that is in the Sun POS. The solar panel illumination constraint dictates that the spacecraft remain close to this plane for most of the mission. Otherwise, the solar panels would have to be turned too far off the Sun-line in order to image nadir. Also, staying in this plane for the first few weeks of the mission will minimize the effect of solar pressure on the trajectory and on the attitude control momentum management. This is particularly important at high altitudes where the solar pressure acceleration is a large contributor to the total acceleration and is much easier to model when the spacecraft is pointed directly at the Sun.

At the time of arrival at Eros, the spin axis of Eros is pointed away from the Sun. Since the Sun POS coordinate frame z axis also points away from the Sun, a retrograde orbit in the Sun POS will also retrograde in the asteroid equator. This is the direction that is established for the initial orbits. As Eros moves in its orbit about the Sun, the Eros spin axis first points away from the Sun, then perpendicular to the Sun-line, and then toward the Sun. The spin axis, which remains essentially fixed in inertial space, appears to rotate in the Sun POS frame. During the time that the spin axis is pointed almost directly away from the Sun, a retrograde orbit in the asteroid equator will be close to being in the Sun POS and thus suitable for imaging Eros without turning far off the Sun-line. Retrograde equatorial orbits are generally very stable and thus the orbit altitude may be lowered to 35 km for gamma ray spectrometer observations. As the Eros spin axis aligns perpendicular to the Sun-line, the Sun POS orbit results in a polar orbit with respect to Eros. This is also a stable orbit. A problem occurs when the Eros spin axis is about 45° off the Sun-line. For these orbits, the node of the orbit plane with respect to the asteroid equator precesses at a fast rate sometimes approaching 5° per day. During this transition zone, the spacecraft orbit must be actively controlled with maneuvers to keep the spacecraft within 30° inclination of the Sun POS. As Eros spin axis rotates from perpendicular to the Sun POS to near alignment with the direction toward the Sun, it passes through another transition region and then is placed in a retrograde equatorial orbit for the second time. This orbit is direct with respect to the Sun POS. Therefore, at the time the spacecraft is in the polar orbit and the Eros spin axis crosses the Sun POS, it is necessary to execute a "plane flip" maneuver sequence to reverse the direction of the Sun POS orbit from retrograde to direct. When this sequence is completed, the Eros equator orbit remains polar.

3.5.3 Targeting Algorithm

The trajectory design is accomplished by transforming the targeting strategy into a specific step-by-step procedure referred to as the targeting algorithm. The general approach is first to translate spacecraft and science constraints to geometrical parameters that may be computed directly from the spacecraft trajectory about Eros. When necessary, propulsive maneuvers are targeted to these trajectory-related geometrical parameters and spacecraft and science constraints are implicitly satisfied. Therefore, the success of the targeting strategy depends on the ability to define geometrical parameters that relate directly to mission constraints. The geometric parameters of interest that may be closely related to mission constraints are distances from Eros and angular positions of various celestial bodies with respect to Eros. A problem with these angles and distances is that they vary rapidly as the spacecraft orbits Eros and are thus difficult to target. For the targeting to be successful, a set of parameters that vary slowly with time need to be defined. A convenient set of parameters are classical orbit elements. The classical elements describe the size, shape, and orientation of a spacecraft orbit about a central body that may be represented as a point mass. As long as the spacecraft acceleration is dominated by the central gravity, the classical orbit elements do not vary significantly. This was true during a large part of the NEAR mission. In high orbits, the solar pressure becomes a significant perturber relative to the central body gravity and in low orbits the gravity harmonics cause the classical orbit elements to osculate. However, we may use the osculating orbit elements as short-term predictors of spacecraft motion and thus control the trajectory and satisfy mission constraints by targeting to these parameters.

The classical orbit parameters of interest for targeting the NEAR trajectory may be separated into several general categories. The first category describes the size and shape of the orbit which relates directly to energy and angular momentum. The radius of periapsis and radius of apoapsis may be used to control the size and shape of the orbit. These parameters also implicitly control the period of the orbit. The second category describes the orientation of the orbit in inertial space. The longitude of the ascending node, argument of periapsis, and inclination orient the orbit in inertial space. These angles may be computed in either the Sun POS or asteroid equator coordinate frames. The solar panel illumination constraint may be satisfied by keeping the inclination in the Sun POS coordinate frame less than 30°. The asteroid equator coordinate frame may be used to target polar or low-inclination orbits. The final category of orbit parameters, obtained by solution of Kepler's equation, are the times that the spacecraft arrives at various points in the orbit. The true anomaly of the spacecraft, which is the angle measured from periapsis, is also included in this category. These points are candidate maneuver placements. The times of periapsis, apoapsis, and crossings of the line of nodes or reference planes are of interest for maneuver placement. In addition to the classical elements, the Cartesian components of position and velocity in various coordinate frames may be used as target parameters. Also, the Cartesian components may be mixed with classical elements to define target parameter sets.

The first step of the targeting algorithm is to determine the time of the next propulsive maneuver. The spacecraft orbit is propagated into the future and the values of the target parameters are computed. If a constraint is violated or a point is reached in the mission where it is desirable to change the orbit size or orientation, a complete set of orbit elements is displayed which describes the local region of the trajectory. We assume that the orbit elements are osculating slow enough that the conic trajectory propagation is sufficiently accurate. The conic orbit propagation is generally good for several orbits. A suitable maneuver placement point is selected which is normally at periapsis, apoapsis, or the crossing of some reference plane. A precision trajectory is propagated to the nominal maneuver time obtained by solution of Kepler's equation. The osculating orbit elements are reevaluated and a few iterations may be required to determine the precise time of maneuver placement.

The next step is to select three target parameters that describe the post maneuver orbit. These parameters must be independent and include the maneuver point on the post maneuver orbit. The independence of the parameters may be verified by determining that the Jacoby matrix is nonsingular. For example, the parameters periapsis radius, apoapsis radius, and eccentricity would not be independent because eccentricity may be determined from the other two. However, the parameters periapsis radius, apoapsis radius, and inclination are independent and would be suitable for targeting. The inclusion of the maneuver point on the post maneuver orbit is a subtle condition to satisfy. An example would be transfer to a 35 km circular orbit from a maneuver placement at 50 km. Clearly, this would not be possible because the 35 km orbit would not contain any point at 50 km. A more subtle example would be transfer to zero inclination. For this target parameter to converge, the maneuver placement point would have to be in the reference plane.

The third step is to determine the time to evaluate the target parameter constraints. Most of the time, this is immediately after the maneuver. However, for some orbits that are osculating severely, the constraints may be evaluated sometime in the future where it is desired to control some parameter of particular interest. For example, if we are trying to control the precise periapsis radius, the constraint may be evaluated at the nominal time of periapsis passage several orbits after the maneuver thus forcing the minimum radius to occur at this time.

The final step is to determine the finite burn velocity correction that satisfies the constraint parameters. The matrix of partial derivatives that relate the target parameters to the maneuver velocity components is computed by finite difference from precision trajectory propagations. This 3 by 3 matrix is inverted and multiplied times the required target parameter correction to obtain the delta velocity correction. The Newton-Raphson iteration is repeated several times until the desired target parameters are achieved.

3.5.4 NEAR Trajectory Design

The design of the NEAR orbit phase trajectory involves repeated application of the above targeting algorithms. The orbit phase begins after a series of rendezvous burns that slow the spacecraft down from an approach speed of about 1 km/s–5 m/s. The NEAR orbit phase trajectory is divided into 27 segments beginning with the Eros Orbit Insertion (EOI) maneuver (segment 0). Each segment begins with an OCM and the final segment ends on February 6, 2000, the nominal end of the mission.

3.5.5 Approach Through 100 km Orbit

An orbit insertion maneuver is performed at closest approach to transfer the spacecraft from an approach hyperbola with a periapsis radius of 1000 km to a highly eccentric ellipse with a periapsis radius of 400 km as shown in Fig. 3.23. When the spacecraft arrives at 400 km radius, a maneuver is executed to transfer the spacecraft to an orbit with a periapsis radius of 200 km. At 200 km radius, the orbit is circularized and science observations are carried out for about 9 days. Two more maneuvers are executed to lower the spacecraft orbit to 100 km. The projection of the spacecraft orbit into the Sun POS coordinate frame is shown in Fig. 3.23 for the initial orbits through 100 km radius. The view is from behind Eros looking toward the Sun. The spacecraft moves in a retrograde (clockwise) direction while the Eros rotation is counterclockwise when viewed from this direction. The initial orbit following the orbit insertion burn is in the Sun POS with a radius of periapsis of 400 km. The approach trajectory is targeted to cross the Sun POS at a radius of 1000 km periapsis on the side of Eros that results in a retrograde orbit. The orbit insertion bum is placed at the point where the approach hyperbola pierces the Sun POS; that is, the plane perpendicular to the Sun-line that passes through the center of Eros. The target parameters are radius of periapsis, true anomaly, and inclination in the Sun POS.

The target parameter constraints are computed on Jan 20, 1999 16:00:00. This targeting strategy provides a 10-day separation between maneuvers and results in the spacecraft arriving at a periapsis radius of 400 km on January 20 in an orbit with an inclination of 178° in the Sun POS. A 178° inclination orbit with respect to the Sun POS is 2° inclination retrograde or very nearly in the Sun POS. By targeting to orbit elements evaluated on January 20, 1999, the effect of solar pressure on the orbit elements was mitigated.

The first orbit correction maneuver (OCM) is placed at periapsis of the initial transfer orbit. This maneuver is targeted to a periapsis radius of 200 km, apoapsis radius of 400 km, and inclination of 177° in the Sun POS. OCM 2 is also performed at periapsis using the same targeting strategy and circularizes the orbit at 200 km radius.

After 9 days, which is about one revolution in the 200 km orbit, a transfer orbit is computed to lower the spacecraft orbit to 100 km. At this time, the Sun POS inclination is about 171°. Recall that the Sun POS coordinate frame rotates as Eros orbits about the Sun and the orbit plane tends to remain fixed with respect to inertial space. In order to keep the spacecraft in the Sun POS, we must perform the orbit circularization maneuver at 100 km in the Sun POS. This may be accomplished by biasing the transfer orbit to a periapsis altitude of 84 km such that the spacecraft is at 100 km radius when the spacecraft crosses the Sun POS. This little trick saves a maneuver and the spacecraft orbit is circularized with another maneuver placed at the point where the spacecraft crosses the Sun POS at 100 km. In order to maintain 1 week separation between maneuvers, the spacecraft remains in the 200 km by 84 km transfer orbit for 2 and one half revolutions about Eros. The period of this transfer orbit is 5 days.

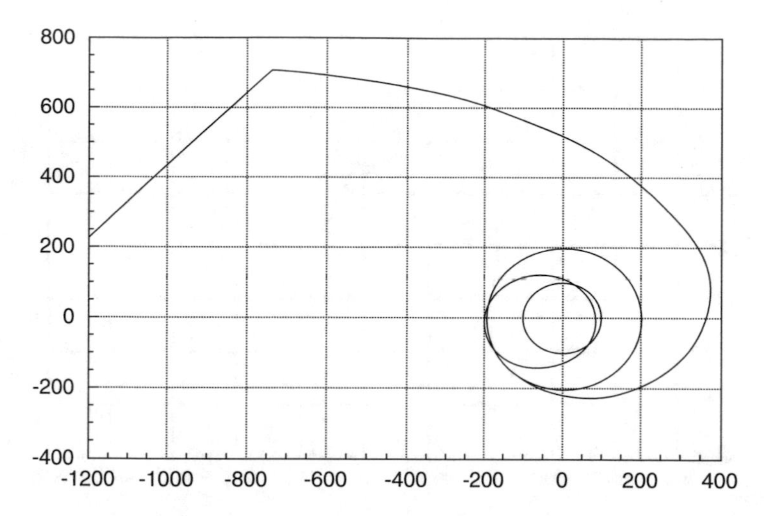

Fig. 3.23 Approach through 100 km orbit

3.5.6 Subsolar Overfly Through 50 km Orbit

An important science objective is to obtain infrared images of Eros at low phase angle. In order to obtain these images, it is necessary to place the spacecraft in an orbit that overflies the subsolar point. This may be accomplished by raising the inclination in the Sun POS coordinate frame to 90°. A convenient time to perform this over flight is early in the mission from the 100 km orbit. After the subsolar overfly, the spacecraft is parked in an elliptical transfer orbit for a week before the orbit is circularized at 50 km radius. Figure 3.24 shows the projection of the spacecraft trajectory in the Sun POS for the subsolar overfly through the initial 50 km circular orbit.

OCM 5 is targeted for overfly of the subsolar point. An orbit that flies over the subsolar point will also fly over the point opposite from the subsolar point or the anti-subsolar point. Therefore, if left in this overfly orbit, the spacecraft will fly through the shadow of Eros and violate an important spacecraft constraint to keep the solar panels illuminated. In order to avoid flying through the shadow and maintain 1 week separation between maneuvers, the spacecraft is placed in an orbit that flies over the subsolar point at a radius of about 150 km and continues out to an apoapsis altitude of about 500 km. The apoapsis radius is selected to give an orbit period of 2 weeks. After 1 week, at the point where the spacecraft crosses the Sun POS, OCM 6 is targeted for an elliptical return trajectory in the Sun POS and at a periapsis radius of 100 km. The periapsis radius of the return trajectory is selected to be 100 km in order to avoid being on an impact trajectory should the maneuver execution error associated with OCM 6 exceed the required accuracy. On return to periapsis, the spacecraft is parked in a 170 km by 50 km elliptical orbit for 1 week. This orbit is a compromise that enables the circular 50 km orbit to return to

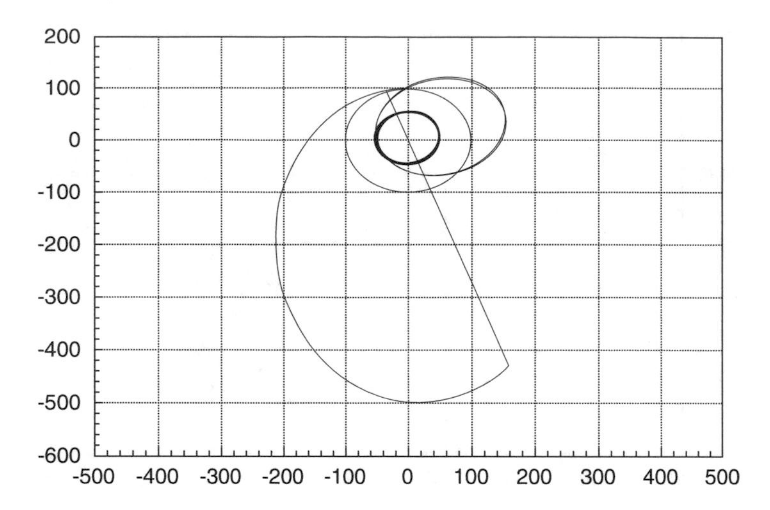

Fig. 3.24 Subsolar overfly through 50 km orbit

the Sun POS and avoids some of the rapid precession of the nodes associated with the 50 km orbit at this time. The spacecraft is then placed in a circular 50 km orbit where it remains until the Eros spin vector comes into favorable alignment for an equatorial orbit about Eros with a low inclination in the Sun POS. OCMs 6 through 9 are all targeted to radius of periapsis, radius of apoapsis, and inclination in the Sun POS.

3.5.7 Transfer to Southern Illuminated 35 km Orbit

A major science objective of the NEAR mission is to obtain low altitude gamma ray and x-ray spectrometer measurements of Eros. From a 35 km orbit, Eros fills the field of view and long integration times are required to obtain the data needed to characterize the composition of Eros. Figure 3.25 shows the projection of the spacecraft orbit on the Sun POS for the transfer orbit from 50 Ian to 35 km and the 35 km orbit.

OCM 9 is targeted for an elliptic transfer orbit from the 50 km circular orbit to a 35 km circular orbit. The actual dimensions of the nominal transfer orbit is 54 km by 32 km. For these low orbits, the conic elements are osculating such that the actual dimensions of the orbit vary in a complicated way. The dimensions of the orbit are contained by targeting the energy and angular momentum. The orbit elements periapsis and apoapsis radius are used to control energy and angular momentum. The orbital distances and velocity are controlled implicitly by managing energy and angular momentum. Even the energy and angular momentum are not conserved with respect to the two-body motion of the spacecraft about Eros. The gravity harmonics

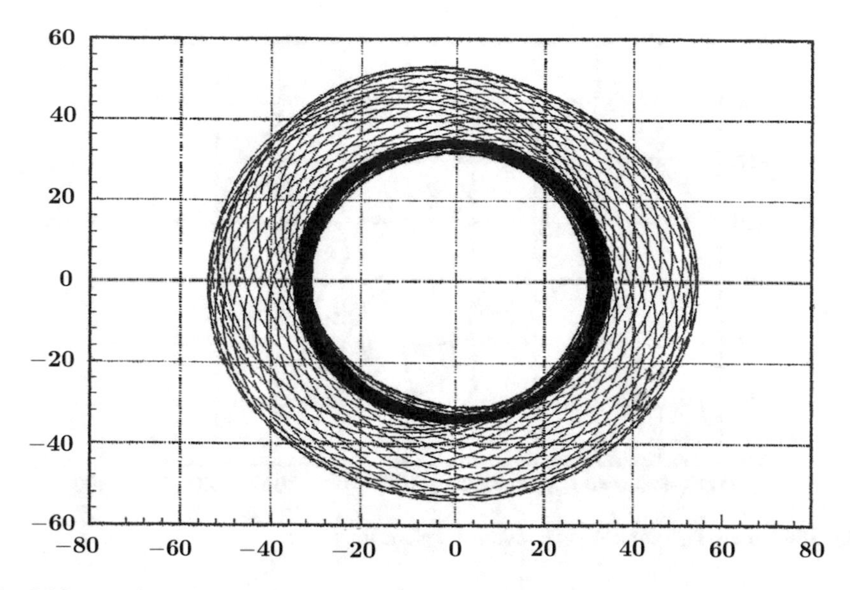

Fig. 3.25 Transfer to southern illuminated 35 km orbit

act to exchange energy between the spacecraft orbit and Eros's rotation. In some cases, the gravity harmonics may act to eject the spacecraft from Eros orbit. The spacecraft actually receives a gravity assist from the gravity harmonics and the rotation of Eros. The basket weave appearance of the orbit shown in Fig. 3.25 is caused by the rapid precession of the spacecraft orbit about Eros in inertial space and not by some artifact of a rotating coordinate system. After one month in the elliptic transfer orbit, the orbit is circularized at 35 km radius with OCM 10. For the transfer orbit and 35 km circular orbit, the target parameters are radius of periapsis, radius of apoapsis, and inclination in the Eros equatorial coordinate frame.

3.5.8 Active POS Control, Polar Orbits, and Plane Flip

The projection of the spacecraft orbit on the Sun POS for the period of active POS control through the polar orbits and plane flip maneuver are shown in Fig. 3.26.

As Eros moves in its orbit about the Sun, the Eros equatorial plane rotates from alignment with the Sun POS to being perpendicular to the Sun POS and back to alignment. This rotation of the planes with respect to one another occurs because the Eros spin axis remains fixed in inertial space while the Sun POS coordinate frame slowly rotates with Eros in its orbit around the Sun. During the time of the 35 km equatorial orbits, these planes are nearly aligned. As they change to an angle greater than about 30°, it is no longer possible to stay in an Eros equatorial orbit and at the same time satisfy the Sun constraint defined in the Sun POS. During the time that the angle between these planes is about 30–60°, the precession of the orbit about

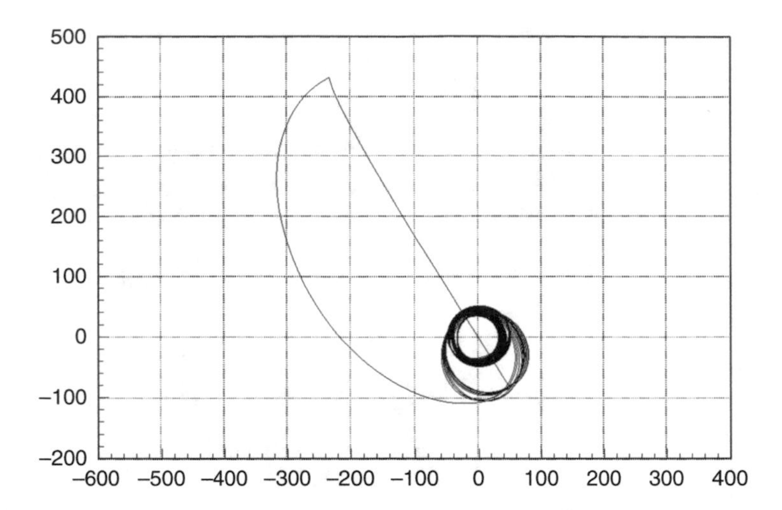

Fig. 3.26 Active POS control, polar orbits, and plane flip

the Eros equator causes the Sun constraint to be violated if left unattended. Active POS control must be executed to prevent the Sun constraint from being violated. Eventually, the angle between the planes approaches 90° and the spacecraft orbit may be transferred to a polar orbit with respect to Eros and at the same time satisfy the Sun constraint. In the polar orbit, the mission is once again interrupted for another overfly of the subsolar point. During this overfly sequence of maneuvers, the direction of the orbit in the Sun POS is reversed from retrograde to direct. This sequence of maneuvers is referred to as the plane flip.

OCM 11 transfers the spacecraft from the 35 km circular orbit to a 55 km by 35 km elliptical transfer orbit. The apoapsis radius of 55 km is set in anticipation of circularizing the orbit at 50 km and returning to the Sun POS. OCM 12 circularizes the orbit a week later at 50 km and the plane of the orbit remains in the Eros equatorial plane. A week later, the orbit plane is transferred to the Sun POS with OCM 13 and the period of active plane of sky control begins. Precession of the line of nodes with respect to Eros equator causes a break in the longitude of ascending node and a gradual increase in inclination as observed in the Sun POS. One week later or at the time the inclination reaches 30°, a maneuver is performed to flip the line of nodes in the Sun POS by 180°. The target parameters for this maneuver are periapsis radius, apoapsis radius, and the z component of velocity in the Sun POS. Reversing the z component of velocity is a device to flip the line of nodes 180°. This strategy will result in the inclination with respect to the Sun POS decreasing from 30° to 0° and then increasing again to 30° as the node precesses with respect to the Eros equatorial plane. This strategy is repeated several times through OCM 17 where the plane flip maneuver sequence is executed. The strategy for the plane flip sequence is the same as executed previously for the subsolar point overfly only the spacecraft returns in a direct orbit. After the plane flip sequence is completed, the spacecraft returns to a polar orbit about Eros.

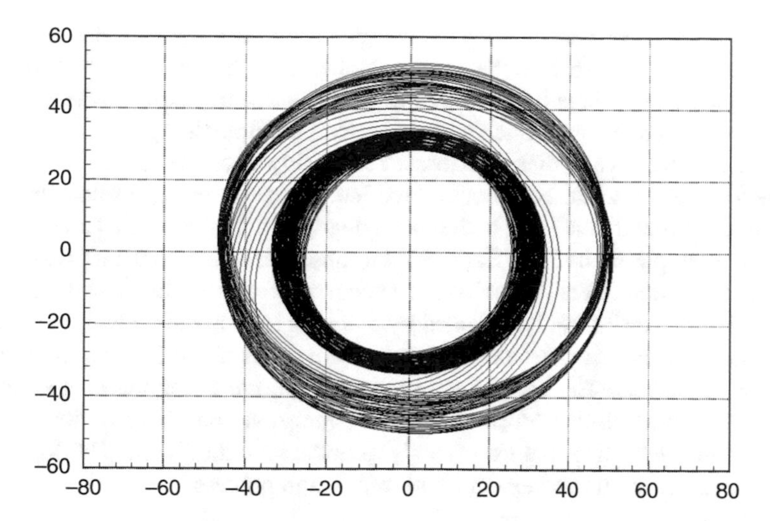

Fig. 3.27 Northern illuminated 50 km and 35 km orbits

3.5.9 Northern Illuminated 50 km and 35 km Orbits

Following the polar orbits, the spacecraft is placed in a 50 km by 45 km orbit for several weeks and active POS control is performed to keep the spacecraft from violating the Sun constraint. The targeting strategy is the same as used previously. When the north pole of Eros comes within 30° of alignment with the Sun direction, a transfer to a 35 km equatorial orbit is executed. The spacecraft remains in the 35 km circular orbit until the end of the mission. The projection of the spacecraft orbit into the Sun POS for these orbits is illustrated in Fig. 3.27.

3.6 Summary

The objective of the design of a spacecraft trajectory is to determine the initial state that will result in satisfying navigation and mission design constraints when the trajectory is integrated to the end of the mission. With experience, the spacecraft initial state can sometimes be guessed and a search conducted to satisfy constraints. In space, the spacecraft motion is usually dominated by the acceleration from the most massive nearby body. This motion was determined by Kepler and is either an ellipse or a hyperbola. Therefore, a very good approximation to the desired trajectory may generally be obtained by patching together the conic solutions with respect to the planet or satellite that is dominant. This is referred to as a two-body solution.

When the spacecraft is in a region of space where the gravitational accelerations from two nearby bodies are nearly equal, such as the Earth and Moon, it is more difficult to patch together two-body conic trajectories. These trajectories are referred to as three-body or restricted three-body if one of the bodies has very little mass. Another example of three-body motion is gravity assist trajectories where the three bodies are the sun, a planet, and the spacecraft. If the time the spacecraft is in limbo between two massive bodies is short, a reasonably accurate initial orbit may be obtained by patching two-body conics together. By extension, a four-body trajectory involves a spacecraft and three bodies with nearly equal gravitational accelerations. The best example is the Sun, Earth, Moon, and spacecraft orbit that was used to navigate the Hiten, Genesis, and other spacecraft from the Earth to Lunar orbit. Four-body trajectories can be obtained by patching together three-body orbits.

Perhaps, the easiest trajectories to design are orbits about planets, asteroids, or satellites of planets. The orbits are easy to design, but satisfying all the mission constraints can result in a lot of complicated maneuvering. The solution is to just stay in orbit a long time until everybody gets the data they want. The Viking and NEAR missions are a good examples of this design process.

Exercises

3.1 For a hyperbolic orbit, the argument F in Kepler's equation can be computed by $F = \ln(\sinh(F) + \cosh(F))$ for positive F and $F = -\ln(\cosh(F) - \sinh(F))$ for negative F. Why are two formulas, which are mathematically the same, needed.

3.2 The flight path angle (γ) is defined as the angle between the velocity vector and the local horizontal plane. Derive the flight path angle for a spacecraft in an elliptical orbit.

3.3 When a spacecraft flies by a target body, errors in the trajectory are amplified by the target body and are corrected by a maneuver performed shortly after encounter. An error in energy after the encounter is particularly troublesome. The energy error is directly related to the error in the magnitude of the B vector (Δb). Show that the ΔV correction is given by

$$\Delta V = \frac{2V_\infty^3\, GM}{GM^2 + b^2\, V_\infty^4}\, \Delta b$$

3.4 Show that the following identity (Eq. 3.38) is true by making use of Eq. (3.8).

$$p\left(\frac{r_1 + r_2}{r_1 r_2}\right) = 2 + 2e\cos\left(\frac{\eta_2 + \eta_1}{2}\right)\cos\left(\frac{\eta_2 - \eta_1}{2}\right)$$

3.5 Show that the following identity (Eq. 3.39) is true.

$$\sqrt{r_1 r_2}\cos\left(\frac{\theta_t}{2}\right) = 2a\sin\frac{\alpha}{2}\sin\frac{\beta}{2}$$

3.6 Determine an equation for the lag angle required for a Hohmann transfer and then compute the time between launch opportunities for an Earth–Mars transfer.

3.7 In Eq. (3.57) and Eq. (3.58), show that

$$\omega^2 x'^2 + \omega^2 y'^2 = 2\omega(x\dot{y} - y\dot{x})$$

Bibliography

Battin, R. H., "An Introduction to the Mathematics and Methods of Astrodynamics", American Institute of Aeronautics and Astronautics, Inc., Reston, VA, 1999.

Belbruno, E. A., J. K. Miller, "Sun-Perturbed Earth-to-Moon Transfers with Ballistic Capture", Vol 16, No. 4, *Journal of Guidance, Control and Dynamics*, July-August 1993.

Egorov, V. A., "Certain Problems of Moon Flight Dynamics," in *The Russian Literature of Satellites*, Part 1, International Physical Index, Inc., New York, 1958.

Ehricke, K. A., "Space Flight", D. Van Nostrand, Princeton, NJ, 1960.

Fesenkov, V. G., *Journal of Astronomy*, 23, No. 1, 1946.

Hintz, G. R., "Orbital Mechanics and Astrodynamics", Springer International Publishing, Switzerland, 2015

Miller, J. K., C. J. Weeks, and L. J. Wood, Orbit Determination Strategy and Accuracy for a Comet Rendezvous Mission. Journal of Guidance, Control and Dynamics 13, 775–784., 1990

Miller, J. K., E. A. Belbruno, "A Method for the Construction of a Lunar Transfer Trajectory Using Ballistic Capture", AAS 91–100, AAS/AIAA Spaceflight Mechanics Meeting, Houston, TX, February 11, 1991.

Miller, J. K., E. Carranza, C. E. Helfrich, W. M. Owen, B. G. Williams, D. W. Dunham, R. W. Farguhar, Y. Guo and J. V. McAdams, "Near Earth Asteroid Rendezvous Orbit Phase Trajectory Design", AIAA 98–4286, AAS/AIAA Astrodynamics Specialist Conference, Boston, MA, August 10, 1998.

Miller, J. K. and C. J. Weeks, "Application of Tisserand's Criterion to the Design of Gravity Assist Trajectories", AIAA 2002–4717, AAS/AIAA Astrodynamics Specialist Conference, Monterey, CA, August 5, 2002.

Miller, J. K., "Lunar Transfer Trajectory Design and the Four Body Problem", AAS 03–144, 13th AAS/AIAA Space Flight Mechanics Meeting, Ponce, Puerto Rico, February 9, 2003.

Miller, J. K. and G. R. Hintz, "Weak Stability Boundary and Trajectory Design", AAS paper 15–297, AAS/AIAA Astrodynamics Specialist Conference, Vail, CO, August 9, 2015.

Roy, A, E., *Orbital Motion*, Adam Hilgar Ltd., Bristol, UK., 1982

Strange, N. J. and J. A. Sims, "Methods for the Design of V-Infinity Leveraging Maneuvers", AAS paper 01–437., 2001.

Strange, N. J. and J. M. Longuski, "Graphical Methods for Gravity-Assist Trajectory Design", *Journal of Spacecraft and Rockets*, Vol. 39, No. 1, January-February 2002.

Yamakawa, H., Kawaguchi, J. and Nakajima, T., "LUNAR-A Trajectory description, " ISTS 94-c-30, 19th International Symposium on Space Technology and Science, Yokohama, Japan, May 15–24, 1994.

Yamakawa, H., Kawaguchi, J., Ishii, N. and Matsuo, H., "A Numerical Study of Gravitational Capture Orbit in the Earth-Moon System," AAS 92–186, 1992.

Chapter 4
Trajectory Optimization

Navigation operations require the refinement of the design trajectory to obtain a high-precision trajectory for flight path control and science operations. The preliminary trajectory design often involves approximate solutions of boundary value problems that provide sufficient accuracy for mission design but are not accurate enough for flight operations. The final precision trajectory is obtained by driving a high-precision trajectory model with targeting and optimization algorithms that yield the final high-precision solution. The preliminary trajectory design provides an initial guess for starting the targeting algorithm. With experience, the preliminary design may sometimes be omitted and the trajectory design obtained directly by targeting.

Targeting algorithms can be separated into two classes: those that involve the solution of a two-point boundary value problem with no optimization and those that target a reduced set of parameters and optimize some performance criterion such as fuel expenditure. The former is sometimes called shooting and the solution is obtained by first computing the partial derivatives of the target parameters with respect to the initial condition or control parameters. These partial derivatives are used to compute a correction to the initial condition and control parameters iteratively using the Newton–Raphson technique. The latter type of algorithm is called an optimizer and performs a similar search for a solution that achieves the target and minimizes a performance criterion.

4.1 Parameter Optimization

When properly formulated, an optimizer may be used to solve a wide variety of problems that extend far beyond navigation of spacecraft. For example, problems of the calculus of variations may be solved with an optimizer by parameterizing the solution and solving for the parameters. Consider the problem of finding the shape of

J. Miller, *Planetary Spacecraft Navigation*, Space Technology Library 37,
https://doi.org/10.1007/978-3-319-78916-3_4

a wire, strung between two points, that a bead will slide down in minimum time. This problem is called the brachistochrone problem and was first posed by John Bernoulli in 1697. The shape of the wire may be represented by a polynomial, and the problem converted to a parameter optimization problem where the independent parameters are the coefficients of the polynomial. The problem in trajectory optimization of finding the optimum programmed thrust direction for a rocket may be solved in a similar fashion.

An optimization algorithm is described that solves the problem of constrained optimization by the method of explicit functions. This method was originally devised to minimize propellant expenditure for the Viking mission to Mars. Additional arbitrary constraint functions are adjoined to the given equations of constraint to completely span the space of the independent parameters. The search is performed on the arbitrary constraint parameters to obtain the values of these parameters that minimizes the performance criterion. First derivatives of the constraint functions with respect to the independent parameters are used to drive the dependent constraint variables or target variables to satisfy the desired constraints, and second partial derivatives of the minimization criterion with respect to the same independent parameters are used to drive the optimization condition to zero. The search is referred to as a second-order gradient search.

The partial derivatives that are required by the optimization algorithm may be obtained analytically or by finite difference. Analytic partial derivatives are often not pursued because of the difficulty in obtaining the partial derivatives, particularly the second derivatives. A problem with exact second derivative finite difference equations is the large number of function evaluations that are required to compute the derivatives for one iteration. These grow as the square of the number of parameters. Approximate techniques may be used to accelerate the computation of the second derivatives, and a method along the lines suggested by Fletcher-Powell-Davidon was investigated. However, these acceleration techniques generally work well only for the problems they were designed to solve and require modification for specific problems making parameter optimization more of an art than a science.

Because of nonlinearity and ill-conditioned problems, a second order gradient search will often diverge. An algorithm is developed to enable inequality constraints to control the search for a solution. Constraining the dependent target variables to an interval permits the optimization algorithm to find a minimum solution within the interval and prevents the search from diverging to a local maximum or inflection point outside the interval.

4.2 Statement of Problem

A performance index (J) is defined that is a function of N independent variables ($\mathbf{U}$). We also have M equations of constraint ($M < N$) that define the target variables ($\mathbf{Z}_C$), and the equations of constraint are also functions of $\mathbf{U}$. Thus we have,

$$J = f(\mathbf{U}) \tag{4.1}$$

$$\mathbf{Z}_C = g(\mathbf{U}) \tag{4.2}$$

and

$$J = f(U_1, U_2, U_3, \ldots U_N)$$

$$Z_{C1} = g_1(U_1, U_2, U_3, \ldots U_N)$$

$$Z_{C2} = g_2(U_1, U_2, U_3, \ldots U_N)$$

$$\cdot \quad \cdot \quad \cdot$$

$$Z_{CM} = g_M(U_1, U_2, U_3, \ldots U_N)$$

The problem is to find a $\mathbf{U}^*$ such that

$$\mathbf{Z}_C(\mathbf{U}^*) = \mathbf{C} \tag{4.3}$$

where $\mathbf{C}$ are constant target parameters and J is a minimum for all $\mathbf{U}$ subject to the constraint $\mathbf{C}$.

4.3 Condition for Optimum Solution

A simple method, in principal, for solving the problem of constrained optimization is to solve the equations of constraint (g) for a selected subset of the independent parameters $(\mathbf{U}_C)$ and substitute these expressions into the objective function (f), thus reducing the number of unknowns from N to $N - M$ where M is the number of constraint functions. The partial derivative of J with respect to the remaining independent parameters is obtained and set equal to zero. These equations are solved in conjunction with the equations of constraint. The selection of which independent control parameters to include in $\mathbf{U}_C$ is arbitrary. However, the choice may have some effect on performance when a numerical solution is sought.

The method of explicit functions carries this concept a step further. In place of the arbitrary selection of control parameters, additional arbitrary constraint functions $(\mathbf{Z}_F)$ are defined to bring the total number of $\mathbf{Z}$ parameters to N. The $\mathbf{Z}_F$ functions are not completely arbitrary in that a one to one mapping must exist between $\mathbf{U}$ and $\mathbf{Z}$. At the solution point, any change in $\mathbf{U}$ holding $\mathbf{Z}_C$ constant will increase J. Since a one to one mapping must exist, any unique change in $\mathbf{Z}_F$ holding $\mathbf{Z}_C$ constant will cause a unique change in $\mathbf{U}$ holding $\mathbf{Z}_C$ constant and consequently increase J. Mathematically, the partial derivative of J with respect to $\mathbf{Z}_F$ holding

$\mathbf{Z}_C$ constant being set equal to zero is a necessary and sufficient condition for a stationary point which is a minimum if J is properly defined and $\mathbf{Z}_C$ is properly constrained. The performance criterion and augmented equations of constraint are given by,

$$J = f(U_1, U_2, U_3, \ldots U_N)$$
$$Z_{C1} = g_1(U_1, U_2, U_3, \ldots U_N)$$
$$Z_{C2} = g_2(U_1, U_2, U_3, \ldots U_N)$$
$$\cdot \quad \cdot \quad \cdot$$
$$Z_{CM} = g_M(U_1, U_2, U_3, \ldots U_N)$$
$$\cdot \quad \cdot \quad \cdot$$
$$Z_{FN} = g_N(U_1, U_2, U_3, \ldots U_N)$$

and the solution is obtained by solving

$$\mathbf{Z}_C = \mathbf{C} \tag{4.4}$$

$$\frac{\partial J}{\partial \mathbf{Z}_F} = 0 \tag{4.5}$$

Observe that the above solution reduces to direct elimination if $\mathbf{Z}_F$ is taken to be identically equal to a subset of U of dimension N minus M.

Because of the difficulty in obtaining the inverse functions analytically, direct solution of the above equations is only practical for relatively simple systems of equations. For complex systems, solutions may be obtained by searching using Newton's method. The theory behind techniques currently in use such as Lagrange multipliers and gradient projection follow directly from the method of explicit functions.

The method of explicit functions involves adjoining to the equations of constraint some additional equations that define the parameters $\mathbf{Z}_F$. The $\mathbf{Z}_F$ parameters replace the independent parameters selected by the method of direct elimination for the purpose of minimizing J. An equation that relates the optimization condition to the independent control parameters, equations of constraint, and performance criterion may be obtained by application of the chain rule.

$$\frac{\partial J}{\partial \mathbf{U}} = \frac{\partial J}{\partial \mathbf{Z}} \frac{\partial \mathbf{Z}}{\partial \mathbf{U}} \tag{4.6}$$

The partial derivatives of $\mathbf{Z}$ with respect to the independent parameters $\mathbf{U}$ are contained in a square matrix of dimension N by N. The partial derivatives of J with respect to $\mathbf{U}$ and $\mathbf{Z}$ are row matrices also of dimension N. Partitioning the above matrices separating the $\mathbf{Z}_C$ dependent elements from the $\mathbf{Z}_F$ dependent elements yields,

$$\left[\frac{\partial J}{\partial \mathbf{U}}\right] = \left[\frac{\partial J}{\partial \mathbf{Z}_C}\ \frac{\partial J}{\partial \mathbf{Z}_F}\right]\left[\begin{array}{c}\frac{\partial \mathbf{Z}_C}{\partial \mathbf{U}} \\ \frac{\partial \mathbf{Z}_F}{\partial \mathbf{U}}\end{array}\right] \tag{4.7}$$

The above partitioned matrices may be factored to further separate those submatrices dependent on $\mathbf{Z}_C$ from those dependent on $\mathbf{Z}_F$ and after rearranging terms the following equation is obtained.

$$\frac{\partial J}{\partial \mathbf{U}} - \frac{\partial J}{\partial \mathbf{Z}_C}\frac{\partial \mathbf{Z}_C}{\partial \mathbf{U}} = \frac{\partial J}{\partial \mathbf{Z}_F}\frac{\partial \mathbf{Z}_F}{\partial \mathbf{U}} \tag{4.8}$$

Equation (4.8) provides a fundamental relationship that may be used to tie together various methods of constrained parameter optimization including the methods of Lagrange multipliers, gradient projection, and explicit functions. Comparison of these methods provides insight into which approach may work best depending on the problem.

4.3.1 Lagrange Multipliers

The classic solution of constrained parameter optimization was derived by the eighteenth-century mathematician Joseph Luis Lagrange. This solution is particularly appealing since a choice of independent parameters is not necessary. Referring to Eq. (4.8), at the solution point the right side is zero because the partial derivative of J with respect to the $\mathbf{Z}_F$ must be zero as required by Eq. (4.5).

$$\frac{\partial J}{\partial \mathbf{U}} - \frac{\partial J}{\partial \mathbf{Z}_C}\frac{\partial \mathbf{Z}_C}{\partial \mathbf{U}} = 0 \tag{4.9}$$

The terms of Eq. (4.9) may be readily obtained from the equations of constraint and the equation for the performance index with the exception of the partial derivative of J with respect to the $\mathbf{Z}_C$. Lagrange's insight was to make the elements of this term parameters to be solved for in conjunction with the equations of constraint. These parameters are called Lagrange multipliers and are defined by

$$\boldsymbol{\lambda} = -\frac{\partial J}{\partial \mathbf{Z}_C} \tag{4.10}$$

The sign of the Lagrange multipliers is arbitrary and it may be conjectured that Lagrange selected the minus sign for convenience. He was certainly aware of Eq. (4.8) but apparently did not consider the right side important since the computer had not been invented in his time. The equations that must be solved to obtain an optimum are thus,

$$\mathbf{Z}_C = \mathbf{C} \quad \text{(M equations)} \tag{4.11}$$

$$\frac{\partial J}{\partial \mathbf{U}} + \boldsymbol{\lambda}\frac{\partial \mathbf{Z}_C}{\partial \mathbf{U}} = 0 \quad \text{(N equations)} \tag{4.12}$$

The method of Lagrange multipliers requires the solution of M+N equations for N **U** parameters and M Lagrange multipliers. This method is well suited for obtaining analytic solutions since the equations of constraint need not be solved for the independent **U** parameters as a function of the **Z** parameters. However, the need to solve for the Lagrange multipliers makes numerical solutions more complicated than necessary.

4.3.2 Explicit Functions

The methods of explicit functions and gradient projection use the right side of Eq. (4.8) to obtain a solution and thus avoid the need to solve for Lagrange multipliers. The method of explicit functions requires an equation for the partial derivative of J with respect to $\mathbf{Z}_F$. Application of the chain rule gives,

$$\frac{\partial J}{\partial \mathbf{Z}} = \frac{\partial J}{\partial \mathbf{U}}\frac{\partial \mathbf{U}}{\partial \mathbf{Z}} \tag{4.13}$$

The partial derivatives of **U** with respect to the dependent target parameters **Z** are obtained by matrix inversion.

$$\left[\frac{\partial J}{\partial \mathbf{Z}_C} \; \frac{\partial J}{\partial \mathbf{Z}_F}\right] = \left[\frac{\partial J}{\partial \mathbf{U}}\right]\begin{bmatrix} \dfrac{\partial \mathbf{Z}_C}{\partial \mathbf{U}} \\ \dfrac{\partial \mathbf{Z}_F}{\partial \mathbf{U}} \end{bmatrix}^{-1} \tag{4.14}$$

where

$$\frac{\partial \mathbf{U}}{\partial \mathbf{Z}} = \left[\frac{\partial \mathbf{Z}}{\partial \mathbf{U}}\right]^{-1} = \begin{bmatrix} \dfrac{\partial \mathbf{Z}_C}{\partial \mathbf{U}} \\ \dfrac{\partial \mathbf{Z}_F}{\partial \mathbf{U}} \end{bmatrix}^{-1}$$

The equations that must be solved to obtain an optimum are the equations of constraint and the last $N - M$ columns of Eq. (4.14).

$$\mathbf{Z}_C = \mathbf{C} \quad \text{(M equations)} \tag{4.15}$$

$$\frac{\partial J}{\partial \mathbf{Z}_F} = 0 \quad \text{(N} - \text{M equations)} \tag{4.16}$$

The method of explicit functions requires the solution of N equations for N control parameters **U**. This algorithm is well suited for obtaining numerical solutions on a computer but not for analytic solutions since it involves matrix inversion of a matrix with analytic functions for elements. Observe that the Lagrange multipliers are obtained as a by-product of Eq. (4.14) (the first M columns).

4.3.3 Gradient Projection

The method of gradient projection is a special case of the method of explicit functions. The independent parameters are partitioned into what are referred to as state parameters ($\mathbf{U}_C$) and decision parameters ($\mathbf{U}_F$). The choice between which independent parameters to designate as decision parameters is not unique. The distinction between state and decision parameters is generally only a matter of convenience. However, the decision parameters must determine the state parameters through the constraint relations. Expanding Eq. (4.14), separating the $\mathbf{U}_C$ dependent elements from the $\mathbf{U}_F$ dependent elements, yields,

$$\begin{bmatrix} \frac{\partial J}{\partial \mathbf{Z}_C} & \frac{\partial J}{\partial \mathbf{Z}_F} \end{bmatrix} = \begin{bmatrix} \frac{\partial J}{\partial \mathbf{U}_C} & \frac{\partial J}{\partial \mathbf{U}_F} \end{bmatrix} \begin{bmatrix} \frac{\partial \mathbf{Z}_C}{\partial \mathbf{U}_C} & \frac{\partial \mathbf{Z}_C}{\partial \mathbf{U}_F} \\ \frac{\partial \mathbf{Z}_F}{\partial \mathbf{U}_C} & \frac{\partial \mathbf{Z}_F}{\partial \mathbf{U}_F} \end{bmatrix}^{-1} \tag{4.17}$$

The $\mathbf{Z}_F$ constraint relationships have yet to be specified. Depending on the choice of which **U** are included in $\mathbf{U}_F$, some reordering of the rows and columns of Eq. (4.17) may be necessary. Since the selection of the $\mathbf{Z}_F$ equations of constraint is arbitrary, $\mathbf{Z}_F$ may be made identically equal to $\mathbf{U}_F$. Equation (4.17) then reduces to,

$$\begin{bmatrix} \frac{\partial J}{\partial \mathbf{Z}_C} & \frac{\partial J}{\partial \mathbf{Z}_F} \end{bmatrix} = \begin{bmatrix} \frac{\partial J}{\partial \mathbf{U}_C} & \frac{\partial J}{\partial \mathbf{U}_F} \end{bmatrix} \begin{bmatrix} \frac{\partial \mathbf{Z}_C}{\partial \mathbf{U}_C} & \frac{\partial \mathbf{Z}_C}{\partial \mathbf{U}_F} \\ 0 & I \end{bmatrix}^{-1} \tag{4.18}$$

Performing the indicated matrix inversion yields,

$$\begin{bmatrix} \frac{\partial J}{\partial \mathbf{Z}_C} & \frac{\partial J}{\partial \mathbf{Z}_F} \end{bmatrix} = \begin{bmatrix} \frac{\partial J}{\partial \mathbf{U}_C} & \frac{\partial J}{\partial \mathbf{U}_F} \end{bmatrix} \begin{bmatrix} \frac{\partial \mathbf{Z}_C}{\partial \mathbf{U}_C}^{-1} & -\frac{\partial \mathbf{Z}_C}{\partial \mathbf{U}_C}^{-1} \frac{\partial \mathbf{Z}_C}{\partial \mathbf{U}_F} \\ 0 & I \end{bmatrix} \tag{4.19}$$

and

$$\frac{\partial J}{\partial \mathbf{Z}_F} = \frac{\partial J}{\partial \mathbf{U}_F} - \frac{\partial J}{\partial \mathbf{U}_C} \frac{\partial \mathbf{Z}_C}{\partial \mathbf{U}_C}^{-1} \frac{\partial \mathbf{Z}_C}{\partial \mathbf{U}_F} = 0 \tag{4.20}$$

Equation (4.20) is solved in conjunction with the equation of constraint to obtain an optimum as is done for the method of explicit functions. Observe that the Lagrange multipliers are obtained as a by-product from both the method of explicit functions and gradient projection.

$$\boldsymbol{\lambda} = -\frac{\partial J}{\partial \mathbf{Z}_C} = -\frac{\partial J}{\partial \mathbf{U}_C}\frac{\partial \mathbf{Z}_C}{\partial \mathbf{U}_C}^{-1} \tag{4.21}$$

Even though the Lagrange multipliers do no enter into the optimal solution, they are useful for determining which bound is appropriate for inequality constraints.

4.4 Sample Problem

A sample problem is solved to illustrate the various methods of constrained parameter optimization. Consider an ellipse with semi-major axis a and semi-minor axis b oriented along the coordinate axes. The problem is to find the greatest rectangle with sides parallel to the coordinate axes that will fit inside the ellipse. The geometry is illustrated in Fig. 4.1. The equation of constraint describes an ellipse and the performance criterion is the area of the rectangle. The area in the first quadrant is multiplied by four and assigned a minus sign since we are seeking a maximum.

$$Z_c = \frac{U_1^2}{a^2} + \frac{U_2^2}{b^2} = C = 1 \tag{4.22}$$

$$J = -4U_1U_2 \tag{4.23}$$

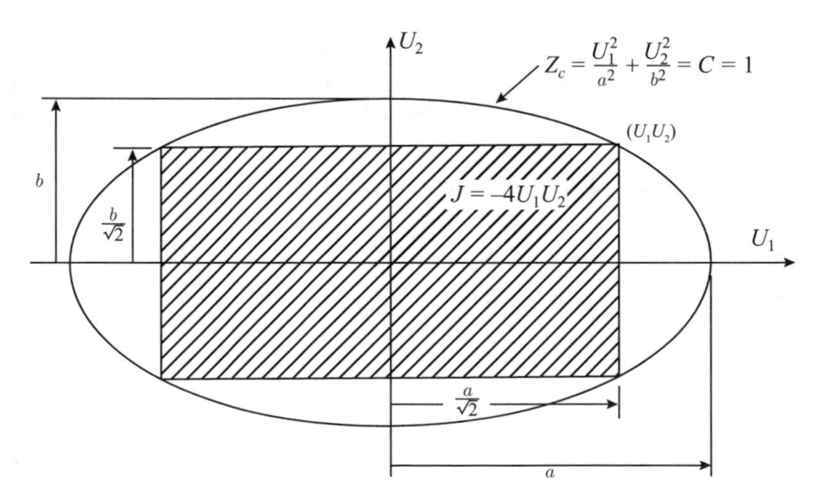

Fig. 4.1 Sample problem

4.4.1 Solution by Method of Lagrange Multipliers

The method of Lagrange multipliers requires a solution of Eq. (4.12) in conjunction with the equation of constraint (Eq. 4.22). For the sample problem

$$\frac{\partial J}{\partial \mathbf{U}} = \begin{bmatrix} -4U_2 & -4U_1 \end{bmatrix} \tag{4.24}$$

$$\frac{\partial \mathbf{Z}_C}{\partial \mathbf{U}} = \begin{bmatrix} \frac{2U_1}{a^2} & \frac{2U_2}{b^2} \end{bmatrix} \tag{4.25}$$

Substituting into Eq. (4.12) gives the following two equations:

$$-4U_2 + \lambda \frac{2U_1}{a^2} = 0 \tag{4.26}$$

$$-4U_1 + \lambda \frac{2U_2}{b^2} = 0 \tag{4.27}$$

which may be solved in conjunction with the equation of constraint (Eq. 4.22) to obtain the solution, $U_1 = \frac{a}{\sqrt{2}}$, $U_2 = \frac{b}{\sqrt{2}}$ and $\lambda = 2ab$.

4.4.2 Solution by Method of Explicit Functions

The method of explicit functions requires a solution of Eq. (4.14) in conjunction with the equation of constraint (Eq. 4.22). Since there are two independent parameters, an additional equation of constraint is needed to square up the system of equations. For numerical solutions, a good choice is a function that is nearly normal to the constraint function. A hyperbola is selected for Z_F.

$$Z_F = \frac{U_1^2}{c^2} - \frac{U_2^2}{d^2} \tag{4.28}$$

For the sample problem, the terms of Eq. (4.14) are given by,

$$\begin{bmatrix} \frac{\partial \mathbf{Z}_C}{\partial \mathbf{U}} \\ \frac{\partial \mathbf{Z}_F}{\partial \mathbf{U}} \end{bmatrix} = \begin{bmatrix} \frac{2U_1}{a^2} & \frac{2U_2}{b^2} \\ \frac{2U_1}{c^2} & -\frac{2U_2}{d^2} \end{bmatrix} \tag{4.29}$$

The required matrix inverse is

$$\begin{bmatrix} \dfrac{\partial \mathbf{Z}_C}{\partial \mathbf{U}} \\ \dfrac{\partial \mathbf{Z}_F}{\partial \mathbf{U}} \end{bmatrix}^{-1} = \frac{1}{4U_1U_2}\left(\frac{a^2b^2c^2d^2}{a^2d^2+b^2c^2}\right)\begin{bmatrix} \dfrac{2U_2}{d^2} & \dfrac{2U_2}{b^2} \\ \dfrac{2U_1}{c^2} & \dfrac{-2U_1}{a^2} \end{bmatrix} \tag{4.30}$$

Substituting Eqs. (4.24) and (4.30) into Eq. (4.14) yields,

$$\begin{bmatrix} \dfrac{\partial J}{\partial \mathbf{Z}_C} & \dfrac{\partial J}{\partial \mathbf{Z}_F} \end{bmatrix} = \frac{-1}{4U_1U_2}\left(\frac{a^2b^2c^2d^2}{a^2d^2+b^2c^2}\right)\begin{bmatrix} \dfrac{8d^2U_1^2+8c^2U_2^2}{c^2d^2} & \dfrac{8a^2U_2^2-8b^2U_1^2}{a^2b^2} \end{bmatrix} \tag{4.31}$$

The equation

$$\frac{8U_2^2}{b^2} - \frac{8U_1^2}{a^2} = 0 \tag{4.32}$$

is solved in conjunction with Eq. (4.22) to obtain $U_1 = \dfrac{a}{\sqrt{2}}$ and $U_2 = \dfrac{b}{\sqrt{2}}$. The Lagrange multiplier, obtained from the first column of Eq. (4.31), is $\lambda = 2ab$. Observe that at the solution point, the constants c and d completely cancel from the solution, as expected, verifying that Eq. (4.28) is arbitrary.

4.4.3 Solution by Method of Gradient Projection

The method of gradient projection requires a solution of Eq. (4.21) in conjunction with Eq. (4.22). For the sample problem, U_1 is selected for $\mathbf{U}_C$ and U_2 for $\mathbf{U}_F$. Because of symmetry, the selection of which independent parameter is a "state" parameter and which is a "decision" parameter is arbitrary.

$$\frac{\partial J}{\partial \mathbf{U}_C} = -4U_2 \tag{4.33}$$

$$\frac{\partial J}{\partial \mathbf{U}_F} = -4U_1 \tag{4.34}$$

$$\frac{\partial \mathbf{Z}_C}{\partial \mathbf{U}_C} = \frac{2U_1}{a^2} \tag{4.35}$$

$$\frac{\partial \mathbf{Z}_C}{\partial \mathbf{U}_F} = \frac{2U_2}{b^2} \tag{4.36}$$

Substituting the above equations into Eq. (4.21) yields

$$[-4U_1] - [-4U_2]\left[\frac{a^2}{2U_1}\right]\left[\frac{2U_2}{b^2}\right] = 0$$

$$-4U_1^2 b^2 + 4U_2^2 a^2 = 0 \tag{4.37}$$

which is solved in conjunction with Eq. (4.22) to obtain $U_1 = \frac{a}{\sqrt{2}}$ and $U_2 = \frac{b}{\sqrt{2}}$. The Lagrange multiplier, which is also obtained as a by-product, is given by substituting into Eq. (4.21).

$$\lambda = -[-4U_2]\left[\frac{2U_1}{a^2}\right]^{-1} = 2ab \tag{4.38}$$

4.5 Second-Order Gradient Search

Parameter optimization problems with constraints, where the dependent parameters are obtained by numerical integration, are difficult if not impossible to solve analytically. Numerical solutions may be obtained by searching using an iterative technique like Newton's method. For the explicit method, the equations that need to be solved are

$$\mathbf{Z}_C = \mathbf{C} \quad (\text{M equations}) \tag{4.39}$$

and from the last $N - M$ columns of

$$\left[\frac{\partial J}{\partial \mathbf{Z}_C} \; \frac{\partial J}{\partial \mathbf{Z}_F}\right] = \left[\frac{\partial J}{\partial \mathbf{U}}\right]\left[\begin{matrix}\frac{\partial \mathbf{Z}_C}{\partial \mathbf{U}} \\ \frac{\partial \mathbf{Z}_F}{\partial \mathbf{U}}\end{matrix}\right]^{-1} \tag{4.40}$$

the following equation is extracted:

$$\frac{\partial J}{\partial \mathbf{Z}_F} = 0 \quad (\text{N} - \text{M equations}) \tag{4.41}$$

From the definition of the derivative, the following difference equations may be written:

$$\Delta \mathbf{Z}_C = \frac{\partial \mathbf{Z}_C}{\partial \mathbf{U}} \Delta \mathbf{U} \tag{4.42}$$

$$\Delta \frac{\partial J}{\partial \mathbf{Z}_F} = \frac{\partial^2 J}{\partial \mathbf{U} \partial \mathbf{Z}_F} \Delta \mathbf{U} \tag{4.43}$$

The search for a solution involves finding a change in the independent control parameters that will move the current values of the constraint parameters and optimization condition to their desired values. The desired changes in the constraint parameters and optimization condition are given by

$$\Delta \mathbf{Z}_C^i = \mathbf{C} - \mathbf{Z}_C^i \tag{4.44}$$

$$\Delta \frac{\partial J^i}{\partial \mathbf{Z}_F} = 0 - \frac{\partial J^i}{\partial \mathbf{Z}_F} \tag{4.45}$$

corresponding to a change in the control parameters from $\mathbf{U}^i$ to $\mathbf{U}^{i+1}$,

$$\Delta \mathbf{U} = \mathbf{U}^{i+1} - \mathbf{U}^i \tag{4.46}$$

Solving for $\mathbf{U}^{i+1}$, an iterative equation is obtained for the ith iteration.

$$\mathbf{U}^{i+1} = \mathbf{U}^i - \begin{bmatrix} \dfrac{\partial \mathbf{Z}_C}{\partial \mathbf{U}} \\ \dfrac{\partial^2 J}{\partial \mathbf{U} \partial \mathbf{Z}_F} \end{bmatrix}^{-1} \begin{bmatrix} \mathbf{Z}_C^i - \mathbf{C} \\ \dfrac{\partial J^i}{\partial \mathbf{Z}_F} \end{bmatrix} \tag{4.47}$$

The partial derivatives required by the second-order gradient search are obtained by finite difference. Computation of these finite difference partial derivatives requires repeated evaluation of the functions f and g for the performance index and constraint parameters at each iteration.

$$\frac{\partial J}{\partial U_i} = \frac{f(\mathbf{U} + \Delta \mathbf{U}_i) - f(\mathbf{U} - \Delta \mathbf{U}_i)}{2\Delta U_i} \tag{4.48}$$

The $\Delta \mathbf{U}_i$ vector is zero except for the ith element that contains the partial derivative step size. ΔU_i is the ith element of $\Delta \mathbf{U}_i$. The partial derivatives of the constraint parameters with respect to the independent control parameters are given by

$$\frac{\partial Z_j}{\partial U_i} = \frac{g_j(\mathbf{U} + \Delta \mathbf{U}_i) - g_j(\mathbf{U} - \Delta \mathbf{U}_i)}{2\Delta U_i} \tag{4.49}$$

The matrix of second partial derivatives in Eq. (4.47) is a mixed tensor that is covariant in $\mathbf{U}$ and contravariant in $\mathbf{Z}$. It serves the same purpose as the covariant Hessian in optimization theory and may be a Hessian depending on the definition. The Hessian matrix was developed by the nineteenth-century mathematician Ludwig Otto Hesse. The elements of the required matrix of second partial derivatives are given by

$$\frac{\partial^2 J}{\partial U_j \partial Z_i} = \frac{1}{2\Delta U_j}\{[f(\mathbf{U} + \Delta \mathbf{U}_j + \Delta \mathbf{U}_i) - f(\mathbf{U} + \Delta \mathbf{U}_j - \Delta \mathbf{U}_i)]$$

$$[g_i (\mathbf{U} + \Delta\mathbf{U}_j + \Delta\mathbf{U}_i) - g_i (\mathbf{U} + \Delta\mathbf{U}_j - \Delta\mathbf{U}_i)]^{-1}$$
$$- [f(\mathbf{U} - \Delta\mathbf{U}_j + \Delta\mathbf{U}_i) - f(\mathbf{U} - \Delta\mathbf{U}_j - \Delta\mathbf{U}_i)]$$
$$[g_i (\mathbf{U} - \Delta\mathbf{U}_j + \Delta\mathbf{U}_i) - g_i (\mathbf{U} - \Delta\mathbf{U}_j - \Delta\mathbf{U}_i)]^{-1}\} \quad (4.50)$$

The partial step size for the first partial derivatives should be as small as possible to achieve linearity but large enough, relative to the machine precision, to maintain accuracy. The partial step size for the second partial derivatives ($\Delta\mathbf{U}_j$) should be about 5–10 times larger than the corresponding ($\Delta\mathbf{U}_i$). The computation of the second partial derivatives (Eq. 4.50) will require $4N^2$ evaluations of the performance index and constraint functions. For six control parameters, 144 function evaluations are needed. Several methods have been explored to accelerate the computation of the second partial derivatives. Since the optimization conditions are not a function of the second partial derivatives, approximations may be used to speed up the search without compromising accuracy. An approximation that worked well for optimization of the Viking orbit insertion maneuver was to set all the terms of Eq. (4.50) where $i \neq j$ to zero. For this approximation, $2N + 1$ function evaluations are required. Another approach was suggested by Fletcher-Powell-Davidon. The matrix of second partial derivatives is primed with an approximate solution. Subsequent changes in the control parameters computed during the search are used to estimate and thus improve the second partial derivative matrix. This bootstrap approach can greatly speed up the search but may lead to instabilities if the search is not properly controlled.

4.6 Inequality Constraints

Sometimes the constraint on a Z parameter is not a specific target value but a range of values. In other situations, the second-order gradient search described above may not converge to the desired minimum if the initial guess required to start the search is too far from the solution but wander off toward a local maximum or inflection point. For these reasons, it is often convenient to specify inequality constraints where the $\mathbf{Z}$ are constrained to a specified range of values.

$$C_{Li} \leq Z_i \leq C_{Ui} \quad (4.51)$$

An algorithm has been devised to transform the problem of optimization with inequality constraints into the problem of optimization with equality constraints described above. At any step in the iteration for a solution, the Z_i parameters are tested and sorted into the $\mathbf{Z}_C$ category or $\mathbf{Z}_F$ category. The algorithm is diagrammed in Fig. 4.2.

The following conditions result in the Z_i target variable being placed in the constrained $\mathbf{Z}_C$ category:

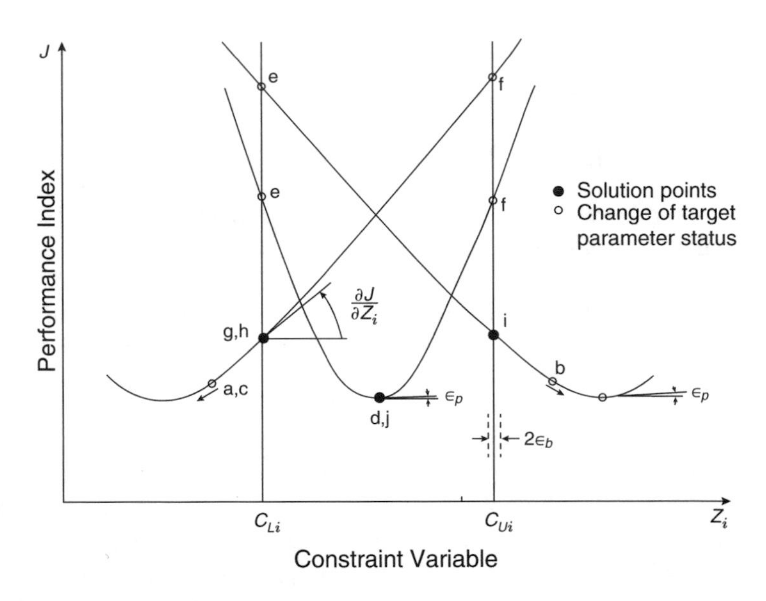

Fig. 4.2 Inequality constraint status determination

(a) If $C_{Li} = C_{Ui}$, C_i is set equal to C_{Li} and Z_i is a hard constraint
(b) If $Z_i > C_{Ui}$, C_i is set equal to C_{Ui} and Z_i is a soft constraint
(c) If $Z_i < C_{Li}$, C_i is set equal to C_{Li} and Z_i is a soft constraint

The following conditions result in the Z_i target variable being placed in the unconstrained $\mathbf{Z}_F$ category:

(d) If $C_{Li} < Z_i < C_{Ui}$ and the constraint is released
(e) If $|Z_i - C_{Li}| < \epsilon_b$ and $\frac{\partial J}{\partial Z_i} < 0$ and the constraint is released
(f) If $|Z_i - C_{Ui}| < \epsilon_b$ and $\frac{\partial J}{\partial Z_i} > 0$ and the constraint is released

The following conditions result in convergence if true for all Z_i.

(g) If $|Z_i - C_{Li}| < \epsilon_b$ and $C_{Li} = C_{Ui}$, hard constraint
(h) If $|Z_i - C_{Li}| < \epsilon_b$ and $\frac{\partial J}{\partial Z_i} > 0$, soft constraint
(i) If $|Z_i - C_{Ui}| < \epsilon_b$ and $\frac{\partial J}{\partial Z_i} < 0$, soft constraint
(j) If $C_{Li} < Z_i < C_{Ui}$ and $\frac{\partial J}{\partial Z_i} < \epsilon_p$, a true minimum satisfying the constraints

A soft constraint applies to the current iteration and may be released as the search progresses. A hard constraint is an equality constraint and applies throughout the search. The tolerance ϵ_b is on the value of the constrained variable and the tolerance ϵ_p is on the partial derivative of J with respect to Z_i. The conditions for control of the search and confirmation of a solution are lettered a-j and shown in Fig. 4.2. There are three possible cases that each constraint variable may describe provided the optimization problem has been properly defined and constrained. The constraint variable may either be an increasing monotone across the constraint interval, achieve a minimum within the constraint interval, or be a decreasing monotone across the constraint interval. If a maximum is sought, the sign of J is changed and the algorithm searches for a minimum. These three cases are illustrated in Fig. 4.2. For the first case, conditions (a) or (c) will select the lower bound and condition (f) will release the constraint from the upper bound. At the solution point (g,h), the partial derivative of J with respect to Z_i, the negative of the Lagrange multiplier, indicates that releasing the constraint will result in an increase in J. The solution is thus held at the lower bound. For the second case, conditions e or f will release the constraint from the lower and upper bounds, respectively and a minimum is obtained (d,j) between the bounds. The third case is simply the mirror image of the first case.

4.7 Mission to Mercury

The MESSENGER spacecraft was launched on August 3, 2004, on a mission to explore the planet Mercury. The trajectory reencountered the Earth a year after launch, to obtain a gravity assist, and then proceeded on to several encounters with Venus and Mercury before being inserted into Mercury orbit in 2011. The initial injection error at Earth launch resulted in a 20 m/s under burn. Two Trajectory Correction Maneuvers (TCMs) were scheduled to make up the energy deficit and place the spacecraft on the proper trajectory. Two TCMs are necessary to achieve the target: The first corrects the energy and the second corrects the orbit plane. Because of the near 360° transfer, the first maneuver, which is performed shortly after launch, is unable to correct the orbit plane. The second maneuver, which is performed about three months after launch, is less efficient in correcting energy or flight time. Since there are only two constraints that need to be satisfied related to the position relative to the Earth at the second encounter and there are six maneuver components available to control the trajectory, the remaining four degrees of freedom may be used to minimize propellant expenditure.

The initial Earth launch injection conditions ($\mathbf{X}_0$) on August 3, 2004, are propagated to the nominal time of Earth return on August 2, 2005. Two TCMs were initially planned for August 18, 2004 and November 19, 2004. The spacecraft state at Earth return is determined by numerical integration.

$$\mathbf{X}_e = g_1(t_0,\ \mathbf{X}_0,\ t_1,\ \Delta\mathbf{V}_1,\ t_2,\ \Delta\mathbf{V}_2,\ t_e)$$

The maneuver velocity components, $\Delta\mathbf{V}_1$ and $\Delta\mathbf{V}_2$, are applied as finite burns at the maneuver start times t_1 and t_2. At the end time (t_e), the Cartesian state vector is transformed into hyperbolic orbit elements.

$$\mathbf{H}_e = g_2(\mathbf{X}_e)$$

$$\mathbf{H}_e = [B\cdot R,\ B\cdot T, t_p, V_\infty, \alpha_\infty, \delta_\infty]$$

The hyperbolic elements $B \cdot R$ and $B \cdot T$ (see Fig. 3.4) are the coordinates of the approach asymptote in the target B-plane, t_p is the time of closest approach, V_∞ is the approach hyperbolic velocity magnitude, and α_∞ and δ_∞ are the right ascension and declination of the approach asymptote. The optimization problem is to find the velocity change components of the two TCMs that will acquire the target and eventually arrive at Mercury and minimize propellant consumption which is related to the sum of the magnitudes of the maneuver velocity change associated with each maneuver.

The optimization problem described above must first be cast into the framework required by the optimization method being used. The following constraint variables, constraint parameters, performance index, and control variables are defined:

$$\mathbf{Z}_C = [B \cdot R,\ B \cdot T]$$

$$\mathbf{C}_C = [-14,463.00\ \text{km},\ -17,793.00\ \text{km}]$$

$$J = |\Delta\mathbf{V}_1| + |\Delta\mathbf{V}_2|$$

$$\mathbf{U} = [\Delta\text{V}_{1x},\ \Delta\text{V}_{1y},\ \Delta\text{V}_{1z},\ \Delta\text{V}_{2x},\ \Delta\text{V}_{2y},\ \Delta\text{V}_{2z}]$$

The B-plane parameters are restored to their nominal prelaunch target values, and all the other hyperbolic parameters at the second Earth flyby including flight time are permitted to float. Experience has revealed that the flight time and approach velocity errors are small enough to be corrected by subsequent maneuvers. For the method of explicit functions, four additional equations of constraint ($\mathbf{Z}_F$) must be defined. A natural choice are the four hyperbolic parameters that are not constrained.

$$\mathbf{Z}_F = [t_p,\ V_\infty,\ \alpha_\infty,\ \delta_\infty]$$

A problem with this choice for $\mathbf{Z}_F$ is the sensitivity of the first maneuver to parameters defined after the second maneuver. The first maneuver must be determined through the second maneuver. For this reason, a preliminary search is conducted with $\mathbf{Z}_F$ defined by t_p and the three velocity components of the second maneuver rotated to along track, cross track, and out of plane components. The inplane velocity components for the second maneuver are constrained to zero, and a solution is obtained that is within 5 m/s of optimum.

The initial guess is input to initialize the optimizer which uses the method of explicit functions. The results after each iteration are given in Table 4.1.
The search algorithm attempts to drive the constraint variables to their desired values at the same time the performance index is being driven to a minimum value. At iteration 2, for example, a substantial reduction in J is achieved at the expense of driving the constraint variables away from their desired values. At iteration 4, a slight increase in performance index is obtained as the constraint variables nearly achieve their objective. From iteration 5 through 9, convergence is achieved as the optimization algorithm drives the optimization conditions to smaller values. The solution achieves an optimum within 0.1 mm/s before machine precision prohibits

Table 4.1 Optimization by explicit functions

	PI	Constraints		Optimization condition			
	J	B · R	B · T	$\frac{\partial J}{\partial t_p}$	$\frac{\partial J}{\partial V_\infty}$	$\frac{\partial J}{\partial \alpha_\infty}$	$\frac{\partial J}{\partial \delta_\infty}$
1	30.065554	−14,462.115	−17,789.846	0.1240E−05	0.3155E+00	−0.1107E+02	−0.3651E+01
2	26.249672	−14,287.217	−17,347.515	−0.4375E−06	−0.1903E+00	0.3855E+01	0.1314E+01
3	25.531653	−14,447.957	−17,750.009	−0.1150E−06	0.8816E−02	0.8356E+00	0.6061E+00
4	25.566200	−14,462.791	−17,792.569	−0.1677E−06	−0.2009E−01	0.1230E+01	0.8770E+00
5	25.520888	−14,462.540	−17,792.179	−0.4402E−07	−0.1596E−02	0.3625E+00	0.3040E+00
6	25.513694	−14,462.889	−17,792.492	0.2466E−07	0.3987E−02	−0.1789E+00	−0.1282E+00
7	25.530977	−14,462.585	−17,792.014	−0.1009E−06	−0.1068E−01	0.7454E+00	0.5284E+00
8	25.512837	−14,462.740	−17,792.466	−0.2351E−08	0.1684E−02	0.3485E−01	0.4391E−01
9	25.512751	−14,462.996	−17,792.953	0.1312E−07	0.1629E−02	−0.9681E−01	−0.6979E−01

any further reduction. The velocity components of the two maneuvers in the Earth mean equator of J2000 are

$$\Delta \mathbf{V}_1 = [12.186085, -13.684292, -8.4862428]\,\mathrm{m/s}$$

$$\Delta \mathbf{V}_2 = [3.6276212, -3.4959270, 1.7057893]\,\mathrm{m/s}$$

The first maneuver was a bit large for the maneuver system that had not been tested in space and was delayed until August 24, 2004, and only about 80% of the required velocity change was executed at this time. A small makeup maneuver was executed on September 24, 2004. The maneuver scheduled for November 19, 2004, was executed as planned.

The same problem may be solved by the method of gradient projection. This method requires an awkward choice of which independent parameters are "state" parameters and which are "decision" parameters. A choice of four $\mathbf{U}_F$ parameters must be made from two sets of maneuver parameters each of dimension three. The following arbitrary partition of maneuver velocity components into the categories required by gradient projection was used for the search:

$$\mathbf{U}_C = [\Delta \mathrm{V}_{1x},\ \Delta \mathrm{V}_{1y}] \tag{4.52}$$

$$\mathbf{U}_F = [\Delta \mathrm{V}_{1z},\ \Delta \mathrm{V}_{2x},\ \Delta \mathrm{V}_{2y},\ \Delta \mathrm{V}_{2z}] \tag{4.53}$$

The gradient projection search algorithm was implemented by making $\mathbf{Z}_F$ equal to $\mathbf{U}_F$ and using the same explicit function algorithm as above. The search was started with the maneuver velocity components set to zero, and the results after each iteration are given in Table 4.2. The first iteration moved the target variables from about 2 million km to within 20,000 km of the desired target. By the third iteration the target variables were within 200 km of their desired value and the performance index was within 1 m/s of optimum. Iterations 4–8 were within the linear region of the second partial derivatives, and quadratic convergence is observed. The indication of quadratic convergence is an order of magnitude reduction in the optimization condition after each iteration until the machine precision limit is reached.

4.8 Multiple Encounter Optimization

For the problem of multiple encounter trajectory design, the independent control parameters are the components of propulsive maneuvers strategically placed along the flight path to enable the spacecraft to attain the target body. The candidate constraint variables are simply the position or some simple function of the position of the spacecraft with respect to the various bodies that the spacecraft encounters or some simple function of the independent parameters such as the magnitude or direction of propulsive maneuvers. For a typical trajectory design problem, the constraint variables may be the two components of the position vector in the final

Table 4.2 Optimization by gradient projection

	PI	Constraints		Optimization condition			
	J	B · R	B · T	$\frac{\partial J}{\partial \mathrm{U}_F} - \frac{\partial J}{\partial \mathrm{U}_C} \frac{\partial \mathrm{Z}_C^{-1}}{\partial \mathrm{U}_C} \frac{\partial Z_C}{\partial \mathrm{U}_F} = 0$			
1	0.000000	810,838.303	2,107,832.409	0.0000E+00	0.0000E+00	0.0000E+00	0.0000E+00
2	26.231575	−41,369.351	−982.876	0.3501E−02	0.1454E+00	−0.2393E+00	−0.7324E−01
3	26.323229	−14,420.254	−17,938.400	−0.2132E−01	0.5700E+00	0.4645E+00	−0.1632E+00
4	25.530799	−14,462.669	−17,805.150	0.1309E−01	−0.3024E+00	0.9834E−01	0.1208E+00
5	25.512553	−14,462.601	−17,791.852	−0.3134E−02	0.4177E−01	−0.5895E−02	−0.1555E−01
6	25.512403	−14,462.998	−17,792.996	0.4131E−03	−0.5040E−02	0.8746E−03	0.1905E−02
7	25.512401	−14,463.000	−17,793.000	−0.4762E−04	0.5815E−03	−0.1010E−03	−0.2197E−03
8	25.512401	−14,463.000	−17,793.000	0.5495E−05	−0.6708E−04	0.1165E−04	0.2535E−04

target body B-plane, the time of closest approach to the final target body and the altitude of closest approach at some of the intervening bodies. The performance criterion is the sum of the magnitudes of the propulsive maneuvers. At least two propulsive maneuvers are placed between each encounter en route to the target body. For example, a multiple encounter mission to Mercury that is launched from Earth and encounters the Earth one additional time, Venus two times, and Mercury three times before arriving at the fourth and final Mercury encounter would have 7 legs with 14 propulsive maneuvers for a minimum total of 42 independent parameters. The constraint variables are the target body B-plane parameters including time of closest approach at the fourth Mercury encounter. This strategy was implemented and a multiple encounter trajectory designed from Earth to Mercury. Since this trajectory design is similar to the actual design of the MErcury Surface Space ENvironment, GEochemistry, and Ranging (MESSENGER) mission to Mercury, the results will be compared.

4.8.1 Multiple Encounter Strategy

Application of parameter optimization algorithms to the problem of multiple planetary encounter trajectory optimization from Earth launch to the target planet encounter is impractical. The number of independent parameters is excessive. For a seven-encounter mission from Earth to Mercury with 18 propulsive maneuvers, four more than the minimum required, the number of independent parameters is 54 corresponding to three velocity components for each maneuver. The number of function evaluations per iteration required by the method of explicit functions is four times the square of the number of parameters or 11,664. Other methods, such as gradient projection, would probably not fair any better. Each function evaluation involves integrating the trajectory from each propulsive maneuver to the final planetary encounter. With current computer technology, the computer processing time would be excessive. Another problem is even more insidious. The sensitivity of a position perturbation at the target to a velocity perturbation near the Earth is about 10^{32} s. In order to successfully target maneuvers, the trajectory calculations would need to be carried out in quadruple precision (116 bits). A velocity perturbation of one Angstrom per 20 billon years at Earth launch would result in a 15 km perturbation at the final planetary encounter.

The strategy for reducing sensitivity to velocity perturbations is to divide the trajectory into several legs and then group the legs into segments for optimization. A trajectory leg starts shortly after a planetary encounter and ends shortly after the next planetary encounter. A trajectory segment consists of two successive trajectory legs. For a given trajectory, the trajectory segments overlap and this results in the number of trajectory segments being one less than the number of legs. This strategy is illustrated on Fig. 4.3.

The ith leg is integrated from the initial time (to_i) to the final time (tf_i) and the integration is stopped at each propulsive maneuver $(ta_i,\ tb_i,\ tc_i,\ \cdots)$ and at

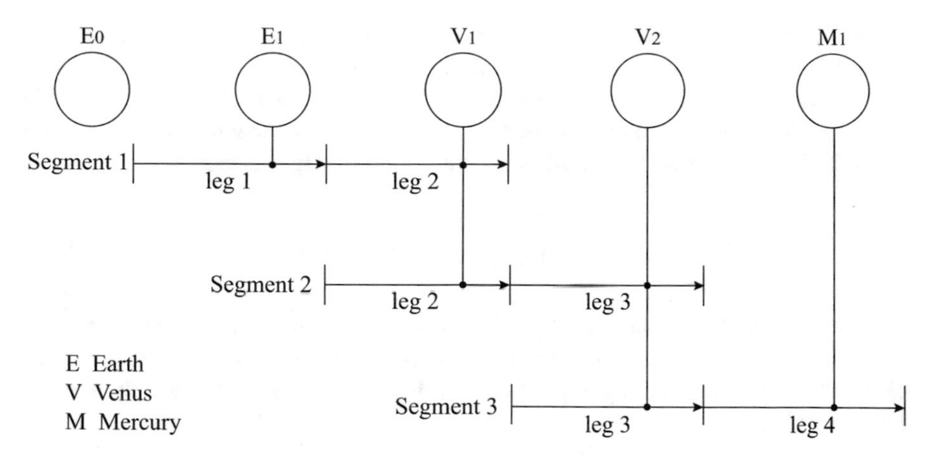

Fig. 4.3 Definition of segments and legs

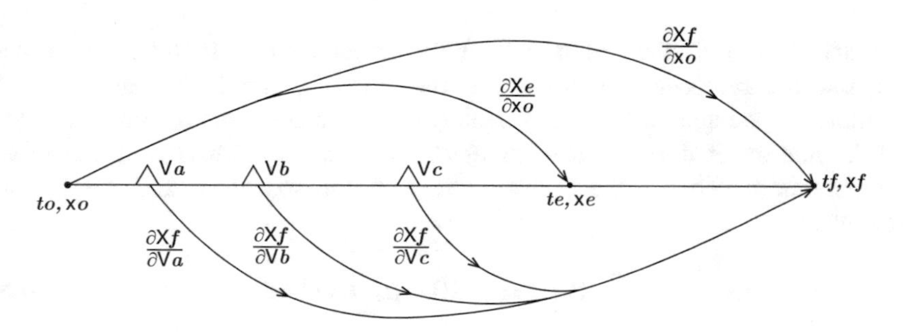

Fig. 4.4 Trajectory leg schematic diagram

the nominal time of the planetary encounter (te_i). The spacecraft states and partial derivatives that are needed by the optimization algorithm are saved. This strategy is schematically represented on Fig. 4.4. The trajectory optimization algorithm is initialized with a preliminary design obtained by patching conic sections using Lambert's theorem and other design techniques including shooting. This preliminary design fixes the time of the maneuvers and trajectory end points. The initial state is obtained for each leg along with nominal values for the deterministic propulsive maneuvers and planetary encounter aim points. At the conclusion of an optimization iteration, the propulsive maneuver velocity components are updated along with the time and aim points at each planetary encounter. Because of nonlinearity and machine precision, the initial state and propulsive maneuvers result in the trajectory missing the desired encounter conditions at the planetary encounter associated with each leg. The intermediate planetary encounter times and aim points are needed to shepherd the trajectory to the final planetary encounter.

For the ith trajectory leg, the initial state and propulsive maneuvers that occur during the ith leg are input to a precision trajectory propagator and the trajectory is integrated from to_i to tf_i. The spacecraft state at the planetary encounter of the

ith leg will differ from the desired B-plane encounter conditions. A correction is computed for one of the propulsive maneuvers in the ith leg to force the trajectory through the desired position in the B-plane. The velocity error is left uncontrolled and is permitted to accumulate. Restoring the position error will tend to also restore the velocity error. Thus, if the first maneuver in the ith leg is selected,

$$\Delta \mathbf{V}a_i = \left[\frac{\partial \mathbf{B}p_i}{\partial \mathbf{V}a_i} \right]^{-1} (\mathbf{B}t_i - \mathbf{B}p_i) \tag{4.54}$$

B is a column matrix containing B-plane parameters that are a simple transformation of the spacecraft state at encounter into a two-body conic. $\mathbf{B}p_i$ contains the first three elements of **B** corresponding to position, and $\mathbf{B}t_i$ are the desired target values at the ith encounter.

$$B = (B \cdot T, B \cdot R, t_p, V_\infty, \alpha_\infty, \delta_\infty)$$

$B \cdot T$ and $B \cdot R$ are the coordinates of the approach asymptote in the target plane normal to the approach asymptote, t_p is the time of periapsis passage, V_∞ is the magnitude of the approach velocity vector, and α_∞ and δ_∞ are the right ascension and declination of the approach asymptote, respectively. There is a one to one correspondence between the Cartesian state of the spacecraft and the B plane parameters.

$$\mathbf{B}_i(t) = f(\mathbf{X}e_i, \mu_i, te_i) \tag{4.55}$$

where μ is the gravitational parameter of the ith planet and te_i is the time of the state vector $\mathbf{X}e_i$ at the ith planetary encounter. The incremental velocity change ($\Delta \mathbf{V}a_i$) is added to $\mathbf{V}a_i$ and the trajectory integrated again from to_i to tf_i. The targeting calculations are repeated iteratively until the B-plane error is nulled to an acceptable tolerance.

The 3×3 matrix of partial derivatives in Eq. (4.54) may be obtained from the first three rows of the full state maneuver matrix that may be computed from the matrices illustrated in Fig. 4.4.

$$\left[\frac{\partial \mathbf{B}_i}{\partial \mathbf{V}a_i} \right] = \left[\frac{\partial \mathbf{B}_i}{\partial \mathbf{X}e_i} \right] \left[\frac{\partial \mathbf{X}e_i}{\partial \mathbf{X}o_i} \right] \left[\frac{\partial \mathbf{X}f_i}{\partial \mathbf{X}o_i} \right]^{-1} \left[\frac{\partial \mathbf{X}f_i}{\partial \mathbf{V}a_i} \right]$$

The state at the end of the ith trajectory leg is used to initialize the state at the beginning of the $i+1$th leg. Starting with the first leg and continuing to the final leg, adjustments are made to the propulsive maneuvers that result in a smooth trajectory from launch to the final encounter that passes through the desired aim point at each planetary encounter with no position or velocity discontinuities. Since the propulsive maneuvers are modeled as finite burns, there is no velocity discontinuity associated with propulsive maneuvers.

4.8.2 Trajectory Segment Optimization

The trajectory optimization algorithm refines the constraint parameters and reduces the total ΔV required for propulsive maneuvers. This algorithm is applied iteratively until no further decrease is obtained. Between each iteration, the trajectory must be retargeted to remove position discontinuities introduced by nonlinearity. Because of the sensitivity of state perturbations at the end of a trajectory leg to velocity perturbations in the previous leg, it will be convenient to break the trajectory into segments that span two successive trajectory legs as illustrated on Fig. 4.3. For a given trajectory segment, the initial state of the first leg and final state of the second leg are constrained to their current values and the only parameters that are permitted to vary are those associated with the propulsive maneuvers that occur during the segment and the planetary encounter of the first leg. Since the legs are defined such that there are no propulsive maneuvers after the planetary encounter, constraining the end state of the second leg is equivalent to constraining the planetary encounter on the second leg.

The constrained parameter optimization algorithm requires as input target values for the constraint parameters, nominal values for the control parameters and constraint parameters along with their partial derivatives obtained by precision numerical integration of the state and variational equations. In addition, tolerances on finite difference partial step sizes and convergence tolerances are needed. A linear correction to the control parameters is output that lowers the performance index and holds the constraint parameters at their target values. This processing is repeated for each trajectory segment starting with the first two legs and proceeding to the final leg. At the completion of each segment, the encounter time and aim point for the first leg in the segment must be updated. This update consists of mapping the velocity change associated with propulsive maneuvers that occur during the first leg of each segment to the first leg encounter.

$$\Delta \mathbf{B}_i = \frac{\partial \mathbf{B}_i}{\partial \mathbf{V}a_i} \Delta \mathbf{V}a_i + \frac{\partial \mathbf{B}_i}{\partial \mathbf{V}b_i} \Delta \mathbf{V}b_i + \frac{\partial \mathbf{B}_i}{\partial \mathbf{V}c_i} \Delta \mathbf{V}c_i + \cdots \tag{4.56}$$

The encounter time and aim point for the second encounter need not be updated since they are constrained. The second encounter of a trajectory segment becomes the first encounter of the next segment and is updated as part of the processing of that segment. The updated control and encounter aim points are targeted as described above to remove position discontinuities between the segments caused by nonlinearity. The targeting and optimization processing is repeated until convergence is obtained.

4.8.3 Multiple Encounter Example

As an example, a seven-encounter trajectory from Earth to Mercury is targeted to minimize propellant expenditure. This example is close to the MESSENGER Mission trajectory. The major differences are in the trajectory propagator, and initial conditions assumed. The initial state after launch is determined by processing several weeks of Doppler and range tracking data, and the resulting optimum trajectory includes removal of actual launch vehicle injection errors. The first leg is a return to Earth trajectory, and the first segment includes the first encounter of Venus leg. Subsequent legs are a return to Venus and four Mercury encounters. In order to prevent the trajectory design from intersecting the surfaces of Venus and Mercury, the second Venus encounter altitude is constrained to be no less than 300 km and the first three Mercury encounter altitudes are constrained to be no less than 200 km.

The velocity change associated with each maneuver is shown in Table 4.3. The initial solution was obtained from the MESSENGER prelaunch trajectory design modified to remove the launch vehicle injection error. Because of minor trajectory model errors, probably associated with solar pressure, the first iteration diverges from the prelaunch reference trajectory and the total propulsive ΔV is about 1187 m/s. The next four iterations reduce the ΔV by about 130 m/s resulting primarily from optimizing allocation between maneuvers on the same leg. Some ΔV is shifted from leg 2 to leg 4 by adjusting the encounter aim point and arrival time at leg 3. The last four iterations refine the propellant allocation among the maneuvers and achieve another 10 m/s reduction in ΔV.

The MESSENGER postlaunch trajectory design results are shown at the bottom of Table 4.3. The total ΔV of 1042 m/s is 3 m/s less than obtained here. The ΔV allocation to the individual legs may differ by 10–20 m/s. These differences may be attributed to modeling errors and curvature of the performance index function near the optimum solution. The performance index function is nearly flat at the solution point. Large changes in the control parameters that satisfy the constraints result in small changes in the performance index or ΔV. This behavior of the performance index function makes it difficult to find the true minimum. However, it does not cost much to be a little suboptimum.

4.9 Summary

Trajectory optimization performed for navigation involves searching for the initial conditions and deterministic propulsive maneuvers that acquire the target, satisfy mission constraints, and minimize some performance criterion such as fuel expenditure. Analytic solutions are only approximate because their does not exist a closed form solution for the trajectory that is accurate enough. The method of Lagrange multipliers, that is useful for an analytic solution, does not perform well when incorporated into a numerical search algorithm. The methods of explicit functions or gradient projection provide accurate numerical solutions. Since the solution does not

Table 4.3 Optimization of seven-leg mission to Mercury

Iteration	LEG 1 (E0-E1) MVR1[a] MVR2	LEG 2 (E1-V1) MVR1 MVR2	LEG 3 (V1-V2) MVR1 MVR2 MVR3	LEG 4 (V3-M1) MVR1 MVR2 MVR3	LEG 5 (H1-M2) MVR1 MVR2	LEG 6 (H2-M3) MVR1 MVR2 MVR3	LEG 7 (M3-H4) MVR1 MVR2 MVR3	Total
1	20.53	80.62	3.02	7.46	33.58	3.49	13.97	
	5.11	304.33	2.11	209.26	73.20	240.86	178.02	
			2.34	4.57		2.76	1.96	1187.16
2	20.47	70.24	2.98	1.21	34.17	1.80	4.10	
	5.14	303.49	1.72	208.92	73.67	241.33	175.66	
			3.29	4.11		0.99	0.20	1153.51
3	20.36	0.54	6.25	11.33	30.75	2.30	2.71	
	5.34	301.65	0.94	123.85	74.39	242.74	175.68	
			4.88	92.20		0.26	0.22	1096.36
4	20.37	0.69	2.45	11.91	0.29	0.41	2.49	
	5.46	304.19	1.71	103.31	69.14	243.13	175.87	
			1.82	119.17		0.26	0.13	1062.79
5	20.74	0.66	2.40	20.74	0.63	0.53	0.27	
	5.33	306.92	1.52	200.12	73.91	243.03	174.96	
			1.37	0.00		1.19	1.13	1055.45
6	19.95	1.60	4.85	0.37	3.65	0.60	4.19	
	5.72	303.04	5.89	216.01	71.88	239.91	175.82	
			0	0		0	0	1053.49

(continued)

Table 4.3 (continued)

Iteration	LEG 1 (E0-E1) MVR1[a] MVR2	LEG 2 (E1-V1) MVR1 MVR2	LEG 3 (V1-V2) MVR1 MVR2 MVR3	LEG 4 (V3-M1) MVR1 MVR2 MVR3	LEG 5 (H1-M2) MVR1 MVR2	LEG 6 (H2-M3) MVR1 MVR2 MVR3	LEG 7 (M3-H4) MVR1 MVR2 MVR3	Total
7	20.75	1.39	1.58	13.43	1.08	0.79	0.31	
	5.34	307.15	1.88	202.27	73.86	242.26	175.01	
			0.58	2.24		0.22	1.00	1051.14
8	20.74	1.82	1.58	9.89	1.52	0.00	0.86	
	5.38	306.67	2.18	199.70	73.09	1.32	0.33	
			1.20	5.78		242.98	174.92	1049.97
9	20.74	0.69	1.03	9.20	0.53	0.64	0.33	
	5.37	306.49	1.62	199.90	71.86	242.91	174.91	
			0.88	5.81	1.25	0.53	1.24	1045.95
MESSENGER	0.64	0.02	1.43	1.59	0.08	0.81	1.74	
	25.82	312.20	3.006	204.82	73.77	240.35	176.35	1042.65

[a]MVR1 is the i'th propulsive maneuver of the indicated leg in meters/second

depend on programming exact partial derivatives, finite difference partial derivatives are computed for the search. This is fortunate because a new set of analytic partial derivatives would have to be derived for every new problem. Once a solution is obtained, the high- precision trajectory can verify optimality by systematically perturbing the constraints.

A problem with trajectory optimization is the large number of function evaluations required when there are many constraint and control parameters. For the MESSENGER mission, it was necessary to segment the trajectory into overlapping segments and optimize each segment. The resulting segments had to be retargeted a small amount after each iteration to remove position discontinuities caused by nonlinearity. The segments also removed the extreme sensitivity of constraints to the propulsive maneuvers.

Exercises

4.1 A cylindrical oil can is being manufactured and, for a given volume, the cost of the steel is proportional to the surface area of the can if it is assumed to have uniform thickness. The problem is to find the height and radius that minimize the amount of steel. This problem can be formulated as a constrained parameter optimization problem where $U_1 = r$, the radius, and $U_2 = h$, the height.

$$Z_c = \pi U_1^2 U_2$$

$$J = V = 2\pi U_1^2 + 2\pi U_1 U_2$$

Using one of the optimization methods described in Sect. 4.4, determine the optimum U_1 and U_2. If the method of explicit functions is selected and $Z_f = \pi U_1 U_2^2$, the matrix inversion is simple and the solution is straightforward.

4.2 A sphere is inscribed inside the oil can of Exercise 4.1. Determine the ratio of the volume of this sphere to the volume of the oil can. This problem was solved by Archimedes and a cylinder and sphere were atop his tomb according to Cicero.

4.3 The Hessian matrix $(\frac{\partial^2 J}{\partial \mathbf{U} \partial \mathbf{Z}_F})$ which is used to obtain a minimum in a Newton–Raphson search can be used to verify that a minimum has been obtained. For the sample problem in Sect. 4.5, determine the Hessian and verify that it is positive indicating a minimum.

4.4 For interplanetary maneuvers, the spacecraft is often targeted to $B \cdot R$ and $B \cdot T$ and the time of flight is not corrected and permitted to float. The optimum maneuver can be found by solving a constrained parameter optimization problem.

$$Z_c = (B \cdot R, B \cdot T)$$

$$Z_f = t_p$$
$$J = |\Delta \mathbf{V}|$$

This maneuver is called a critical plane maneuver and can be computed from the K matrix defined by

$$\left[\frac{\partial B \cdot R,\ B \cdot T,\ t_p}{\partial \mathbf{V}}\right]$$

Determine the critical plane maneuver directly from the K matrix and the B-plane miss.

4.5 A spacecraft that is approaching Mars is inserted directly into orbit with a periapsis altitude of 1500 km and an orbit period of 24 h. The approach velocity (V_∞) is 2.54 km/s, GM is 42,828 km^3/s^2, and the radius of Mars is 3310 km. A second spacecraft with the same approach velocity is inserted into an orbit with a periapsis altitude of 1000 km and the same apoapsis radius as the first spacecraft. A maneuver is executed at apoapsis to raise the periapsis altitude to 1500 km. Assuming the orbit insertion maneuvers are impulsive, determine the total Δv required for each strategy. Which strategy is most fuel efficient?

4.6 Show that a Hohmann transfer between two circular orbits minimizes the launch energy from the first orbit and the orbit insertion energy at the second orbit. The minimum energy transfer orbit will be tangential at the second orbit and r_a will equal the radius of the second orbit. Crossing the second orbit requires more energy. The problem is thus reduced to minimizing v with respect to the flight path angle γ subject to the constraint that r_a is the radius of the second orbit and r is the radius of the first orbit where

$$v = \frac{\sqrt{GM\,p}}{r \cos\gamma}$$

4.7 If γ happens to be zero in Exercise 4.6, minimizing v also minimizes the velocity change since v and the orbital velocity are in the same direction. Compute the launch energy from Earth for a mission to Mars and the orbit insertion energy at Mars where $r_e = 0.149 \times 10^9$ km, $r_m = 0.227 \times 10^9$ km, $GM = 0.132 \times 10^{12}$ km^{3}s^{-2}

Bibliography

Bryson, A. E. and Yu-Chi Ho, *Applied Optimal Control*, Ginn Blaisdell Pub. Co., Waltham, MA., 1969.

Fletcher, R. and M.J.D. Powell, “A Rapidly Convergent Descent Method for Minimization”, *Computer Journal*, July, 1963.

Hoffman, L. H. and R. N. Green,"Thrusting Trajectory Minimization Program for Orbital Transfer Maneuvers", NASA TN d-1620, June 1971.

Miller, J. K., "Determination of an Optimal Control for a Planetary Orbit Insertion Maneuver", internal JPL document TM 392-94, July 20,1972.

Miller, J. K., "A Parameter Search and Optimization Algorithm for the Viking Mars Orbit Insertion Maneuver", internal JPL document IOM 392.5-747, April 11,1974.

Miller, J. K., "Determination of an Optimal Control Subject to Constraints by the Method of Explicit Functions", internal JPL document EM 314–412, February 20,1987.

Miller, J. K., "Constrained Parameter Optimization by the Method of Explicit Functions: Application to MESSENGER Mission", Paper AAS 05–174, 15th AAS/AIAA Space Flight Mechanics Conference, Copper Mountain,CO, January 23–27, 2005.

Miller, J. K., "A Multiple Encounter Targeing and Optimization Algorithm", AAS 05–336, AAS/AIAA Astrodynamics Specialist Conference, Lake Tahoe, CA, August 7, 2005.

Pines, S., "An accelerated Version of the JPL-MOIOP Numerical Search Optimization Method", Rep.No. 73–32, Analytical Mechanics Associates Inc., June, 1973

Pines, S., "A Direct Gradient Projection Method for Parameter Optimization with Nonlinear Constraints", AMA Rep. No. 75-8, Analytical Mechanics Associates Inc., April, 1975

Chapter 5
Probability and Statistics

Navigation of planetary spacecraft requires determining a nominal design trajectory that obeys the laws of physics and has a high probability of achieving mission success within the constraints of the mission objectives and the cost of the spacecraft design and mission operations. It is relatively easy to design a trajectory that satisfies all the physical laws but cannot be flown. For example, a trajectory describing the path of a coin that is tossed on the floor and rolls to a stop remaining on its edge is easy to design. However, the perturbations that the coin encounters as it rolls on the floor almost guarantees that it will not remain on its edge. Spacecraft trajectory design encounters this same problem in many forms. Statistical perturbations of the trajectory along the flight path may result in failure to meet mission objectives if not complete failure as in the case of the coin. Therefore, the trajectory designer and navigator must give as much attention to the mathematics of probability and statistics as to the laws of physics.

5.1 Normal Probability Distribution Function (PDF)

An arrow or unguided rocket would score a direct hit on the target provided they were launched with the proper initial conditions and there were no perturbing forces acting during transit. However, a body moving through the atmosphere will encounter dust particles or random hits from gas molecules that will deflect it from the target. If a large number of arrows or rockets are launched, a pattern emerges for the distribution of the impacts around the target. It is the characterization of this distribution that is of interest. Consider the case of a body that is launched and encounters dust particles on the way to the target. Assume that the dust particles deflect the body a fixed amount either to the right or to the left with equal probability. The geometry is illustrated on Fig. 5.1. The first impact is identified by $m = 0$ and the body moves an equal amount to the right or to the left. Both paths are shown

J. Miller, *Planetary Spacecraft Navigation*, Space Technology Library 37,
https://doi.org/10.1007/978-3-319-78916-3_5

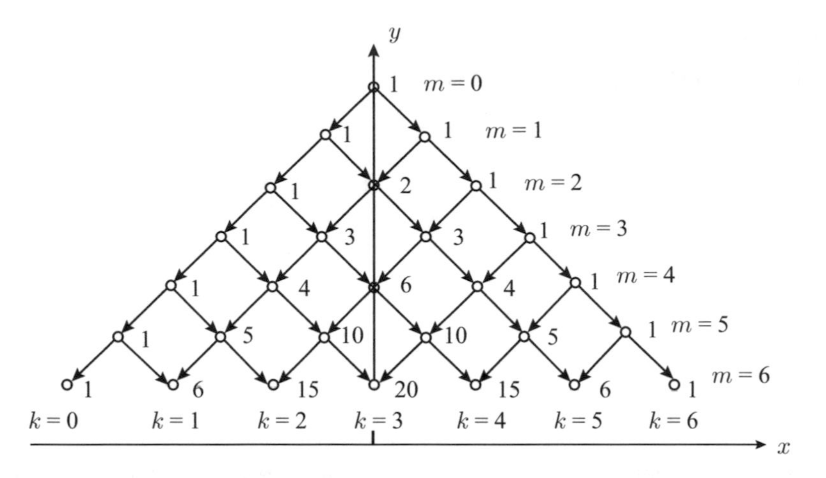

Fig. 5.1 Pascal's triangle

in Fig. 5.1. Each generation of impacts, corresponding to increasing values of m, double the number of possible paths. At the end of m generations, there are 2^m possible paths. The number beside each of the nodes is the total number of paths that pass through the node. Therefore, the probability that a path will pass through a given node is the number beside the node divided by the total number of possible paths since all paths are equally likely. The triangle illustrated on Fig. 5.1 is called Pascal's triangle, and the number of paths passing through each node is given by

$$p(m, k) = \binom{m}{k}$$

$$\binom{m}{k} = \frac{m!}{(m-k)!\,k!}$$

where k is the node numbered from the left starting at $k = 0$. For any given node of Pascal's triangle, the number is the sum of the two numbers immediately above.

$$\binom{m+1}{k+1} = \binom{m}{k} + \binom{m}{k+1}$$

The rows of Pascal's triangle are the coefficients of the binomial expansion. It will be shown later that the mth row of Pascal's triangle may be approximated by the normal probability distribution function (PDF) given by

$$\frac{dP(x)}{dx} = \frac{1}{\sigma\sqrt{2\pi}}\, e^{-\left(\frac{x^2}{2\sigma^2}\right)} \tag{5.1}$$

after x is properly scaled. In the limit as m goes to infinity, the approximation is exact. The probability that x is in the interval from x_1 to x_2 is given by

$$P(x_1, x_2) = \int_{x_1}^{x_2} \frac{1}{\sigma\sqrt{2\pi}} e^{-\left(\frac{x^2}{2\sigma^2}\right)} dx$$

and

$$\int_{-\infty}^{\infty} \frac{1}{\sigma\sqrt{2\pi}} e^{-\left(\frac{x^2}{2\sigma^2}\right)} dx = 1$$

A convenient measure of the spread of a PDF is given by the second moment about the y axis, which is the moment of inertia, and when applied to a PDF is called the variance. The variance is simple to compute and has the property of giving increased weight to the tails of the distribution just as the moment of inertia gives more weight to mass that is further from the axis of rotation. The variance is given by

$$V = \int_{-\infty}^{\infty} x^2\, p(x) dx$$

Since the integral of the normal PDF from minus infinity to infinity is one, the following is obtained after differentiating with respect to σ.

$$\int_{-\infty}^{\infty} \frac{x^2}{\sigma^4\sqrt{2\pi}} e^{-\left(\frac{x^2}{2\sigma^2}\right)} dx - \int_{-\infty}^{\infty} \frac{1}{\sigma^2\sqrt{2\pi}} e^{-\left(\frac{x^2}{2\sigma^2}\right)} dx = 0$$

After multiplying by σ^3, the variance is given by

$$V = \int_{-\infty}^{\infty} \frac{x^2}{\sigma\sqrt{2\pi}} e^{-\left(\frac{x^2}{2\sigma^2}\right)} dx = \sigma^2 \int_{-\infty}^{\infty} \frac{1}{\sigma\sqrt{2\pi}} e^{-\left(\frac{x^2}{2\sigma^2}\right)} dx = \sigma^2 \qquad (5.2)$$

The variance (σ^2) provides a measure of the error associated with a random variable and the reciprocal provides a measure of the accuracy. Since the application of variance is often in its minimization, the square or quadratic form is mathematically convenient since the minimum of a function of σ^2 is also the minimum of a function of σ. The simple result obtained for the variance was no accident but followed directly from the scaling assumed in Eq. (5.1).

5.2 n-Dimensional Normal PDF

The joint PDF of n independent normally distributed random variables $(y_1, y_2, \dots y_n)$ is defined as the probability that y is in all of the intervals from y_i to $y_i + \Delta y_i$. The PDF is obtained by multiplying together n normal PDFs.

$$p(y_1, y_2, \dots y_n) = \frac{1}{(2\pi)^{\frac{n}{2}} (\sigma_{y_1} \sigma_{y_2} \sigma_{y_3} \dots \sigma_{y_n})} \exp\left(-\frac{1}{2}\sum_{i=1}^{n} \frac{y_i^2}{\sigma_{y_i}^2}\right)$$

In matrix notation, the normal joint PDF becomes

$$p(Y) = \frac{1}{(2\pi)^{\frac{n}{2}} |A|^{-\frac{1}{2}}} \exp\left(-\frac{1}{2} Y^T A Y\right) \tag{5.3}$$

where

$$Y = \begin{bmatrix} \sigma_{y_1} \\ \sigma_{y_2} \\ \vdots \\ \sigma_{y_1} \end{bmatrix}$$

and

$$A = \begin{bmatrix} \frac{1}{\sigma_{y_1}^2} & 0 & 0 & \dots & 0 \\ 0 & \frac{1}{\sigma_{y_2}^2} & 0 & \dots & 0 \\ 0 & 0 & \frac{1}{\sigma_{y_3}^2} & \dots & 0 \\ \vdots & \vdots & \vdots & \ddots & \vdots \\ 0 & 0 & 0 & \dots & \frac{1}{\sigma_{y_1}^2} \end{bmatrix}$$

If a new random variable X is defined that is a transformation or mapping of Y then

$$p(X) = \frac{1}{(2\pi)^{\frac{n}{2}} |B|^{-\frac{1}{2}}} \exp\left(-\frac{1}{2} X^T B X\right) \tag{5.4}$$

where

$$X = R\,Y$$

$$B = RAR^{-1}$$

The matrix B is called the information matrix and the inverse of B is called the covariance matrix. Each diagonal element of B^{-1} is the variance of the associated random variable. The covariance matrix of the new multidimensional normal PDF of the random variables X is given by

$$B^{-1} = \begin{bmatrix} \sigma_{x_1}^2 & \rho_{12}\sigma_{x_1}\sigma_{x_2} & \rho_{13}\sigma_{x_1}\sigma_{x_3} & \cdots & \rho_{1n}\sigma_{x_1}\sigma_{x_3} \\ \rho_{21}\sigma_{x_2}\sigma_{x_1} & \sigma_{x_2}^2 & \rho_{23}\sigma_{x_2}\sigma_{x_3} & \cdots & \rho_{2n}\sigma_{x_2}\sigma_{x_n} \\ \rho_{31}\sigma_{x_3}\sigma_{x_1} & \rho_{32}\sigma_{x_3}\sigma_{x_2} & \sigma_{x_3}^2 & \cdots & \rho_{3n}\sigma_{x_3}\sigma_{x_n} \\ \vdots & \vdots & \vdots & \ddots & \vdots \\ \rho_{n1}\sigma_{x_n}\sigma_{x_1} & \rho_{n2}\sigma_{x_n}\sigma_{x_2} & \rho_{n3}\sigma_{x_n}\sigma_{x_3} & \cdots & \sigma_{x_n}^2 \end{bmatrix}$$

Given B^{-1}, the mapping matrix R can be found by extracting the eigenvalues of B^{-1}. The matrix R is the matrix of eigenvectors, and the diagonal matrix A^{-1} has the eigenvalues on the diagonal.

5.3 Bivariate Normal PDF

The multidimensional normal PDF for $n = 2$ is called the bivariate normal PDF. The probability that x is in the interval from x_1 to $x_1 + \Delta x_1$ and in the interval from x_2 to $x_2 + \Delta x_2$ is the probability that x is in the region defined by these intervals. If we assume that x_1 and x_2 are the Cartesian coordinates x and y, respectively, then the bivariate PDF gives the probability density associated with areas in the $x - y$ plane. Let X and Y be joint normal random variables. The covariance and determinate of B^{-1} are given by,

$$B^{-1} = \begin{bmatrix} \sigma_x^2 & \rho\sigma_x\sigma_y \\ \rho\sigma_x\sigma_y & \sigma_y^2 \end{bmatrix}$$

$$|B^{-1}| = (1 - \rho^2)\sigma_x^2\sigma_y^2$$

and

$$B = \frac{1}{1-\rho^2} \begin{bmatrix} \frac{1}{\sigma_x^2} & -\frac{\rho}{\sigma_x\sigma_y} \\ -\frac{\rho}{\sigma_x\sigma_y} & \frac{1}{\sigma_y^2} \end{bmatrix}$$

The bivariate normal PDF is thus given by

$$p(x, y) = \frac{1}{2\pi\sigma_x\sigma_y\sqrt{1-\rho^2}} \exp\left(-\frac{1}{2(1-\rho^2)}\left[\frac{x^2}{\sigma_x^2} - \frac{2\rho xy}{\sigma_x\sigma_y} + \frac{y^2}{\sigma_y^2}\right]\right) \tag{5.5}$$

Contours of constant $p(x, y)$ plotted in the $x - y$ plane are ellipses given by

$$\frac{x^2}{\sigma_x^2} - \frac{2\rho xy}{\sigma_x \sigma_y} + \frac{y^2}{\sigma_y^2} = \mathrm{C}$$

where C is a constant. These ellipses have semi-major and semi-minor axes $\mathrm{C}\lambda_1$ and $\mathrm{C}\lambda_2$, respectively, given by

$$\lambda_1^2 = \frac{1}{2}(\sigma_x^2 + \sigma_y^2) + \sqrt{\left(\frac{\sigma_x^2 - \sigma_y^2)}{2}\right)^2 + \rho^2 \sigma_x^2 \sigma_y^2}$$

$$\lambda_2^2 = \frac{1}{2}(\sigma_x^2 + \sigma_y^2) - \sqrt{\left(\frac{\sigma_x^2 - \sigma_y^2)}{2}\right)^2 + \rho^2 \sigma_x^2 \sigma_y^2}$$

The major axis of the error ellipse is inclined to the x axis at an angle (θ) given by

$$\theta = \frac{1}{2} \tan^{-1} \left(\frac{2\rho \sigma_x \sigma_y}{\sigma_x^2 - \sigma_y^2} \right)$$

It can be shown that the probability that the random variables (x, y) are inside the error ellipse is

$$p(x, y) = 1 - e^{-\left(\frac{\mathrm{C}^2}{2}\right)}$$

For integer values of C the corresponding error ellipses are often referred to as the C sigma error ellipses. Thus, the probability of X being in the one sigma or three sigma error ellipse is 0.393 and 0.989, respectively.

An example of the application of the bivariate PDF to navigation was provided by the Viking mission to Mars. The Viking lander was targeted to a landing site defined by target coordinates x_t and y_t. Analysis of the accuracy of the orbit of the Viking orbiter and the lander descent trajectory revealed a footprint centered at the targeted landing site and oriented as shown schematically in Fig. 5.2. The footprint is jargon for a bivariate PDF. Orbiter reconnaissance images of the landing site region revealed a large crater just outside of the footprint. The crater is also shown in Fig. 5.2 overlaid with a grid of rectangles of width Δx and Δy. The center of each rectangle within the crater has coordinates (x_i, y_j). The probability that the lander will land in the crater is obtained by integrating the associated bivariate PDF over the crater area. This integral, in the limit as Δx and Δy approach zero, is given by

$$P = \sum p(x_i, y_j) \Delta x \Delta y$$

The numerical integration revealed a probability of less than 10^{-5}, and this was judged to be small enough that the lander was not retargeted.

Fig. 5.2 Lander footprint

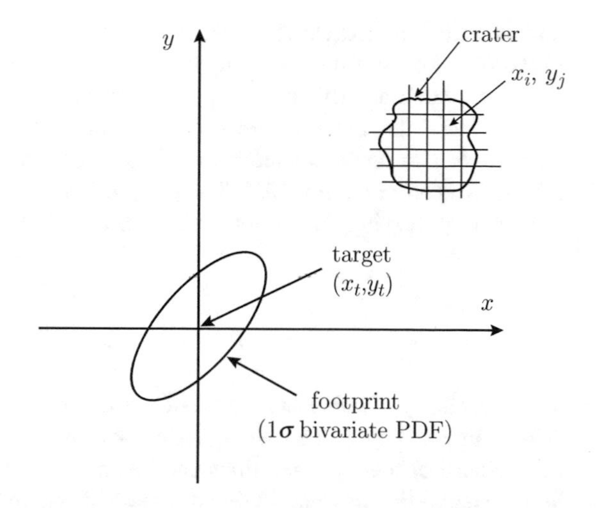

5.4 Rayleigh PDF

A special case of the bivariate PDF occurs when σ_x and σ_y are equal and x and y are independent ($\rho = 0$). The resulting error ellipse is a circle and the PDF reduces to

$$p(x, y) = \frac{1}{2\pi\sigma^2} e^{-\left(\frac{x^2+y^2}{2\sigma^2}\right)}$$

The probability of the random variable X being in the circle is obtained by integrating the PDF over the circle. A change of variable to polar coordinates simplifies the integration and

$$P(r, \theta) = \frac{1}{2\pi\sigma^2} \int_0^{2\pi} \int_0^r e^{-\left(\frac{r^2}{2\sigma^2}\right)} r dr d\theta$$

Performing the θ integration first,

$$P(r) = \frac{1}{\sigma^2} \int_0^r e^{-\left(\frac{r^2}{2\sigma^2}\right)} r dr$$

and the PDF associated with r is given by

$$p(r) = \frac{r}{\sigma^2} e^{-\left(\frac{r^2}{2\sigma^2}\right)}$$

and is called the Rayleigh PDF. The Rayleigh PDF has many applications in probability and statistics. One application is in setting accuracy requirements for armament such as artillery and rockets. Several rounds are fired, and a circle is drawn around the target that encompasses the impacts of exactly half of the rounds. This circle and its associated probability is called the Circular Error Probable in military jargon or simply CEP. The probability of impact within the circle may be obtained by carrying out the integration of the Rayleigh PDF with respect to r.

$$P(r) = 1 - e^{-\left(\frac{r^2}{2\sigma^2}\right)} \tag{5.6}$$

The probability of being in a circle of radius one sigma where the PDF is Rayleigh is the same as being in a one sigma error ellipse where the PDF is bivariate. The equivalence of these probabilities can be shown by an appropriate scaling of the x and y axes of the bivariate PDF. For the CEP, where the probability is one half, the corresponding error ellipse of the bivariate is at 1.17741 sigma.

5.5 Central Limit Theorem

The artifically contrived problem of an arrow or unguided rocket being deflected to the left or to the right with equal probability is equivalent to the problem of flipping a coin to decide the path to follow. An interesting result has been obtained. The sum of many trials involving a probability distribution function that has only two states of equal probability results in the normal distribution function. The central limit theorem considers the problem of a sum drawn from a large number of probability distribution functions of arbitrary distribution. The central limit theorem states, in essence, that this sum also has a normal distribution. The proof is nontrivial and involves some complex issues. However, the central limit theorem is of such importance to probability and statistics that a simplified outline of this proof, provided by Harry Lass, is given.

Consider an arbitrary probability density function of zero mean and variance σ. The Fourier transform is obtained from the moment generating function and

$$F_x(\omega) = 1 - \frac{\sigma^2\omega^2}{2} + K_3\omega^3 + K_4\omega^4 + \dots$$

The PDF of the sum of two samples drawn from F_x is obtained by evaluation of the convolution integral associated with the probability distribution of each sample. The convolution of probability density functions involves a double integral that sums the probabilities associated with all the ways that two numbers can sum to a third number. The Fourier transform of the convolved functions is simply the product of the Fourier transform associated with each function. Therefore, the

Fourier transform of the sum of n samples from F_x is obtained by raising F_x to the nth power. As the number of convolutions approaches infinity, the variance of the resulting probability distribution function will also approach infinity. In order to bound the resulting variance, the variance of F_x must be scaled down and this is accomplished by scaling the variable ω to $\frac{\omega}{\sqrt{n}}$. The scaling preserves the variance without changing the shape of the probability distribution function. The Fourier transform of the resulting probability distribution becomes,

$$F_y(\omega) = \left[1 - \frac{\sigma^2\omega^2}{2n} + K(\frac{\omega}{\sqrt{n}})\frac{\omega^3}{n^{\frac{3}{2}}}\right]^n = f(n)^n$$

Making use of the relationship

$$F_y = [f(n)]^n = \exp[n \ln f(n)] = \exp\left(\frac{\ln f(n)}{n^{-1}}\right)$$

The limit of F_y, as n approaches infinity, is obtained by applying L'Hospital's rule.

$$\lim_{n\to\infty} F_y = \lim_{n\to\infty} \exp\left[\frac{\frac{\sigma^2\omega^2}{2}n^{-2} - \frac{3}{2}K(\frac{\omega}{\sqrt{n}})\omega^3 n^{-\frac{5}{2}} - \frac{1}{2}\frac{\partial K(\frac{\omega}{\sqrt{n}})}{\partial(\frac{\omega}{\sqrt{n}})}\omega^4 n^{\frac{-5}{2}}}{-n^{-2} f(n)}\right]$$

$$\lim_{n\to\infty} F_y = \exp\left(\frac{-\sigma^2\omega^2}{2}\right)$$

In the limit as n approaches infinity, the inverse Fourier transform of F_y is the normal PDF.

5.6 Monte Carlo Methods

Navigation requirements are generally based on an analysis of various errors that affect the determination and control of a spacecraft trajectory. The errors in parameters that are of interest are generally expressed in the elements of the covariance matrix associated with these parameters. Instrumentation and spacecraft execution errors are evaluated to determine the covariance matrix that describes the errors in design parameters such as closeness of the spacecraft to the target or the amount of fuel that may be consumed in performing the mission. Conversely, the desired maximum value for the errors in mission design parameters drives the accuracy requirements on navigation instrumentation and spacecraft system design.

For a spacecraft approaching a planet or in a well-determined orbit, the covariance matrix is often used to describe the errors of interest. Since the errors in trajectory parameters in deep space far from a planetary body are small compared to the magnitude of the parameters being determined, the statistics tend to be normally distributed owing to the central limit theorem. Also, for small perturbations from the nominal value of these parameters, the design parameters are linear, the linear theory that orbit determination is based on is validated, and the normal PDF associated with the covariance matrix provides an excellent description of the error distribution.

As a spacecraft approaches a planetary body and is inserted into orbit or descends from an orbit to land on a body, the errors in knowledge of trajectory grow and may exceed the region where linear theory may be applied. The resulting probability distribution is often distorted from the normal curve. Unfortunately, the probability distribution of the spacecraft state on achieving orbit or at touchdown of a lander is of considerable interest to trajectory design. The probability distribution may be determined by application of a simple but powerful technique called Monte Carlo mapping named after the gambling casinos of Monte Carlo. The Monte Carlo technique consists of defining a mathematical model of the system being investigated including all the error sources that affect the outcome and performing a statistical analysis of the outcome of many executions of the mathematical model.

For application of the Monte Carlo method to navigation analysis, a precision model of the spacecraft trajectory must be defined. This model involves numerical integration of the equations of motion, including propulsive events, from initial conditions determined, for example, on approach to a planetary body to final conditions computed in orbit. The error sources include initial spacecraft state and propulsion system execution errors. Error sources are described by their associated PDF which is generally normal or uniform. A random number generator is used to generate samples of the PDF associated with each error source. A separate random number generator is required for each PDF and may not be readily available. Most computers have a random number generator that will generate random numbers between 0 and 1 that are uniformly distributed. One method for obtaining these numbers might be to take the first 10 digits of pi and put a decimal point in front of them. The next random number may be formed from the next 10 digits of pi. The uniform random number generators for most computers are more sophisticated than this simple example but using pi would suffice since the digits are random. The normal PDF can be formed from uniform random numbers by use of the central limit theorem. For example, 6 random numbers uniformly distributed between -1 and 1 are obtained and added together. Since the variance of a single sample is one third, the PDF obtained by adding 6 samples has a variance of 2 and standard deviation $\sqrt{2}$. The resultant PDF may be scaled to have unity variance by dividing each sum of 6 samples by $\sqrt{2}$. The result is a PDF that approximates the normal PDF and is bounded by plus and minus $3\sqrt{2}$-sigma. Figure 5.3 shows a histogram with a bin width of 0.05 generated from 2 million samples of this normal PDF approximation. The normal PDF (dashed line) is also shown for comparison.

Random numbers generated from a normal PDF with a variance of one are multiplied times the standard deviation of all the independent parameters that contribute to the final result. Included in this set of independent parameters are initial conditions and random parameters associated with propulsive events. The trajectory is propagated from the initial condition to the final condition n times,

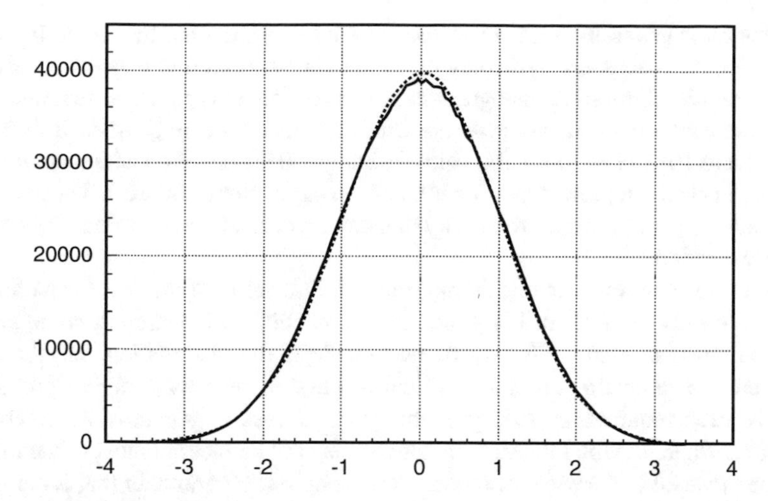

Fig. 5.3 Histogram of normal approximation

and the parameters of interest are saved in n random vectors X_i. For each of the n trajectory propagations, new random numbers are drawn for each of the independent parameters. The number of Monte Carlo samples (n) generated is limited by the computer time required to generate each sample. The more samples, the better, and the minimum required is generally around 500. Once the samples are generated, the results may be displayed in a form that is useful for navigation analysis.

The sample mean is obtained by summing the samples and dividing by the total number of samples.

$$X_\mu = \frac{1}{n}\sum_{i=1}^{n} X_i$$

The sample variance is obtained by summing the squares of the difference between the samples and the true sample mean and dividing by the number of samples. The sample covariance is obtained by summing both the squares and cross-products of the sample differences and dividing by the number of samples (n). The square and cross-product of a sample column vector is obtained by taking the outer product or post multiplying by its transpose. If the sample mean is used for the true mean, it can be shown that the best estimate of the covariance is obtained by dividing by $n-1$ and the best estimate of the sample covariance is given by,

$$Cov = \frac{1}{n-1}\sum_{i=1}^{n}(X_i - X_\mu)(X_i - X_\mu)^T$$

The Monte Carlo method is generally used when the uncertainties of the parameters of interest are greater than can be determined using linear theory. As a result, the distribution is generally not normal and cannot be represented by simple functions. Therefore, it is necessary to display the results in a format that permits observation of the true distribution. A histogram is useful for this purpose. The maximum and minimum values of each parameter are determined, and the range of each variable is divided into 10–25 intervals called bins. A bar graph is generated with the number of samples in each bin plotted as a function of the parameter of interest. The histogram gives some insight into the probability that certain critical design values may or may not be exceeded.

As a rule of thumb, the one sigma probability level is of interest for parameters that are loosely controlled. For example, a probability of obtaining some science observation of around 50% may be acceptable if the observation may be easily repeated. For parameters that are critical to mission success, a higher probability level is often required. A 99% probability of success is generally acceptable for situations where the total mission objectives may not be met but most of the mission may be salvaged in the event of failure. An example of a parameter that is controlled to 99% probability is the amount of fuel required to do the mission. If the spacecraft runs out of fuel and most of the mission objectives are satisfied, a 99% success rate is generally acceptable. For design parameters that can result in catastrophic loss of the mission, a much higher probability of success is required. A failure rate less than 10^{-5} is often specified. Examples of requirements that must be met with very high probability of success are planetary quarantine and unintended planetary impact.

The histogram is generally not a satisfactory tool for evaluating probability levels at the high or low end or tails of the PDF. One approach is to display the actual samples on a graph and inspect for samples that may exceed design limits. Figure 5.4 shows a plot of 250 Monte Carlo samples obtained for analysis of the NEAR landing. Plotted is the sub-latitude point of the NEAR spacecraft as a function of time from the beginning of the descent to landing on the surface of Eros. The variations in latitude on the way down are caused by a series of braking maneuvers designed to slow the spacecraft's descent and the initial orbit determination error. The spread in latitude for all the sample trajectories reveals one component of the landed footprint. Other parameters may be displayed on similar plots to gain insight into the landing site dispersions. The spread in latitude shown in Fig. 5.4 indicates a one sigma error of about 2°. The actual landing was within about one degree of the intended target.

Another approach is to order the Monte Carlo samples from low to high and estimate the probability level directly from the sample. For example, the cumulative 99% probability level may be estimated from a sample size of 500 by determining the fifth largest sample. On the average, 1% or about 5 samples will be above the 99% cumulative probability level. A problem with estimating probability levels from the tails of a distribution is the uncertainty or confidence associated with these estimates. By application of order statistics using the binomial distribution, the confidence level associated with estimates of the cumulative probability may be determined.

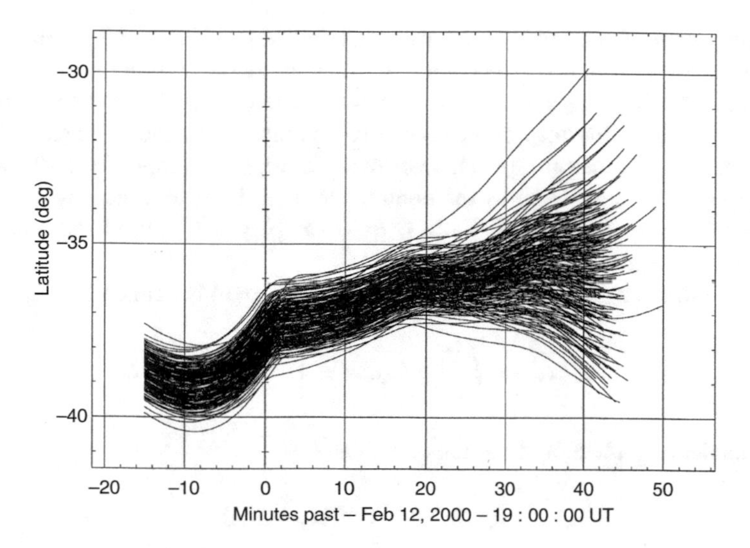

Fig. 5.4 NEAR landing Monte Carlo samples

5.7 Binomial Theorem

It has been said that all problems in probability and statistics can be solved by flipping coins. This may not be true, but it seems to be true for problems associated with navigation of spacecraft. The binomial theorem is the basis for solving problems associated with flipping coins. The normal probability distribution function, the gamma function, and the mathematics associated with computing gambling odds and predicting elections are all based on the binomial expansion. For navigation, the prediction of where a spacecraft will be in the future and the amount of fuel that will be needed is an example of application of the binomial theorem.

5.7.1 Confidence Limits

In determining estimates of cumulative probability levels, the mean and variance of the statistics do not provide a useful measure of the error in these estimates. The closeness of the estimate to the true value is less interesting than the probability that the estimate is bounded by some value. For the cumulative probability, a "best" estimate may be obtained by determining the value that bounds the probability for 50% of the sample sets obtained. The actual cumulative probability is above the estimate for 50% of the sample sets and below the estimate for the other 50%. Thus, the estimate of the 99% cumulative probability from a single sample set of 500 samples would bound this probability for 50% of all of the sample sets obtained.

The usefulness of the 99% cumulative probability level obtained in this manner may be called into question since the actual cumulative probability may exceed this value for half the sample sets. An estimate of the probability that the estimated cumulative probability may exceed some value may be obtained by examining the values of samples that are greater than the fifth highest sample for 500 samples. This probability, referred to as the confidence, may be determined by computing the probability that exactly 4, 3, 2, 1, or 0 samples will exceed the cumulative probability level for a particular sample set.

The cumulative probability function for the PDF $p(x)$ is defined by,

$$p_c(x_q) = \int_{-\infty}^{x_q} p(x)dx = Prob(x < x_q)$$

The confidence probability function is defined by,

$$w(q, x_w) = Prob(x_q < x_w)$$

For a Monte Carlo sample set of n samples, the above functions are only defined on integer values and may be made continuous by linear interpolation between these values. Consider a sample X_k from a set of n samples that are ordered in a decreasing monotone for increasing k. X_1 is the highest sample and X_n is the lowest. The problem of determining w may be cast as an application of Bernoulli trials using the binomial theorem. If the probability w_k, for the kth sample, that X_k is greater than x_q is defined as failure and X_k being less than x_q is defined as success, then the probability of failure may be obtained from the binomial theorem and

$$w_k = 1 - \sum_{i=0}^{k} \binom{n}{i} q^i \, (1-q)^{n-i}$$

where $q = 1 - p_c$ is the probability that a particular sample will exceed x_q.

The quantities being summed are Bernoulli trials that give the probability that exactly i samples will exceed x_q. The summation is needed to compute the probability that exactly 0 or 1 or ... k samples exceed x_q and this result is subtracted from 1 to obtain the desired probability. The probability that X_k exceeds the p_c probability level is w_k. An interesting property of w_k is that the probability is independent of the distribution.

k	Sample number	w_k	X_k
0	500	0.992	3.39
1	499	0.966	3.08
2	498	0.883	2.69
3	497	0.742	2.45
4	496	0.566	2.35
5	495	0.391	2.28

An application of confidence limits to navigation was in the determination of the amount of fuel required for the Viking and Galileo missions. The amount of fuel loaded at launch is the amount required for deterministic maneuvers, an additional amount for statistical maneuvers and project reserve for contingencies. The statistical component was determined by Monte Carlo analysis and was sized to give 99% probability of not running out of fuel. In order to account for uncertainties in determining the statistical component associated with sample size, the statistical fuel budget for Viking was increased by 20%.

5.7.2 Normal PDF from Binomial Coefficients

For a given row (m) of Pascal's triangle, the probability ($P(k)$) that the body will pass through a given node of number k is given by the number of paths that reach node $p(k)$ divided by the total number of possible paths (2^m). All paths are assumed to be equally likely.

$$P(k) = \frac{1}{2^m} p(k) = \frac{1}{2^m} \frac{m!}{(m-k)!\,k!}$$

Connecting the integer values of $p(k)$ with straight lines, a continuous function may be defined. The derivative of $p(k)$ with respect to k is given by

$$\frac{d}{dk} p(k) = \frac{p(k+1) - p(k)}{k+1-k}$$

The derivative with respect to k is discontinuous at integer values of k, so the derivative is defined approaching k from the right. Extracting common factors from $p(k+1)$, the following differential equation is obtained:

$$\frac{d}{dk} p(k) = p(k) \frac{m - 2k - 1}{k+1}$$

Since the function $p(k)$ is symmetrical about the y axis, as shown in Fig. 5.1, it will be convenient to shift the y axis and define k_2 such that it is zero in the middle of the distribution function. Also, for convenience, let m be even.

$$k_2 = k - \frac{m}{2}$$

and

$$\frac{dp(k_2)}{dk_2} = \frac{-4k_2 - 2}{2k_2 + m + 2} p(k_2)$$

In the limit as both m and $k2$ approach infinity and for $k2 <<< m$

$$\frac{dp(k_2)}{dk_2} \doteq \frac{-4k_2}{m} \, p(k_2)$$

The differential equation for $p(k_2)$ has the solution

$$p(k_2) = p(0)\, e^{\frac{-2k_2^2}{m}}$$

$$p(0) = \binom{m}{\frac{m}{2}}$$

The above formula provides approximate values for the mth row of Pascal's triangle. The approximation is good for small values of k_2 near the middle and for large values of m. Some typical values for the numbers in Pascal's triangle are compared with the approximate formula in the table below

k_2	m	k	$\binom{m}{k}$	$p(0)e^{\frac{-2k_2^2}{m}}$
0	20	10	184,756	184,756
1	20	11	167,960	167,174
5	20	15	15504	15,165
5	40	25	4.02E10	3.95E10
9	40	29	2.31E09	2.40E09

The probability that the body defined above will pass through a node at location k_2 is the value from Pascal's triangle divided by 2^m. The probability of passing through an interval between two nodes is obtained by integrating or summing $p(k_2)$ from $k_2(1)$ to $k2(2)$ and the integral of $p(k_2)$ from $-\frac{m}{2}$ to $+\frac{m}{2}$ is one. The probability is given by

$$P = \int_{k_2(1)}^{k_2(2)} \frac{1}{2^m} \, p(k_2)\, dk_2$$

A more convenient method of evaluating the same probability may be obtained by scaling k_2 by an appropriate factor and normalizing the integral. A change of variable from k_2 to x, where x is defined by,

$$x = \sqrt{\frac{4}{m}}\, \sigma\, k_2$$

yields

$$p(x) = C_1\, e^{-\left(\frac{x^2}{2\sigma^2}\right)}$$

Observe that for a given probability the ratio of x to σ is constant,

$$\frac{x}{\sigma} = \frac{2k_2}{\sqrt{m}} = C_2$$

In the limit as m approaches infinity, the ratio of k_2 to m is

$$\lim_{m\to\infty} \frac{k_2}{m} = \lim_{m\to\infty} \frac{\sqrt{m}\, C_2}{2m} = 0$$

and the assumption that k_2 is much smaller than m is valid. Since the integral of p(x) from minus infinity to plus infinity must be one, the final form of p(x) is obtained after evaluating the constant C_1.

$$\int_{-\infty}^{\infty} \exp^{-\left(\frac{x^2}{2\sigma^2}\right)} dx = \sigma\sqrt{2\pi}$$

$$p(x) = \frac{1}{\sigma\sqrt{2\pi}}\, e^{-\left(\frac{x^2}{2\sigma^2}\right)}$$

The function p(x) is called the normal probability distribution function (PDF) and has wide application in the field of probability and statistics. Many applications of the normal PDF involve solving for system parameters that minimize the error or spread of the PDF associated with random variables that are of interest. These random variables may describe the distance of a spacecraft from its intended target or the amount of fuel consumed during the mission.

5.7.3 Approximate Binomial Coefficients from Normal PDF

Recall that the difference equation for the binomial coefficients is given by

$$\frac{d}{dk2} B_m(k2) = \frac{-4\,k2 - 2}{2\,k2 + m + 2}\, B_m(k2)$$

$$k2 = k - \frac{m}{2}$$

In the limit as m becomes much greater than $k2$, the central part of the distribution function corresponding to $k2$ much smaller than m is given by the following differential equation:

$$\frac{dB_m(k2)}{dk2} = \frac{-4\,k2}{m}\,B_m(k2)$$

where B_m are the binomial coefficients for the mth row of Pascal's triangle. The differential equation for $B_m(k2)$ has the solution

$$B_m(k2) = B_m(0)\,e^{\frac{-2\,k2^2}{m}}$$

$$B_m(0) = \binom{m}{\frac{m}{2}}$$

The probability of being in the interval defined by $B_m(0)$ of width one is obtained by integrating the PDF from $k_2 = -\frac{1}{2}$ to $k_2 = +\frac{1}{2}$.

$$\frac{B_m(0)}{2^m} = \int \frac{1}{\sigma\sqrt{2\pi}}\,dx \;=\; \int \frac{1}{\sigma\sqrt{2\pi}}\sqrt{\frac{4}{m}}\,\sigma\,dk2$$

The binomial coefficients are thus given by the following formula:

$$B_m(k2) = \frac{m!}{(m-k)!\,k!} \approx \sqrt{\frac{2}{\pi}}\,\frac{2^m}{\sqrt{m}}\exp^{\frac{-2\,k2^2}{m}}$$

5.7.4 *Stirling Approximation*

The binomial coefficients may also be obtained by direct application of Stirling's approximation to the factorial functions defining the binomial coefficients. The Stirling approximation to the gamma function is given by

$$\ln(n!) \approx n\ln(n) - n + \frac{1}{2}\ln(2\pi n) = \ln\left(\sqrt{2\pi n}\left(\frac{n}{e}\right)^n\right)$$

$$n! \approx \sqrt{2\pi n}\left(\frac{n}{e}\right)^n$$

The results of approximating the binomial coefficients using the Stirling approximation and the normal approximation derived in Sect. 5.7.3 are shown in Fig. 5.5. The Stirling approximation for the binomial coefficients is exact for large values of m but does not do as well in the interior of the PDF. The normal distribution approximation is also exact as m approaches infinity. Observe that the normal and Stirling approximations are equal at the peak of the distribution corresponding to

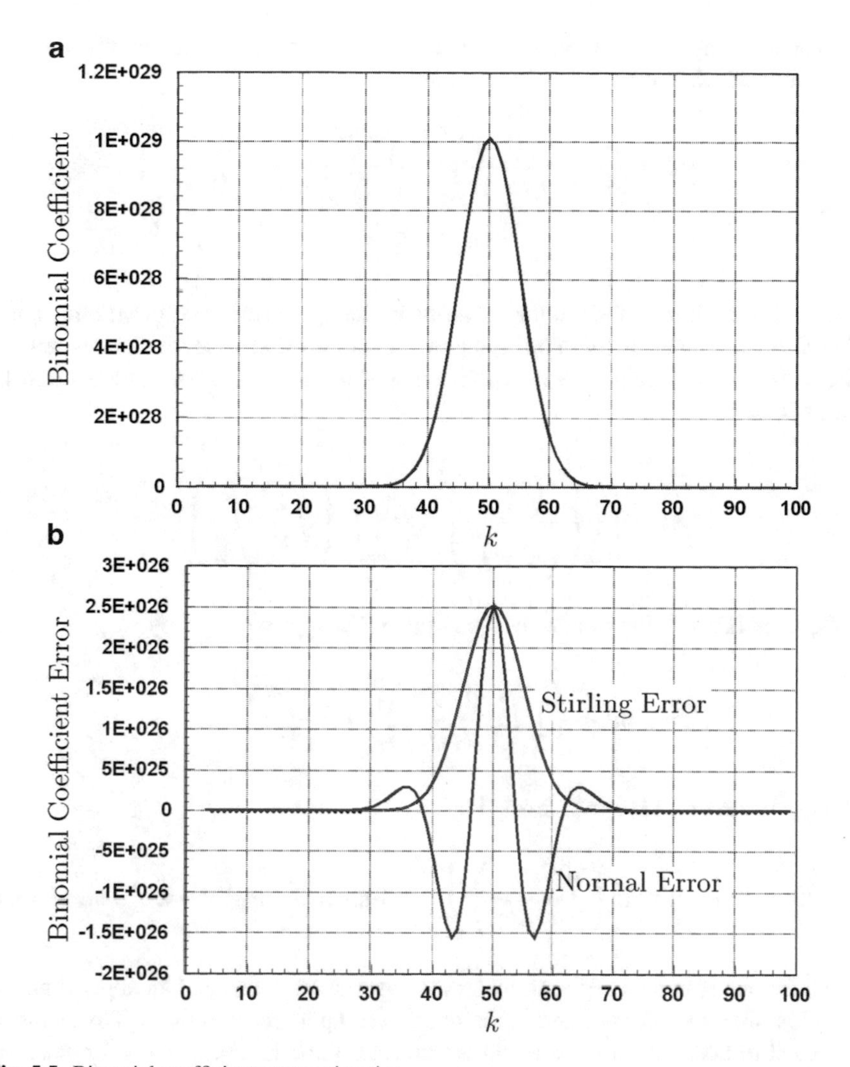

Fig. 5.5 Binomial coefficients approximation

$k = 50$. This suggests that the Stirling approximation has the normal Gaussian distribution embedded. This is indeed the case as the following identity shows:

$$B_m(0) = \frac{m!}{\left(\frac{m}{2}!\right)^2} \approx \sqrt{\frac{2}{\pi}}\,\frac{2^m}{\sqrt{m}} = \frac{\sqrt{2\pi m}\left(\frac{m}{e}\right)^m}{\left[\sqrt{\pi m}\left(\frac{m}{2e}\right)^{\frac{m}{2}}\right]^2}$$

The above identity suggests that the Stirling approximation may be extracted directly from the binomial coefficients. In the following equation, all the numerators are canceled by the denominators in the preceeding term except for the first numerator which is m!.

$$m! = \frac{m!}{\left(\frac{m}{2}!\right)^2}\left(\frac{\left(\frac{m}{2}!\right)}{\left(\frac{m}{4}!\right)^2}\right)^2\left(\frac{\left(\frac{m}{4}!\right)}{\left(\frac{m}{8}!\right)^2}\right)^4\cdots\left(\frac{2!}{1!^2}\right)^N$$

$$N = 2^{\frac{\ln m}{\ln 2}-1}$$

where N is 2 raised to the number of terms in the approximation for $m!$ minus one. N is also the power of two corresponding to the $m!$ that exceeds the desired $m!$. The desired $m!$ is obtained by interpolation. The terms in the above product may be approximated by

$$m! = \prod_{n=0}^{N}\left(\frac{\left(\frac{m!}{2^n}!\right)}{\left(\frac{m}{2^{n+1}}!\right)^2}\right)^{2^n} \approx \prod_{n=0}^{N}\left(\sqrt{\frac{2}{\pi}}\frac{2^{\frac{m}{2^n}}}{\sqrt{\frac{m}{2^n}}}\right)^{2^n}$$

If $B_m(0)$ is factored out of each of the terms, $m!$ is approximated by

$$m! \approx \prod_{n=0}^{N}\left(\sqrt{\frac{2}{\pi}}\frac{2^m}{\sqrt{m}}\right)^{2^n}\left(\frac{2^{\frac{m}{2^n}-m}}{2^{\frac{-n}{2}}}\right)^{2^n}$$

The logarithm of $m!$ is approximated by

$$\ln m! \approx \Gamma(m+1) \approx \ln\left(\sqrt{\frac{2}{\pi}}\frac{2^m}{\sqrt{m}}\right)\sum_{n=0}^{N}2^n + \ln(2)\sum_{n=1}^{N}(n2^{n-1} - m2^n + m) \quad (5.7)$$

The finite series can be replaced by closed form expressions and an approximation for $m!$ is obtained that is exact in the limit as m approaches infinity. The last term is needed to account for the error in the normal binomial coefficients for small m. Thus, we have a closed form expression for $\ln(m!)$ as a function of only m.

$$\ln m! \approx \ln\left(\sqrt{\frac{2}{\pi}}\frac{2^m}{\sqrt{m}}\right)\left[2^{N+1} - 1\right]$$
$$+ \ln(2)\left[1 - 2^N + N2^N - m(2^{N+1} - 2) + mN\right] - 0.08105638054266\, m$$
$$N = 2^{\frac{\ln m - \ln 2}{\ln 2}} \quad (5.8)$$

The constant in the above equation was obtained by forcing the approximation to equal the actual value of $m!$ at $m = 128$. Over the range of m from 1 to

158, the above approximation is twice as accurate as the Stirling approximation. If $5.0862526726 \times 10^{-6}\, m$ is added to the Stirling approximation, both approximations agree to 14 decimal places over the same range. This agreement indicates that a simple correction term can be found to make both approximations exactly equal and the deep underlying mathematical basis for both approximations is the normal PDF or the binomial coefficients.

5.8 Maxwell–Boltzmann Probability Distribution

Significant trajectory perturbations have been observed on interplanetary missions that have been attributed to expulsion of gas from the spacecraft. In 1969, the Mariner 6 spacecraft experienced a large trajectory perturbation and loss of attitude control lock on the star Canopus because of hydrogen gas expelled during a scan platform unlatching event. The scan platform was held in the latched position by pressure from hydrogen gas. Hydrogen gas was expelled by firing a squib and venting to a tee. The tee was supposed to vent the gas in opposite directions, resulting in no net forces on the spacecraft and the observed trajectory and attitude perturbations were a mystery. A complete third spacecraft was built and was sitting on display in Von Karman Auditorium at the Jet Propulsion Laboratory. After some searching, the tee was located in a large cavity under the scan platform. The cavity served as a big rocket engine and the thrust, computed from kinetic theory, matched the velocity change observed in the orbit determination solution. The applied moment was also consistent with the attitude recovered from sun sensor telemetry. The result of this analysis was used to predict the velocity change that Mariner 7 experienced a few weeks later. On the Viking Mission in 1975, venting of gas trapped in the parachute, at least this was the theory, resulted in a perturbation of the spacecraft for about a week after launch making determination of the spacecraft orbit difficult. Kinetic theory was used to predict the velocity change associated with several gas venting events during cruise. Perhaps the most significant application of kinetic theory was in analyzing the acceleration of the MESSENGER spacecraft after launch in 2004. The orbit could not be accurately determined for a couple of weeks. This acceleration was attributed to liquid evaporating from various surfaces on the spacecraft with various time constants, depending on solar illumination. It would take about one cubic inch of water distributed over the spacecraft to produce enough water vapor to cause the observed acceleration.

Several models of particle collisions have been developed for the purpose of observing the kinetics. A simple model involving only particle velocity yields the Maxwell–Boltzmann PDF. The position of the particles is resolved into a number of probability assumptions, and the simulation uses Monte Carlo techniques to obtain the velocity distribution. Depending on the assumption used for the probability of impact, two PDFs referred to as the Maxwell–Boltzmann distribution and velocity-dependent distribution are obtained. Another model includes both position and velocity and is completely deterministic. No probability assumptions or Monte Carlo sampling is included, and this model verified the velocity-dependent model.

The equations of motion in the position/velocity-dependent model are simply Newton's laws of motion. These are the equations of motion for an ideal gas, the so-called "billiard ball" theory. The particles move in a straight line until they impact another particle or a wall of the container. Energy and momentum are conserved. When the particles collide, the velocity change of the two particles, which are assumed to be spheres, is along the line connecting the centers. The angle between the line connecting the spheres and the relative velocity vector is called the "scattering" angle or "Rutherford scattering" angle for molecules. The scattering angle for molecules is a bit more complicated than for billiard balls. It can be shown by arguments of symmetry that the scattering angle has no effect on the final PDF. If the scattering angle affected the PDF, then different gases would have different PDFs and many of the laws of physics pertaining to gasses would be invalidated.

The results of a computer simulation of 400,000 particles and 8,000,000 collisions are shown in Fig. 5.6. Figure 5.6a shows the component velocity distribution in some arbitrary direction. Because of symmetry, all directions have the same distribution. The velocity component is sorted into 50 bins. The abscissa is velocity and zero velocity is bin 25. The scale is not important, because only the shape of the curve is of interest. For simplicity, the root mean square velocity magnitude is initialized with an average value of one. The particles are also assigned a radius of one and all particles have the same mass, also one. The container is sized such that the space between particles averages 30 radii. After proper scaling, it can be shown that the particles have the same temperature and pressure as an ideal gas. The ordinate is the number of molecules in each bin. The probability is obtained by dividing by the total number of molecules.

Also shown by the fine line in Fig. 5.6a is the normal distribution predicted from Maxwell–Boltzmann theory. At zero component velocity, the velocity-dependent model is about 3.0% above the Maxwell–Boltzmann PDF. The difference between the velocity-dependent model and the Maxwell–Boltzmann PDF is plotted in Fig. 5.6b. The peak value of 800 above the 30,000 particles in the associated bin in Fig. 5.6a gives an overshoot of 800/30,000 or 2.67%. The difference may be attributed to the assumed probability of impact. The Maxwell–Boltzmann computer simulation, which assumes equal probability of impact, gives the same result as the theoretical Maxwell–Boltzmann PDF (a normal curve). The velocity-dependent model assumes the probability of impact is dependent on the particle velocity and is confirmed by computer simulation that includes position. It appears that the only way to obtain a Maxwell–Boltzmann PDF by computer simulation is to randomly select the collision participants and ignore their position.

Figure 5.7 shows the same computer simulation results for the velocity magnitude. The velocity magnitude distribution is obtained by mapping three orthogonal velocity component distributions into velocity magnitude and is the well-known Maxwell–Boltzmann PDF of Maxwell–Boltzmann theory. Also shown for com-

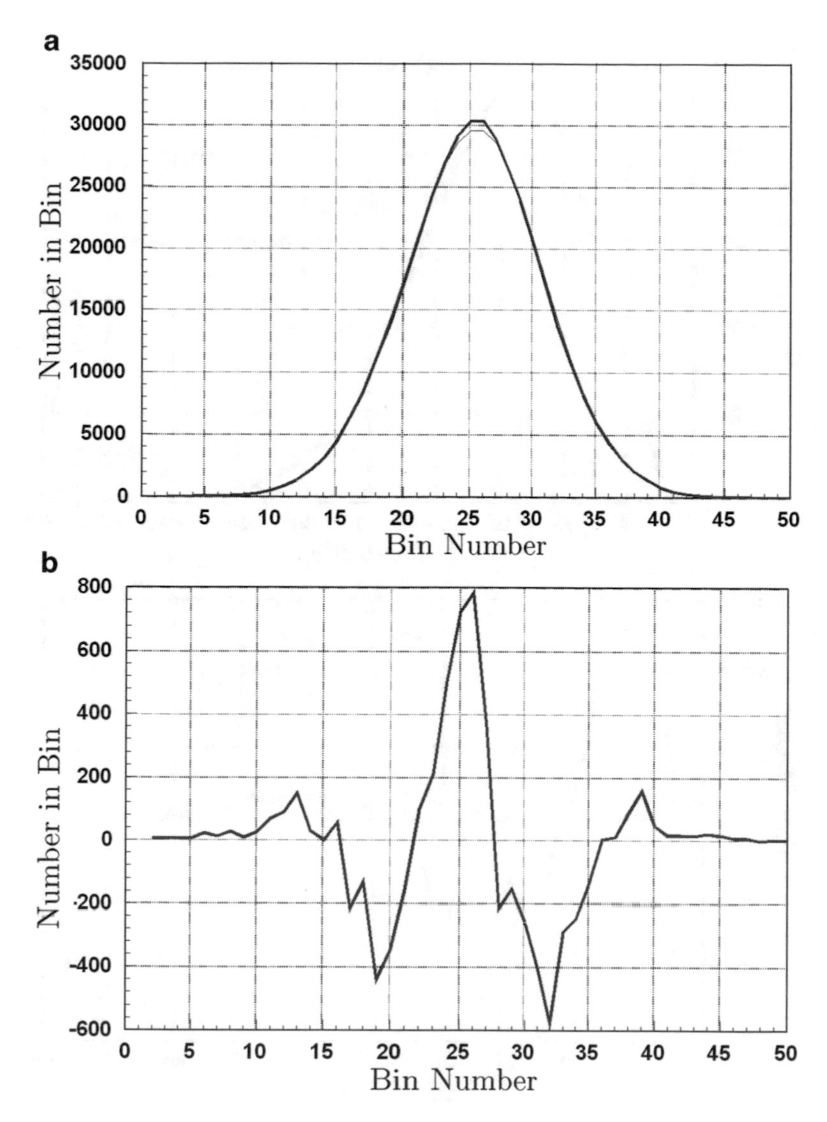

Fig. 5.6 Velocity component distribution comparison

parison is the computer simulation velocity-dependent model results. Here, the velocity-dependent model is about 4% below the Maxwell–Boltzmann PDF (shown by a fine line) at the peak which occurs around bin 13. In Fig. 5.7a, the left side of bin 1 is zero velocity magnitude. Figure 5.7b shows the difference.

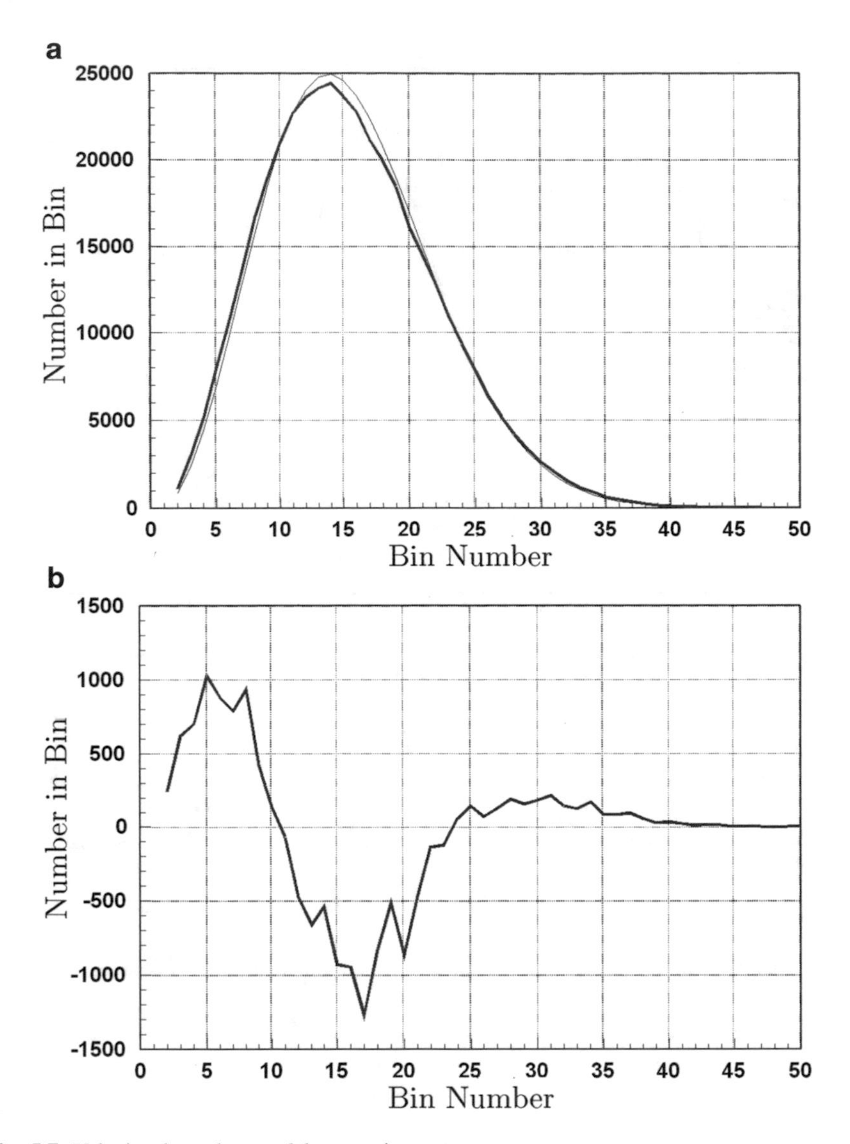

Fig. 5.7 Velocity-dependent model comparison

5.8.1 Experimental Results

An experimental verification of the Maxwell–Boltzmann theory was performed in 1955. The experiment consisted of heating potassium and thallium in an oven to 900°C and venting the atoms through a velocity selector into a detector. The velocity selector was a cylinder with a slot on one side and a curved slot on the other that rotated at 4000 rpm such that atoms would cross the cylinder and enter the curved slot at various transit times depending on their velocity and the rotation rate of the cylinder. The detector would measure the intensity, or number of atoms, at the velocity corresponding to the angle that the cylinder rotates dependent on the rotation rate. The device was quite sophisticated, being cooled with liquid nitrogen and sealed to provide a high vacuum.

The results of the potassium and thallium experiments are shown in Fig. 5.8a. Also plotted is the theoretical Maxwell–Boltzmann PDF. The agreement is very good and verifies the overall veracity of the Maxwell–Boltzmann theory. The agreement with the position/velocity- dependent model does not, at first, appear very good. Inspection of this comparison shown in Fig. 5.7a reveals a gap between Maxwell–Boltzmann theory and the computer model result at the peak intensity or most probable velocity. This gap, if it exists, would not show up in the experimental results because the scaling between intensity and the actual number of molecules is not known precisely. Therefore, the experimental results were scaled to force them to equal the Maxwell–Boltzmann theory at the peak. Also, the temperature was adjusted so that the low-velocity experimental results matched Maxwell–Boltzmann theory. The problem with these adjustments, which do not affect the experiment since only the shape of the distribution is of interest, is that the PDF is a single parameter theory. Once the temperature is fixed, the probabilities and shape of the curve are also fixed. If the gap shown in Fig.5.7a is real, the area between the curves at the peak will be distributed elsewhere. This redistribution is evident in Fig. 5.8a. Close inspection of the high-velocity side of the curve reveals that the experimental results are above the Maxwell–Boltzmann theory prediction. If these results are accurate, they present a problem, since the integral from zero to plus infinity is one and the experimental results are greater than one. R. C. Miller and P. Kusch acknowledged this problem. In their words, “It is seen that the largest discrepancies occur on the high velocity side of the maximum, where there is a small excess of atoms in the experimental distribution. It should be noted that the experimental points could be plotted with the high-velocity side matched to the theoretical curve. The intensities at the maximum velocity would no longer coincide and the experimental distribution would then appear to be deficient of atoms on the low velocity side.” In other words, the theoretical curve could be shifted to the right by adjusting the temperature, but the area under the curve would still be greater than one.

Another approach, that is relatively simple to implement, would be to scale the computer model results to force the peak to coincide with the Maxwell–Boltzmann PDF and adjust the temperature to match the curves on the low-velocity side. These results could then be compared directly with the experimental results. Figure 5.8 shows the results of this procedure. The experimental results in Fig. 5.8a depart a small amount from the theoretical Maxwell–Boltzmann PDF at a reduced velocity of 1.4. The computer model results in Fig. 5.8b also depart a small amount. One may conclude that the experimental results confirm the velocity-dependent model.

Another experiment performed by J. F. C. Wang and H. Y. Wachman in 1976 involved venting molecules from an oven into a detector. The observed flight time is a measure of velocity magnitude. The results of this experiment are plotted in

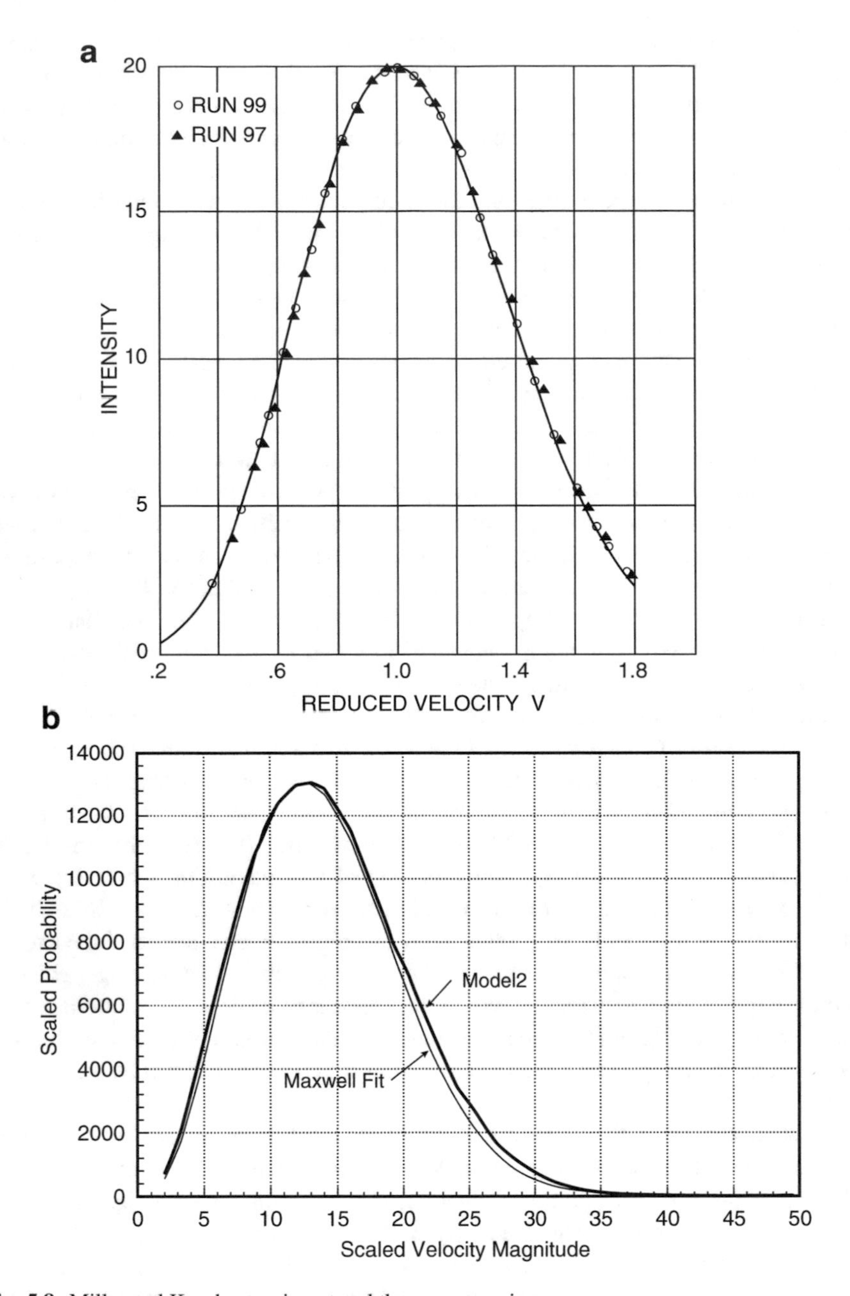

Fig. 5.8 Miller and Kusch experiment and theory comparison

Fig. 5.9. The experimental results fall below the theory prediction by about the same amount and in the same place as the Miller/Kusch results are above the theoretical curve. Both experimenters forced the experimental results to agree at the peak

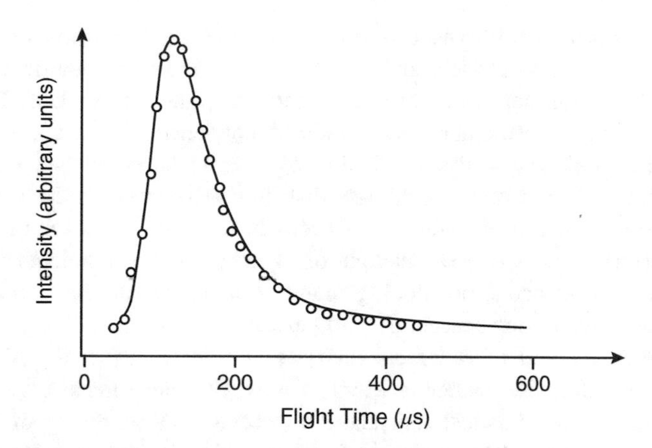

Fig. 5.9 Wang and Wachman experiment and theory comparison

of the PDF. Miller and Kusch scaled up the experimental results and Wang and Wachman scaled down the theoretical prediction. Since the theoretical prediction must integrate to one from minus infinity to plus infinity, the theory must be higher on the tails of the distribution function as can be seen in Fig. 5.9. Observe that the experimental results trend back to the best fit to the theory farther out on the tail of the PDF. The position/velocity computer model results shown in Fig. 5.8b also trend back on the tail only from the opposite side consistent with the assumptions on how the data was fit to the theory for the two experiments. An alternative explanation for the above observations is that the experimental apparatus used for the experiments had equal and opposite systematic errors. There appears to be no other alternative.

5.9 Summary

Statistical analysis of the outcome of a navigation strategy is performed to determine the probability that the spacecraft will safely arrive at the target or the end of the mission and satisfy mission design constraints. These constraints may be related to the probability of impacting the target body, running out of fuel, or acquiring science data. For example, if it is desired to occult a planet, the occultation may be mapped to a region in the B-plane and the probability that the spacecraft will pass through this region would constrain the amount of data that would need to be acquired and maneuver placement. These simple constraints may be satisfied by linear mapping of the orbit determination error. Since the orbit determination error is a mapping of measurement errors that may be characterized with high precision, the confidence in the results of this analysis is high.

Examination of data residuals confirms the accuracy of the measurements and that the probability distribution is normal or white noise. This result was not easy to obtain but resulted from years of analysis by the Jet Propulsion Laboratory to develop high precision models and calibrations of the radio metric data, and a similar analysis was performed to model and calibrate optical data. The normal distribution of the measurement errors and independence of measurements greatly simplifies the statistical analysis. We do not have to be concerned with colored noise or biases. The normal probability distribution function is derived from the binomial coefficients. The binomial theorem is thus the basis for the statistical analysis performed for spacecraft navigation. The binomial coefficients describe the results that are obtained from flipping many coins or rolling dice, and thus have many applications beyond spacecraft navigation.

A powerful method of statistical analysis is referred to as the Monte Carlo method named after the casinos in Monte Carlo. Weather forecasters refer to this method as ensemble statistical analysis. An astronaut sitting on top of a rocket or citizens in a hurricane shelter probably feel better if their life depends on ensemble statistical analysis rather than rolling dice. Monte Carlo analysis enables precision mapping of measurement and model errors when the system is nonlinear. Civil engineers use safety factors to design bridges and buildings. In the past, the safety factors were rather large. The Brooklyn bridge and Empire State building have been around for a long time and it does not appear they will fall down soon. Rockets and spacecraft are built with much lower safety factors. Statistical analysis permits operations much closer to the edge. A critical statistical determination is the amount of extra fuel to put in the fuel tanks to complete the mission. When the Viking spacecraft arrived in orbit about Mars, there was twice as much fuel as was needed to complete the mission. Every extra pound of fuel reduced the science payload by the same amount. Because of problems in selecting launch vehicles and a rocket motor burn failure, the Galileo and Near spacecraft completed their missions with little fuel remaining in the tanks.

Exercises

5.1 Determine the probability of drawing a 5-card royal straight flush (ace, king, queen, jack, and ten of the same suit) from a 52-card deck.

5.2 Determine the probability that the first 5 samples of 500 Monte Carlo samples will be the largest.

5.3 A cannon with a CEP of 50 yards is fired into the Collisiem aiming for the center. Caesar's box is 9 feet by 9 feet and located 100 yards from the center of the Collisiem. Assuming that the PDF is constant within the box, determine the probability that the cannon ball will land in Caesar's box.

5.4 If the cannon in Exercise 5.3 was aimed at Caesar's box, what is the probability of hitting Caesar's box?

5.5 Show that the sum of the binomial coefficients (k) for a given row of Pascal's triangle (m) is equal to 2^m.

5.6 Show that the binomial coefficients are given by

$$B(m,k) = \binom{m}{k} = \frac{m!}{(m-k)!\,k!}$$

5.7 Show that the derivative of $p(k)$ with respect to k holding m constant and assuming linear interpolation between values of k is given by

$$\frac{d}{dk}p(k) = p(k)\frac{m-2k-1}{k+1}$$

where

$$p(k) = \frac{m!}{(m-k)!\,k!}$$

Bibliography

Lass, H. and P. Gottlieb, "Probability and Statistics", Addison-Wesley, Reading, MA, 1971.

Miller, J. K., "Viking Non-Gravitational Accelerations Resulting From Gas Leaks and Venting", internal JPL document IOM 392.5-585, December 11, 1974.

Miller, J. K., "Monte Carlo Analysis of Kinetic Theory of Gases", 16th AAS/AIAA Spaceflight Mechanics Conference, January 2006.

Miller, J. K., G. R. Hintz and P. J. Llanos, "A New Kinetic Theory of Particle Collisions", AAS 16–338, Astrodynamics Specialist Conference, August, 2016.

Miller, R. C. and Kusch, P.,"Velocity Distributions in Potassium and Thallium Atomic Beams", *Physical Review*, Vol 99. No 4. August 15, 1955.

Wang, J. F. C., Wachman, H. Y. and Goodman, F. O., *Dynamics of Gas-Surface Scattering*, Academic Press, New York, 1976.

Chapter 6
Orbit Determination

Determination of the orbit of a spacecraft and all the constant and dynamic parameters that affect the orbit is an application of estimation theory. A model of the spacecraft motion as a function of initial conditions and certain constant and dynamic parameters is developed that may be used to predict the flight path and compute the value of measurements that are obtained during the space flight. Orbit determination involves adjusting the value of the independent parameters that need to be determined until the computed measurements are close to the actual measurements. Since there are many more measurements than independent parameters, the solution is found that minimizes the error in the measurements. This solution will also minimize the error in the estimated parameters.

A measure of orbit determination accuracy is the square of the difference between the computed measurement from the model and the actual measurement summed over all the measurements. An orbit determination filter processes the measurements and computes an update to the estimated parameters starting from an initial guess referred to as the *a priori*. An orbit determination filter does not actually determine an orbit, but takes an initial guess and finds another orbit that is closer to the actual orbit. The actual orbit is never known because we do not have perfect measurements.

In searching for a solution, an orbit determination program will sometimes have problems converging. Before computers, astronomers determined orbits with a few measurements. This process is referred to as deterministic orbit determination and an algorithm is devised for determining an orbit about a body with six Doppler measurements strategically placed around the orbit.

J. Miller, *Planetary Spacecraft Navigation*, Space Technology Library 37,
https://doi.org/10.1007/978-3-319-78916-3_6

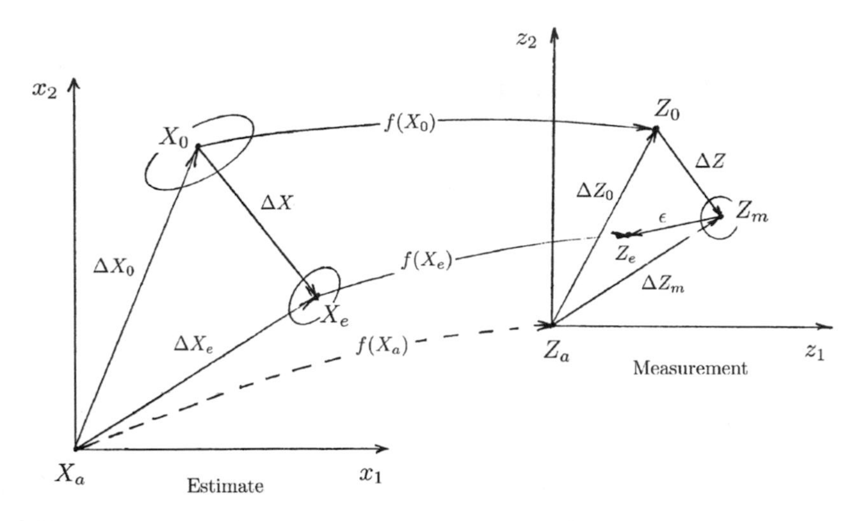

Fig. 6.1 Relationship of measurements to estimated parameters

6.1 Kalman Filter Algorithm

The Kalman filter algorithm computes updates to the *a priori* estimated parameters one measurement at a time. The solution after the current update is used as *a priori* for the next update. This simple one step algorithm, devised by R. E. Kalman in the early 1960s, provides a simple method of computing parameter estimates on a digital computer and is well suited for many applications.

Given an *a priori* estimate of a set of parameters (X_0) and the associated covariance P_0, a measurement Z_m and its covariance and a model of the dynamics, the relationships are illustrated in Fig. 6.1 in two dimensions. Selecting any two parameters, the values of the *a priori* estimated parameters are shown on the left along with the actual value and new estimate of X. The coordinate system is centered at the actual value (X_a) which is unknown. However, the mapping function to the measurement space on the right is known for all values of X. A two-dimensional measurement is shown corresponding to, for example, lines and pixels of an optical measurement. For most measurement updates, the measurements are uncorrelated and Z is one-dimensional. Indeed, lines and pixels are generally processed as two one-dimensional measurements since they are generally assumed to be independent. The actual measurement Z_m is shown along with the measurement(Z_0) computed from the model based on the *a priori* estimate (X_0). The measurement covariance P_m is indicated by the ellipse drawn around the measurement. The estimated measurement (Z_e) is also shown along with the post fit measurement error (ϵ).

The computed measurement is obtained from the model and is given by

$$Z_0 = f(X_0)$$

The data residual (ΔZ) is the difference between the actual measurement and the computed measurement and the Kalman gain matrix (K) multiplied times the data residual and combined with the *a priori* (X_0) provides a new estimate (X_e).

$$\Delta Z = Z_m - Z_0$$

$$\Delta X = K \Delta Z$$

$$X_e = X_0 + \Delta X$$

The desired solution for X_e is X_a. Since X_a is not known, and can never be known exactly, a gain matrix K is desired that moves the solution as close to X_a as can be determined from the information available. Therefore, a solution is sought that minimizes the distance of X_e, shown in Fig. 6.1, from the origin. Minimizing the square of the magnitude of a random column vector, or variance, also minimizes the magnitude and since the column vectors shown in Fig. 6.1 are all examples drawn from a random probability distribution function, the solution that minimizes the variance of X_e is the desired solution. In the terminology used here, a matrix with one column is referred to as a column vector or data vector because of the similarity to real vectors. The random column vectors of interest are defined to have zero mean and are given by

$$\Delta X_0 = X_0 - X_a$$

$$\Delta X_e = X_e - X_a$$

$$\Delta Z_m = Z_m - Z_a$$

$$\Delta Z_0 = Z_0 - Z_a$$

From the model of the equations of motion and data, the partial derivatives of the measurement with respect to the estimated parameters may be computed and these are given by

$$A = \frac{\partial Z}{\partial X}$$

The partial derivatives may be used to map column vectors defined in the estimate space to data vectors in the measurement space. The column vector ΔX_0 thus maps to the data vector ΔZ_0 and

$$\Delta Z_0 = A \, \Delta X_0$$

Some other relationships, shown in Fig. 6.1, are needed to tie X and Z together.

$$\Delta X_e = \Delta X_0 + \Delta X$$

$$\Delta Z_m = \Delta Z_0 + \Delta Z$$

The solution for ΔX_e is then given by

$$\Delta X_e = K\Delta Z_m + (I - KA)\Delta X_0 \tag{6.1}$$

and this result can be extended to higher dimension in X and Z even though the dimension of Z is usually taken to be one. Exact values for ΔX_0, ΔX_e, ΔZ_0, and ΔZ_m cannot be determined. However, we can obtain from prior experience estimates of their errors. The expected value of ΔX_e is given by

$$P_e = E\left(\Delta X_e \Delta X_e^T\right)$$

Assuming the measurement is independent of the estimated parameters,

$$E\left(\Delta X_e \Delta X_e^T\right) = E\left(K\ \Delta Z_m \Delta Z_m^T\ K^T + (I - KA)\ \Delta X_0 \Delta X_0^T\ (I - KA)^T\right)$$

and

$$P_e = KP_mK^T + (I - KA)P_0(I - KA)^T$$

where

$$P_m = E\left(\Delta Z_m \Delta Z_m^T\right)$$

$$P_0 = E\left(\Delta X_0 \Delta X_0^T\right)$$

are the measurement and estimated parameters *a priori* covariances, respectively. The minimum variance estimate is found by taking the variation of P_e with respect to the gain matrix K and setting this result to zero.

$$\delta P_e = \delta K B + B^T \delta K^T$$

$$B^T = KP_m - P_0A^T + KAP_0A^T$$

For a minimum, B is zero and the Kalman gain matrix (K) is given by

$$K = P_0A^T\left[P_m + AP_0A^T\right]^{-1} \tag{6.2}$$

Substituting K into the estimated covariance gives

$$P_e = (I - KA)P_0 \tag{6.3}$$

6.2 Weighted Least Squares

The same result as obtained for the Kalman Filter may be obtained by a different approach to parameter estimation. Since the classical approach grew out of attempts to fit data, the minimization of the measurement residual is the objective rather than the error in the estimated parameters. It will be shown that both approaches yield the same result. Referring to Fig. 6.1, the error in the measurement (ϵ_m) is given by

$$\epsilon_m = Z_m - Z_e = \Delta Z - A\Delta X$$

Since some measurements may be more accurate than others, the square of the residual error may be weighted by dividing by its variance. Thus, a minimum variance estimate may be obtained by dividing the measurement errors by their standard deviation. Dividing by the standard deviation is obtained by multiplying by the square root of the inverse of the measurement covariance. The weighting matrix is defined by

$$W = P_m^{-1}$$

and the weighted measurement error is

$$\epsilon = W^{\frac{1}{2}}(\Delta Z - A\Delta X)$$

The sum of the squares of the weighted residuals (J) is the scalar parameter that is to be minimized.

$$J = \epsilon^T\epsilon = (W^{\frac{1}{2}}\Delta Z - W^{\frac{1}{2}}A\Delta X)^T(W^{\frac{1}{2}}\Delta Z - W^{\frac{1}{2}}A\Delta X)$$

The minimum J may be found by taking the variation with respect to ΔX and setting this result equal to 0.

$$\delta J = (W^{\frac{1}{2}}\Delta Z - W^{\frac{1}{2}}A\Delta X)^T(-W^{\frac{1}{2}}A\delta\Delta X) - \delta\Delta X^T A^T(W^{\frac{1}{2}})^T(W^{\frac{1}{2}}\Delta Z - W^{\frac{1}{2}}A\Delta X)$$

For a minimum, δJ is set equal to 0 and the solution for ΔX is

$$\Delta X = (A^T W A)^{-1}A^T W\Delta Z$$

Assuming that the *a priori* error covariance is infinite, the estimated error covariance is given by

$$P_e = E(\Delta X\Delta X^T)$$

From the equation for the data update,

$$A^T W A \Delta X \Delta X^T (A^T W A)^T = A^T W \Delta Z \Delta Z^T W^T A$$
$$A^T W A P_e (A^T W A)^T = A^T W P_m W^T A$$

and

$$P_e = (A^T W A)^{-1}$$

The weighted least square solution, as derived here, assumes all the data is included in A and there is no *a priori*. The A matrix may be partitioned separating the *a priori* data from the new data and the filter equations put in the same form as for the Kalman filter. The data update for the weighted least squares sequential filter becomes

$$K = (A^T W A + P_0^{-1})^{-1} A^T W \tag{6.4}$$

$$P_e = (A^T W A + P_0^{-1})^{-1} \tag{6.5}$$

The two equations for the Kalman gain (Eq. (6.2) and Eq. (6.4)) solve the same problem but do not appear to be equal. The equivalence of these two solutions may be shown by invoking Shur's identity as was done by Anderson and others in the early 1960s. Shur's identity gives the inverse of a partitioned matrix.

$$\begin{bmatrix} A & B \\ C & D \end{bmatrix} \begin{bmatrix} (A - BD^{-1}C)^{-1} & -A^{-1}B(D - CA^{-1}B)^{-1} \\ -D^{-1}C(A - BD^{-1}C)^{-1} & (D - CA^{-1}B)^{-1} \end{bmatrix} = \begin{bmatrix} I_1 & 0 \\ 0 & I_2 \end{bmatrix}$$

Consider the following matrix product obtained by application of Shur's identity.

$$\begin{bmatrix} P_0^{-1} & A^T \\ A & -W^{-1} \end{bmatrix} \begin{bmatrix} (A^T W A + P_0^{-1})^{-1} & P_0 A^T (W^{-1} + A P_0 A^T)^{-1} \\ W A (A^T W A + P_0^{-1})^{-1} & -(W^{-1} + A P_0 A^T)^{-1} \end{bmatrix} = \begin{bmatrix} I_1 & 0 \\ 0 & I_2 \end{bmatrix}$$

Since the inverse of a symmetric matrix must also be symmetric,

$$\begin{bmatrix} P_0^{-1} & A^T \\ A & -W^{-1} \end{bmatrix} \begin{bmatrix} (A^T W A + P_0^{-1})^{-1} & P_0 A^T (W^{-1} + A P_0 A^T)^{-1} \\ (W^{-1} + A P_0 A^T)^{-1} A P_0 & -(W^{-1} + A P_0 A^T)^{-1} \end{bmatrix} = \begin{bmatrix} I_1 & 0 \\ 0 & I_2 \end{bmatrix}$$

The equation for the I_1 identity submatrix is

$$I_1 = P_0^{-1} (A^T W A + P_0^{-1})^{-1} + A^T (W^{-1} + A P_0 A^T)^{-1} A P_0$$

and

$$P_0 - P_0 A_T (W^{-1} + A P_0 A^T)^{-1} A P_0 = (A^T W A + P_0^{-1})^{-1} \tag{6.6}$$

The $A^T W$ term is factored out of the Kalman gain given by Eq. (6.1).

$$K = P_0 A^T \left[P_m + A P_0 A^T \right]^{-1} [(W^{-1} + A P_0 A^T) W - A P_0 A^T W]$$

$$I = [(W^{-1} + A P_0 A^T) W - A P_0 A^T W]$$

and

$$K = [P_0 - P_0 A^T (W^{-1} + A P_0 A^T)^{-1} A P_0] A^T W$$

Substituting the result obtained from Eq. (6.6) results in the weighted least square formula for the Kalman gain.

$$K = (A^T W A + P_0^{-1})^{-1} A^T W$$

The equivalence of the estimated covariance solutions is simpler to show. Substitute the Kalman gain from the least squares solution into the estimated covariance given by the Kalman filter solution and factor out the desired result.

$$P_e = (I - K A) P_0$$

$$P_e = P_0 - (A^T W A + P_0^{-1})^{-1} A^T W A P_0$$

$$P_e = (A^T W A + P_0^{-1})^{-1} \left[(A^T W A + P_0^{-1}) P_0 - A^T W A P_0 \right]$$

$$P_e = (A^T W A + P_0^{-1})^{-1}$$

6.3 Square Root Information Filter (SRIF)

The SRIF discrete data update algorithm follows directly from the least square data update. The least square solution is given by

$$\hat{X} = \left[A_n^{\mathrm{T}} \Delta W_n A_n \right]^{-1} A_n^{\mathrm{T}} \Delta W_n \hat{Z}_n$$

The measurements can be normalized by factoring ΔW_n into

$$\Delta W_n = \sqrt{\Delta W_n}^{\mathrm{T}} \sqrt{\Delta W_n}$$

and

$$\hat{X} = \left[A_n^{\mathrm{T}} \sqrt{\Delta W_n}^{\mathrm{T}} \sqrt{\Delta W_n} A_n \right]^{-1} A_n^{\mathrm{T}} \sqrt{\Delta W_n}^{\mathrm{T}} \sqrt{\Delta W_n}\, \hat{Z}_n$$

By inspection we can see that

$$R_n = \sqrt{\Delta W_n}\, A_n$$

so after substitution we have

$$\hat{X} = (R_n^{\mathrm{T}} R_n)^{-1} R_n^{\mathrm{T}} \sqrt{\Delta W_n} \hat{Z}_n$$

For the first m measurements, the number of estimated parameters (m) is equal to the number of measurements and R_n is square.

$$\hat{X} = R_n^{-1} \sqrt{\Delta W_n} \hat{Z}_n$$

Multiplying through by R_n gives what is called the data equation.

$$R_n \hat{X} = \sqrt{\Delta W_n} \hat{Z}_n = \hat{\eta}_n \tag{6.7}$$

where $\hat{\eta}_n$ is the normalized measurement. A new measurement can be appended to the data equation resulting in

$$\begin{bmatrix} R_n \\ \\ \sqrt{\Delta W_{n+1}}\, A_{n+1} \end{bmatrix} \hat{X} = \begin{bmatrix} \hat{\eta}_n \\ \\ \hat{\eta}_{n+1} \end{bmatrix}$$

Adding additional measurements results in the row dimension of R exceeding the column dimension. The information matrix would then be given by

$$\Lambda_m = R_{nm}^{\mathrm{T}} R_{nm}$$

where the row dimension n exceeds the column dimension m. Since R_{nm} is not unique, it can be replaced by an upper triangular R_m of dimension m by m.

$$\Lambda_m = R_m^{\mathrm{T}} R_m$$

The Householder algorithm enables one to obtain the matrix R_m without explicitly computing Λ_m. If T is an orthogonal matrix which has the property

$$T^{\mathrm{T}} T = I \tag{6.8}$$

then we have

$$\Lambda_m = R_{nm}^{\mathrm{T}} T^{\mathrm{T}} T R_{nm}$$

The Householder algorithm finds a T that gives R_m when multiplied times R_{nm}. The right side of the data equation $(\hat{\eta})$ is also multiplied by T to obtain a new data equation in upper triangular form. The Householder algorithm thus serves the same purpose in updating the SRIF matrix and right side as the Kalman update algorithm serves to update the covariance and state estimate. An updated state estimate can be obtained from the data equation by simply inverting the SRIF matrix and multiplying times the right side.

6.3.1 Discrete Process Noise Update

For the simple case of exponentially time correlated process noise, the differential equation may be solved by performing a discrete update over a fixed time interval referred to as a batch.

$$\frac{dp}{dt} = \left(\frac{-1}{\tau}\right) p + \omega(t)$$

The solution is

$$p(t) = e^{\frac{t - t_0}{\tau}} p(t_0) + \int_{t_o}^{t} e^{\frac{-(t-\zeta)}{\tau}} \omega(\zeta) d\zeta$$

The variance of $p(t)$ is given by

$$\sigma_p^2(t) = e^{\frac{-2(t - t_0)}{\tau}} \sigma_p^2(t_0) + \int_{t_o}^{t} e^{\frac{-2(t-\zeta)}{\tau}} \sigma_\omega^2(\zeta) d\zeta$$

The process noise variance may also be obtained by solution of the following differential equation,

$$\frac{d\sigma_p^2(t)}{dt} = \left(\frac{-2}{\tau}\right) \sigma_p^2(t) + \dot{q}(t)$$

where

$$\dot{q} = \frac{2\sigma_s^2}{\tau}$$

and σ_s^2 is the steady state noise variance. In difference equation form these give

$$p_{j+1} = Mp_j + \omega_j$$
$$\sigma^2_{p\,j+1} = M^2\sigma^2_{p\,j} + \Delta q$$

where

$$\Delta q = (1 - M^2)\sigma_s^2 \approx \left(\frac{2\Delta t}{\tau}\right)\sigma_s^2$$
$$M = e^{\frac{-\Delta t}{\tau}}$$
$$\Delta t = t_{j+1} - t_j$$

The data equation obtained as a result of processing data from t_j to t_{j+1} is given by

$$\begin{bmatrix} R_p & R_{px} & R_{py} \\ R_{xp} & R_x & R_{xy} \\ 0 & 0 & R_y \end{bmatrix} \begin{bmatrix} p_j \\ x_{j+1} \\ y \end{bmatrix} = \hat{\eta}_j$$

For the discrete process noise data update, the value of the stochastic parameters (p_j) are held constant over the interval t_j to t_{j+1} while the SRIF matrix is mapped forward. At the time t_{j+1}, the process noise variance accumulated over this same time interval is introduced via the following data equation as a discrete impulse.

$$R_\omega\,\hat{\omega}_j = \hat{\eta}_\omega$$

where

$$R_\omega = \frac{1}{\sigma_\omega}$$

Replacing $\hat{\omega}_j$ by the equation in terms of p_j and p_{j+1} we have

$$R_\omega\,\hat{p}_{j+1} - R_\omega M\,\hat{p}_j = \hat{\eta}_\omega = 0$$

The updated data equation is obtained by partitioning and combining with the above noise data equation.

$$\begin{bmatrix} -R_\omega M & R_\omega & 0 & 0 \\ R_p & 0 & R_{px} & R_{py} \\ R_{xp} & 0 & R_x & R_{xy} \\ 0 & 0 & 0 & R_y \end{bmatrix} \begin{bmatrix} p_j \\ p_{j+1} \\ x_{j+1} \\ y \end{bmatrix} = \begin{bmatrix} 0 \\ \\ \hat{\eta}_j \end{bmatrix}$$

The data equation is partially triangularized over the first columns corresponding to the process noise terms to obtain

$$\begin{bmatrix} R^*_{pj} & R^*_{ppj} & R^*_{pxj} & R^*_{pyj} \\ 0 & R^+_p & R^+_{px} & R^+_{py} \\ 0 & R^+_{xp} & R^+_x & R^+_{xy} \\ 0 & 0 & 0 & R^+_y \end{bmatrix} \begin{bmatrix} p_j \\ p_{j+1} \\ x_{j+1} \\ y \end{bmatrix} = \begin{bmatrix} \hat{\eta}^*_j \\ \\ \hat{\eta}_{j+1} \end{bmatrix}$$

where the plus superscript is introduced to indicate a change in the numerical values after the process noise update. The stochastic parameter update is completed by stripping off the top rows corresponding to p_j, those containing the asterisk, and these may be saved along with the right side $(\hat{\eta}^*_j)$ for smoothing.

6.3.2 Solution Epoch

The SRIF formulation computes an estimate of the spacecraft state, target body ephemeris, and target body attitude at the initial epoch. This is purely a matter of convenience since the state and covariance at any future epoch may be obtained by simply mapping the epoch state solution. An epoch state formulation enables the trajectory and variational partial derivatives to be computed before data is processed by the filter. When stochastic parameters are present, the current state cannot be determined by mapping the epoch state solution. The true epoch state solution is obtained by smoothing. Since the final state solution is the solution of most interest and it is desirable to avoid smoothing, the contribution of stochastic parameters is mapped back to epoch creating what is called a pseudo epoch state solution. The pseudo epoch state solution is an exact least squares fit to the data when it is mapped to the current or final state. The pseudo epoch state filter adopted by the Jet Propulsion Laboratory is really a current state filter when mapped to epochs in the future and used for prediction. In the early 1970s orbit determination was a very challenging problem. The high-precision modeling of the solar system, media and station locations made possible by VLBI has reduced the need for stochastic parameters to cover modeling errors. Indeed, most missions are now flown without the need for stochastic parameters and all the data could be processed in a single batch.

6.3.3 Computed and Consider Covariance

After all the data has been processed and a solution obtained, the covariance matrix of the estimated parameters is computed and mapped to an epoch of interest. The SRIF matrix R is inverted to obtain the square root covariance matrix (S) and this matrix is multiplied by its transpose to obtain the covariance matrix (P).

$$R = \begin{bmatrix} R_x & R_{xc} \\ 0 & R_c \end{bmatrix}$$

$$S = R^{-1}$$

$$P = SS^T$$

The R matrix has been partitioned into (R_x) and R_c. The R_x SRIF matrix contains the state, stochastic parameters, and most of the constant parameters. The R_c matrix contains parameters that are suspect because they represent models that are not as accurate as the data can measure. In the early days of the space program station location errors fell into this category. These are called consider parameters. The estimate of consider parameters obtained above can be too optimistic and the modeling errors associated with these parameters can infiltrate into the other estimated parameters which include spacecraft state. One strategy is to just ignore these parameters. This can easily be done by truncating columns pertaining to these parameters from (R). The R_c parameters have been conveniently placed at the end of R for this purpose. The result is called the computed covariance and is given by

$$S_x = R_x^{-1}$$

$$P_x = S_x \, S_x^T$$

The reduced set of estimated parameters associated with the computed covariance can enable orbit determination operations to proceed smoothly, but the infiltration of model errors can result in overly optimistic error estimates. In order to determine the effect of consider parameters on other estimated parameters, the consider covariance is computed. The consider covariance reveals the sensitivity of orbit estimation to modeling errors, but has no affect on the orbit solution. The consider covariance is computed by discarding the rows of R corresponding to R_c and replacing them with the *a priori* consider covariance. The correlation of the consider parameters with the other estimated parameters determined by processing data is preserved. We simply add some negative information to the SRIF matrix that gives us the desired square root covariance for the consider parameters. This can be done by employing some mathematical trickery. Since the triangularization of a SRIF matrix is not unique, if matrix can be partially triangularized. The triangularization is stopped at the row corresponding to the first consider parameter and the square root consider *a priori* consider covariance is inserted. This process is mathematically equivalent to the following.

$$R_c = \begin{bmatrix} R_x & R_{xc} \\ 0 & R_c^* \end{bmatrix}$$

where $R_c^* = S_{con}^{-1}$ and for independent consider parameters the square root covariance S_{con} is a diagonal matrix of consider *a priori* sigmas. Making use of the Schur identity given in Sect. 6.2, the consider square root covariance is given by

$$S_c = R_c^{-1} = \begin{bmatrix} R_x^{-1} & -R_x^{-1}\, R_{xc}\, S_{con}^{-1} \\ 0 & S_{con} \end{bmatrix}$$

and the consider covariance is

$$P_c = S_c\, S_c^T$$

6.3.4 *Smoothing*

A current state orbit determination filter provides a best estimate of the spacecraft state at the end of the data arc. An estimate of the state at the beginning of the data arc or at some time in between may be obtained by mapping the final state back to the epoch of interest. This mapping may also be obtained by integrating the trajectory backwards from the final state. If there are no stochastic parameters, these mappings will also provide a best estimate. When stochastic parameters are included, a best estimate may be obtained by processing the data backwards and this is referred to as a smoothed best estimate. If only a few estimation epochs are of interest a smoothed best estimate may be obtained while processing the data forward. This processing is referred to as single point smoothing and was used by the Galileo project for obtaining an estimate of the probe state at separation.

Data is processed from launch to the separation epoch. At the time of separation or the time that a smoothed best estimate is desired the data processing is halted. The covariance at this time is given by $P(t_s)$

$$P(t_s) = \begin{bmatrix} P(p) & P(p, x1) & 0 & P(p, y) \\ P(x1, p) & P(x1) & 0 & P(x1, y) \\ 0 & 0 & P(xs) & 0 \\ P(y, p) & P(y, x1) & 0 & P(y) \end{bmatrix}$$

The smoothed *a prior* covariance ($P(xs)$) is a place holder and has no affect on the computed covariance ($P(t_s)$). The data partials associated with $P(xs)$ are all zero up to the time t_s. At the smoothing epoch, the computed covariance is updated with

$$P(t_s)^+ = \begin{bmatrix} P(p) & P(p,x1) & P(p,x1) & P(p,y) \\ P(x1,p) & P(x1) & P(x1) & P(x1,y) \\ P(x1,p) & P(x1) & P(x1) & P(x1,y) \\ P(y,p) & P(y,x1) & P(y,x1) & P(y) \end{bmatrix}$$

This update forces the smoothed spacecraft state to be equal to the computed spacecraft state and be perfectly correlated. The smoothed spacecraft state also assumes the same correlation with stochastic parameters ($P(p)$) and constant parameters ($P(y)$). The data filter continues to the end of the data arc. The data partials for xs remain at zero and the smoothed best estimate is updated through its correlation with the other estimated parameters. For the implementation of single point smoothing in the orbit determination program used for Galileo, the update at the smoothing epoch was performed by processing six artificial measurements that forced the spacecraft state to equal the smoothed state at the smoothing epoch.

6.4 Continuous Filter Equations

A data filter processes data in order to obtain an estimate of parameters that are related to the data by a mathematical model. Data filters exist in many forms and use the covariance of the state parameters, or some equivalent representation, along with the measurements and a simulation of the measurements including partial derivatives, to obtain the desired estimate. Data filters may be separated into two categories depending on how the state covariance is evolved as a function of time. Continuous data filters evolve the state covariance by integration of a matrix differential equation or Riccati equation and discrete data filters evolve the state covariance by mapping over a finite time interval. Discrete filters are thus obtained by solving the continuous equations over some finite time interval. The covariance matrix of the state may be represented by its inverse or information matrix or square root factorizations of either of these matrices.

The system dynamics may be described as a linear perturbation of a reference function of the state variables. Given the nominal values of the state variables described by the function $\bar{x}(t)$ and a perturbation of the state (δx) at the initial epoch (t_0), the perturbed state variables are described by

$$x(t) = \bar{x}(t) + \Phi(t, t_0)\, \delta x(t_0)$$

where the state transition matrix (Φ) is given by

$$\Phi = \frac{\partial x(t)}{\partial x(t_0)}$$

The state transition matrix may be obtained as a solution of the following differential equation or by numerical integration.

$$\frac{\partial \dot{x}(t)}{\partial x(t_0)} = \frac{\partial \dot{x}(t)}{\partial x(t)} \frac{\partial x(t)}{\partial x(t_0)}$$

$$\dot{\Phi}(t, t_0) = F\, \Phi(t, t_0)$$

where

$$F = \frac{\partial \dot{x}(t)}{\partial x(t)}$$

The above differential equation, describing the evolution of the state variation, may be generalized to include other parameters and process noise.

$$\dot{X} = F\,X\; +\; G\Omega$$

where G is the mapping of Ω, the process noise. Here, the δ's have been dropped and the variation δx is represented by X. The state vector variation X may be generalized to include constant parameters (y) and stochastic parameters (p) as well as the dynamic state variables (x). The process noise (Ω) contains white noise (ω) on the stochastic parameters. Thus we have

$$X = \begin{bmatrix} p \\ x \\ y \end{bmatrix} \quad \Omega = \begin{bmatrix} \omega \\ 0 \\ 0 \end{bmatrix}$$

The stochastic parameters (p) provide a means of introducing process noise into the state variables. These are defined by scalar differential equations of the form

$$\dot{p}_i = -\frac{1}{\tau_i}\; p_i + \omega_i$$

where τ_i is the correlation time and ω_i is the white noise associated with the i'th stochastic parameter. Thus, white noise is introduced directly to the parameter p and indirectly to the state via the mapping matrix F.

An estimate of the state is obtained from a mathematical model of the system dynamics that include measurements processed by a data filter. The "best" estimate of the variation of the state $(\hat{X})$ is described by the following equations,

$$\dot{\hat{X}} = F\hat{X} + G\hat{\Omega} + K\hat{Z}$$

$$\hat{Z} = Z - A\hat{X}$$

$$A = \frac{\partial Z}{\partial X}$$

where K is the Kalman gain, $\hat{\Omega}$ represents an estimate of the process noise, Z are the actual measurements, and A is the matrix of data partials. The Kalman gain is computed as a function of the measurement error, the data partials, and the state error covariance (P). Thus, in order to obtain a complete set of equations that would enable the computation of the estimated state we need an equation for the Kalman gain and an equation for evolving P as a function of time.

The covariance of the state estimate is defined by the expected value represented by

$$P = E\left\{XX^{\mathrm{T}}\right\}$$

As an alternative, we may compute the information matrix (Λ), the square root of the covariance (S), or the square root of the information matrix (R). The equations that define these matrices are given by

$$P = \Lambda^{-1}$$

$$P = SS^{\mathrm{T}}$$

$$P^{-1} = R^{\mathrm{T}}R$$

Thus, we are interested in obtaining differential equations of the form

$$\dot{P} = \dot{P}_m + \dot{P}_q + \dot{P}_d \tag{6.9}$$

$$\dot{\Lambda} = \dot{\Lambda}_m + \dot{\Lambda}_q + \dot{\Lambda}_d \tag{6.10}$$

$$\dot{S} = \dot{S}_m + \dot{S}_q + \dot{S}_d \tag{6.11}$$

$$\dot{R} = \dot{R}_m + \dot{R}_q + \dot{R}_d \tag{6.12}$$

where the subscript m refers to the mapping terms, the subscript q refers to process noise terms, and the subscript d refers to the data update terms.

The evolution of the covariance as a function of time may be obtained by mapping the state covariance obtained at some epoch (t_0) to some time in the future (t) with the state transition matrix.

$$P(t) = \Phi(t, t_0)\, P(t_0)\, \Phi(t, t_0)^{\mathrm{T}}$$

Taking the derivative with respect to time we obtain

$$\dot{P}(t) = \dot{\Phi}(t, t_0)\, P(t_0)\, \Phi(t, t_0)^{\mathrm{T}} + \Phi(t, t_0)\, P(t_0)\, \dot{\Phi}(t, t_0)^{\mathrm{T}}$$

Since the state transition matrix is obtained by integrating

$$\dot{\Phi}(t, t_0) = F(t)\, \Phi(t, t_0)$$

we obtain after substitution

$$\dot{P}_m = F P \; + \; P F^{\mathrm{T}} \tag{6.13}$$

6.4.1 *Process Noise Term*

In the covariance matrix differential equation, process noise enters as an addition to the covariance. Thus we have

$$P(t + \Delta t) = P(t) \; + \; G \, \Delta Q \, G^{\mathrm{T}}$$

where ΔQ is the covariance of the process noise admitted over the time interval Δt and

$$\Delta Q = Q \; \Delta t$$

where Q is the rate of accumulation of process noise. Thus, in the continuum we have

$$\dot{P}_q = \lim_{\Delta t \to 0} \left\{ \frac{P(t + \Delta t) - P(t)}{\Delta t} \right\} = G \; Q \; G^{\mathrm{T}} \tag{6.14}$$

6.4.2 *Data Update Term*

The discrete covariance update may be obtained assuming an additional measurement A_{n+1} is added to a previously determined estimate based on measurements A_n with covariance P_n.

$$P_{n+1} = \left[A_n^{\mathrm{T}} \Delta W_n A_n + A_{n+1}^{\mathrm{T}} \Delta W_{n+1} A_{n+1} \right]^{-1}$$

In the notation used here, A_n is a matrix with n rows corresponding to the measurements and m columns corresponding to the state parameters. A_{n+1} is a row matrix of dimension m. We also have for the covariance update,

$$P_{n+1}^{-1} = P_n^{-1} + A_{n+1}^{\mathrm{T}} \, \Delta W_{n+1} \, A_{n+1}$$

and since

$$\Lambda = P^{-1}$$

$$\Lambda_{n+1} = \Lambda_n + A_{n+1}^{\mathrm{T}} \, \Delta W_{n+1} \, A_{n+1}$$

Over the time interval Δt between measurements, information accumulates at a rate W and

$$\Delta W_{n+1} = W \, \Delta t$$

$$\Lambda_{n+1} - \Lambda_n = A_{n+1}^{\mathrm{T}} \, W \Delta t \, A_{n+1}$$

Dividing by Δt and taking the limit as Δt approaches zero,

$$\dot{\Lambda}_m = A^{\mathrm{T}} \, W \, A \tag{6.15}$$

we obtain a differential equation for the evolution of the information matrix due to addition of data.

6.4.3 Continuous Filter Differential Equations

Collecting the terms derived above, we have the following matrix differential equation or Riccati equation for the covariance filter,

$$\dot{P} = F P + P F^{\mathrm{T}} + G \, Q \, G^{\mathrm{T}} + \dot{P}_d$$

$$K = P \, A^{\mathrm{T}} \, W$$

and for the information filter,

$$\dot{\Lambda} = \dot{\Lambda}_m + \dot{\Lambda}_q + A^{\mathrm{T}} \, W \, A$$

$$K = \Lambda^{-1} \, A^{\mathrm{T}} \, W$$

The data update term ($\dot{P}_d$) is missing from the covariance equation and the mapping ($\dot{\Lambda}_m$) and process noise ($\dot{\Lambda}_q$) terms are missing from the information filter equation and these may be obtained by transformation using matrix identities. For the covariance and information equations, we need the following matrix identities.

$$P \Lambda = I$$

$$\dot{P} \Lambda + P \dot{\Lambda} = 0$$

$$\dot{P} = -P \dot{\Lambda} \Lambda^{-1} = -P \, \dot{\Lambda} \, P$$

$$\dot{\Lambda} = -P^{-1} \, \dot{P} \, \Lambda = -\Lambda \, \dot{P} \, \Lambda$$

Applying these identities to the above matrix differential equations, we have

$$\dot{P} = F P + P F^{\mathrm{T}} + G \, Q \, G^{\mathrm{T}} - P A^{\mathrm{T}} \, W \, A P$$

$$K = P \, A^{\mathrm{T}} \, W$$

The covariance filter in this form is called the continuous form of the Kalman-Bucy filter. For the information filter, we have

$$\dot{\Lambda} = -\Lambda F - F^{\mathrm{T}}\Lambda - \Lambda\, G Q G^{\mathrm{T}}\, \Lambda + A^{\mathrm{T}}\, W\, A$$

$$K = \Lambda^{-1}\, A^{\mathrm{T}}\, W$$

A similar set of matrix identities, derived by Scheeres, may be developed for the square root covariance filter (SRCF) and the square root information filter (SRIF) that may be used to transform the covariance time derivative.

$$P = S S^{\mathrm{T}}$$

$$\dot{P} = \dot{S} S^{\mathrm{T}} + S \dot{S}^{\mathrm{T}}$$

$$\left[S\dot{S}^{\mathrm{T}} - \frac{1}{2}\dot{P} \right] + \left[\dot{S} S^{\mathrm{T}} - \frac{1}{2}\dot{P} \right] = 0$$

Because of symmetry associated with the above terms in the brackets, both terms in the brackets must be zero and

$$\dot{S} = \frac{1}{2}\dot{P}\, S^{-\mathrm{T}}$$

A similar derivation for the SRIF matrix gives the identity

$$\dot{R} = -\frac{1}{2} R\dot{P}\, R^{\mathrm{T}} R$$

Applying these identities to the covariance and information filter equations gives the following matrix differential equations for the SRCF and SRIF matrices.

$$\dot{S} = \frac{1}{2}\left[F S + S S^{\mathrm{T}} F^{\mathrm{T}} S^{-\mathrm{T}} \right] + \frac{1}{2} G\, Q\, G^{\mathrm{T}} S^{-\mathrm{T}} - \frac{1}{2} S S^{\mathrm{T}} A^{\mathrm{T}} W A S$$

$$\dot{R} = -\frac{1}{2}\left[R F + R^{-1} F^{\mathrm{T}} R^{\mathrm{T}} R \right] - \frac{1}{2} R G Q G^{\mathrm{T}} R^{\mathrm{T}} R + \frac{1}{2} R^{-\mathrm{T}} A^{\mathrm{T}}\, W\, A$$

The mapping terms for both the SRCF and SRIF contain matrix inverses. These may be eliminated by introducing a different factorization of the square roots. Consider the mapping of the square root covariance from an initial epoch t_0 to the epoch t.

$$P(t) = \Phi(t, t_0)\, S(t_0) S(t_0)^{\mathrm{T}}\, \Phi(t, t_0)^{\mathrm{T}}$$

The mapped square root is simply

$$S(t) = \Phi(t, t_0)\, S(t_0)$$

Taking the derivative with respect to time,

$$\dot{S}(t) = \dot{\Phi}(t, t_0)\, S(t_0)$$

$$\dot{S}(t) = \dot{\Phi}(t, t_0)\, \Phi(t, t_0)^{-1} S(t)$$

$$\dot{S}(t) = F(t) S(t)$$

For the SRIF matrix we have

$$S(t)R(t) = I$$

$$\dot{S}(t)R(t) + S(t)\dot{R}(t) = 0$$

$$\dot{R}(t) = -R(t)\dot{S}(t)S(t)^{-1}$$

$$\dot{R}(t) = -R(t)\, F(t)$$

Making the above substitutions for the mapping terms, the matrix differential equations and Kalman gain for the covariance, information, square root covariance, and square root information filters are summarized below.
Covariance (Kalman-Bucy) Filter

$$\dot{P} = FP + PF^{\mathrm{T}} + G\,Q\,G^{\mathrm{T}} - PA^{\mathrm{T}}\,W\,AP \tag{6.16}$$

$$K = P\,A^{\mathrm{T}}\,W \tag{6.17}$$

Information Filter

$$\dot{\Lambda} = -\Lambda F - F^{\mathrm{T}}\Lambda - \Lambda\, GQG^{\mathrm{T}}\,\Lambda + A^{\mathrm{T}}\,W\,A \tag{6.18}$$

$$K = \Lambda^{-1}\,A^{\mathrm{T}}\,W \tag{6.19}$$

Square Root Covariance Filter (SRCF)

$$\dot{S} = FS + \frac{1}{2} G\,Q\,G^{\mathrm{T}} S^{-\mathrm{T}} - \frac{1}{2} SS^{\mathrm{T}} A^{\mathrm{T}} WAS \tag{6.20}$$

$$K = SS^{\mathrm{T}} A^{\mathrm{T}}\,W \tag{6.21}$$

Square Root Information Filter (SRIF)

$$\dot{R} = -RF - \frac{1}{2} RGQG^{\mathrm{T}} R^{\mathrm{T}} R + \frac{1}{2} R^{-\mathrm{T}} A^{\mathrm{T}}\,W\,A \tag{6.22}$$

$$K = R^{-1} R^{-\mathrm{T}}\,A^{\mathrm{T}}\,W \tag{6.23}$$

The data update and process noise terms of the above filter equations exhibit a symmetry or duality when the information filters are compared with the covariance filters. For example, the data update term of the information filter may be obtained by replacing Q with W and G with A^{T} in the process noise term of the covariance filter. Also, the process noise update term of the information filter may be obtained by making similar replacements in the data update term of the covariance filter. These same dual relationships exist for the filters in their square root form. The existence of duality enables algorithms designed for data updating to be used for process noise updating and vice versa. For example, the Potter square root covariance data update algorithm may be used to update process noise in the SRIF.

6.5 Continuous SRIF with Discrete Data Update

The selection of a filter algorithm depends on many competing criteria related to accuracy, computational efficiency, memory utilization, and simplicity of design. Consideration of accuracy seems to favor factorized or square root filters and computational efficiency seems to favor discrete filters. With the proliferation of personal computers, computational efficiency has become less important since computer processing time is now relatively cheap. Simplicity of design and memory utilization favor a continuous approach to filtering. The system dynamics and data partial derivatives enter directly into the filter and the need to compute a state transition matrix is completely eliminated. However, data is generally in the form of discrete data points and may not be easily transformed to the continuous form. This suggests a hybrid approach which allows system dynamics and process noise to be treated continuously and data to be treated as a discrete update.

The continuous SRIF, with discrete data update, is selected for development of a filter algorithm. Information filters have the advantage that *a priori* on the constant parameters does not have to be placed on the filter until after all the data is processed. During filtering, the information arrays may be sparse resulting in less computation.

6.5.1 Process Noise Duality

The continuous process noise update enables one to introduce process noise directly as a differential equation to the filter. This form is convenient for describing process noise and enables the investigation of a wide variety of process noise models without explicitly solving the differential equation. The continuous process noise update term in the information filter has the same form as the data update term in the covariance filter. The Potter square root covariance data update algorithm [6] provides a means of performing a scalar data update to the square root covariance filter. Because of duality, the discrete Potter data update algorithm can be adapted to the SRIF for a discrete scalar process noise update. Taking the limit as Δt approaches zero enables one to convert the discrete process noise update to a continuous process noise update.

Starting with the process noise update term in the information filter (Eq. (6.18)), which is the dual of the data update term of the covariance filter (Eq. (6.16)), we have

$$\dot{\Lambda}_q = -\Lambda\, G Q G^{\mathrm{T}}\, \Lambda = \lim_{\Delta t \to 0} \frac{\Lambda_{n+1} - \Lambda_n}{\Delta t}$$

and in the discrete form,

$$\Lambda = \tilde{\Lambda} - \tilde{\Lambda}\, G \Delta Q G^{\mathrm{T}}\, \tilde{\Lambda}$$

where the notation for Λ_n, the information matrix before the update or priori, is replaced by $\tilde{\Lambda}$ and Λ_{n+1} is replaced by Λ. Since

$$\Lambda = R^{\mathrm{T}} R$$

we have for the POTTER approximation

$$R^{\mathrm{T}} R = \tilde{R}^{\mathrm{T}} \left[I - v\, \Delta Q\, v^{\mathrm{T}} \right] \tilde{R}$$

where

$$v = \tilde{R}\, G$$
$$\Delta Q = Q \Delta T$$

If ΔQ and G are assumed to be diagonal (i.e., uncorrelated process noise parameters), then each diagonal element of ΔQ is given by a scalar Δq_i. Dropping the i subscript, we have for the i'th row of R and diagonal element of ΔQ,

$$I - \Delta q v v^{\mathrm{T}} = (I - \Delta\alpha v v^{\mathrm{T}})^2$$
$$I - \Delta q v v^{\mathrm{T}} = I - 2\Delta\alpha\; v v^{\mathrm{T}} + \Delta\alpha^2 v v^{\mathrm{T}} v v^{\mathrm{T}}$$

Since $v^{\mathrm{T}} v$ is a scalar, the solution of the above quadratic equation is given by

$$\Delta\alpha = \frac{1 - \sqrt{1 - v^{\mathrm{T}} v \Delta q}}{v^{\mathrm{T}} v}$$

and

$$R^{\mathrm{T}} R = \tilde{R}^{\mathrm{T}} (I - \Delta\alpha\; v v^{\mathrm{T}})^{\mathrm{T}}\, (I - \Delta\alpha\; v v^{\mathrm{T}}) \tilde{R}$$
$$R = (I - \Delta\alpha\; v v^{\mathrm{T}}) \tilde{R}$$
$$R = \tilde{R} - \Delta\alpha\; \tilde{R}\, G G^{\mathrm{T}}\, \tilde{R}^{\mathrm{T}} \tilde{R}$$

In the continuum we have

$$\dot{R} = -\Delta\dot{\alpha}\tilde{R}\,GG^{\mathrm{T}}\,\tilde{R}^{\mathrm{T}}\tilde{R}$$

and

$$\Delta\dot{\alpha} = \frac{1}{2}(1 - v^{\mathrm{T}}v\Delta q)^{-\frac{1}{2}}\,\Delta\dot{q}$$

In the limit as Δq and Δt go to zero we have

$$\Delta\dot{\alpha} = \frac{1}{2}\,\Delta\dot{q}$$

and

$$\dot{R} = -\frac{1}{2}\,\Delta\dot{q}\;R\,GG^{\mathrm{T}}\,R^{\mathrm{T}}R$$

If we have more than one stochastic parameter, the $\Delta\dot{q}$ *'s* can be assembled into a diagonal matrix Q and we have

$$\dot{R} = -\frac{1}{2}\;R\,GQG^{\mathrm{T}}\,R^{\mathrm{T}}R \tag{6.24}$$

This is the same equation as derived above for the continuous SRIF process noise update only we have assumed diagonal Q and G.

6.5.2 Numerical Integration of SRIF Matrix

The continuous SRIF data processing algorithm involves mapping the SRIF matrix from the time of a discrete data or process noise update to the time of the next data point or process noise update. The mapping is accomplished by numerical integration of the SRIF matrix differential equation. The numerical integration is performed with a suitable algorithm. The fifth order Runge-Kutta-Fehlberg method with error control has been successfully employed. Recall the matrix differential equation derived above for the SRIF (Eq. (6.22)) and discard the data update term.

$$\dot{R} = -RF - \frac{1}{2}RGQG^{\mathrm{T}}R^{\mathrm{T}}R$$

Consider the following partition.

$$\dot{R} = -\begin{bmatrix} R_d & R_{dy} \\ 0 & R_y \end{bmatrix} \begin{bmatrix} F_d \\ 0 \end{bmatrix} - \begin{bmatrix} R_p & R_{px} & R_{py} \\ 0 & R_x & R_y \\ 0 & 0 & R_y \end{bmatrix} \begin{bmatrix} \frac{1}{2} GQG^{\mathrm{T}} R_p^{\mathrm{T}} R_p \\ 0 \\ 0 \end{bmatrix}$$

where R_d corresponds to the dynamic parameters and the matrix F_d contains only the rows of F corresponding to the dynamic parameters. This equation simplifies to

$$\dot{R} = -R_d \, F_d - \frac{1}{2} R_p \, GQG^{\mathrm{T}} \, R_p^{\mathrm{T}} R_p$$

We only have to integrate the top rows of the SRIF matrix corresponding to the dynamic parameters and the derivative is a function of only the R_d partition of the SRIF matrix. For the simple case of exponentially correlated process noise we have

$$F_d = \begin{bmatrix} \dfrac{\partial \dot{p}}{\partial p} & 0 & 0 \\ \dfrac{\partial \dot{x}}{\partial p} & \dfrac{\partial \dot{x}}{\partial x} & \dfrac{\partial \dot{x}}{\partial y} \end{bmatrix}$$

$$\frac{\partial \dot{p}}{\partial p} = \begin{bmatrix} -\dfrac{1}{\tau_1} & & & \\ & -\dfrac{1}{\tau_2} & & \\ & & \cdots & \\ & & & -\dfrac{1}{\tau_i} \end{bmatrix}$$

$$Q = \begin{bmatrix} \dfrac{2\sigma_{s1}^2}{\tau_1} & & & \\ & \dfrac{2\sigma_{s2}^2}{\tau_2} & & \\ & & \cdots & \\ & & & \dfrac{2\sigma_{si}^2}{\tau_i} \end{bmatrix}$$

and G is the identity matrix.

6.6 Direct Orbit Determination

Direct orbit determination is a method for determining a spacecraft orbit directly from measurements. For Doppler data, the measurement recorded on a tracking data file contains the motion of the Earth, the central body, media and other orbit perturbations. After removing known components from the Doppler signature, the portion of the signature due only to the spacecraft motion relative to the central body can be parameterized. A Newton-Raphson method may be applied to obtain the solution. Parameters which yield a nearly diagonally dominant partials matrix are defined. This method is particularly important for orbits around asteroids and comets where a velocity error of only a few meters per second can place the spacecraft in an entirely unknown orbit.

Another example of direct orbit determination was devised by Hamilton and Melbourne for determining the position of a spacecraft far from the Earth. A single range measurement gives the distance to the spacecraft but provides no information of the position normal to the line-of-sight or the right ascension and declination of the spacecraft. By observing the rotation of a tracking station about the Earth's spin axis, which appears in the Doppler data like a spacecraft orbiting a planet, the right ascension and declination may be determined. The accuracy of this measurement is about 0.25 μrad or less than one tenth of an arc second which is about the same accuracy as the Hubble space telescope.

The determination of a spacecraft orbit can be hampered by the lack of convergence of an orbit determination program, in the case of poor *a priori* information or highly nonlinear measurements. The convergence problem is likely to occur following a large propulsive maneuver where execution errors may result in a poor prediction of the post maneuver orbit. However, the problem may also arise for a spacecraft that is left in orbit unattended for some time. The most common occurrences result from launch into Earth orbit and planetary orbit insertion. In either of these cases, a guidance or propulsion system failure can leave the spacecraft on a trajectory far from the nominal that is generally used as the *a priori* position estimate for the orbit determination solution. Due to nonlinearity, a perturbation of a few degrees in the orbit elements describing orientation or a few percent in period may result in the failure of an orbit determination program to converge.

For a spacecraft orbiting a planet, the energy required to perturb the spacecraft orbit far enough away from its original path to cause an orbit convergence problem is considerable. A planetary orbit insertion maneuver does have the potential for large anomalous orbit perturbations. This subject received much attention on the Viking mission to Mars and the Magellan mission to Venus. If a motor burn anomaly large enough to cause an orbit convergence problem had occurred on either of these missions, it could have been very difficult to recover.

For a spacecraft orbiting a comet or asteroid, a relatively small propulsive maneuver can radically alter the orbit. Only a few meters per second of velocity change are required to reverse the direction of the spacecraft velocity vector. If this were to happen due to an anomalous burn, the spacecraft might be left in a completely unknown orbit. The available data would be Earth-based Doppler tracking since the signal could be recovered by pointing the DSN antennas at the asteroid, and it is assumed that the spacecraft would be able to acquire celestial references and point its antenna at Earth even though the relative direction of the central body is not known. Conventional methods for redetermining the spacecraft orbit based solely on processing Doppler data in an orbit determination program may not be adequate.

The Doppler observable provides a direct measure of the spacecraft velocity projected onto the line-of-sight between the tracking station antenna and the spacecraft. The Doppler measurement is a measure of the difference in range over some time interval scaled by a constant determined by the transmitted frequency, the speed of light and the length of the time interval or count time. The range from the tracking antenna to the spacecraft may be expressed as the sum of the directed distances of the spacecraft to the central body, the central body position relative to the Sun, the Earth position relative to the Sun and the location of the tracking station relative to the center of the Earth, including a small correction for media. For the orbit determination problem, all of the distances referred to above are well known except for the spacecraft position relative to the central body. The contribution of this component may be isolated by subtracting the effect of the known distances and the media delay from the raw Doppler measurement. A multidimensional Newton-Raphson method is then applied for the solution of the spacecraft orbital elements.

6.6.1 Model of Doppler Data Signature

The spacecraft orbit may be determined in terms of six orbit elements in the plane-of-sky coordinate system shown in Fig. 6.2. After deriving a mathematical model for the spacecraft central body component of the Doppler signature, target positions on the signature are chosen that approximately define six target parameters, each of which is directly related to one of the orbital elements to be determined.

Figure 6.2 displays the spacecraft orbital plane with respect to the plane-of-sky coordinate system (X,Y,Z). The plane-of-sky is the plane perpendicular to the line-of-sight between the Earth and the central body of the spacecraft orbit. The X and Y axes lie in the plane-of-sky, and the Z axis points from the central body to the Earth. The plane-of-sky coordinate system is particularly useful for Doppler data, because the line of sight is along the Z axis.

The spacecraft orbit is oriented with respect to the plane-of-sky as follows: the node Ω is the smallest positive angle between the X axis and the line of nodes, which is the intersection of the plane-of-sky with the spacecraft orbital plane. The inclination i is the angle from the plane-of-sky to the spacecraft orbital plane, and the argument of periapsis ω is the angle from the line of nodes to the semi-major axis through periapsis. To convert plane-of-sky coordinates to coordinates in the orbital plane of the spacecraft, rotations through the three angles Ω, i, and ω must be performed.

$$\begin{bmatrix} X_3 \\ Y_3 \\ Z_3 \end{bmatrix} = T \begin{bmatrix} X \\ Y \\ Z \end{bmatrix}$$

and

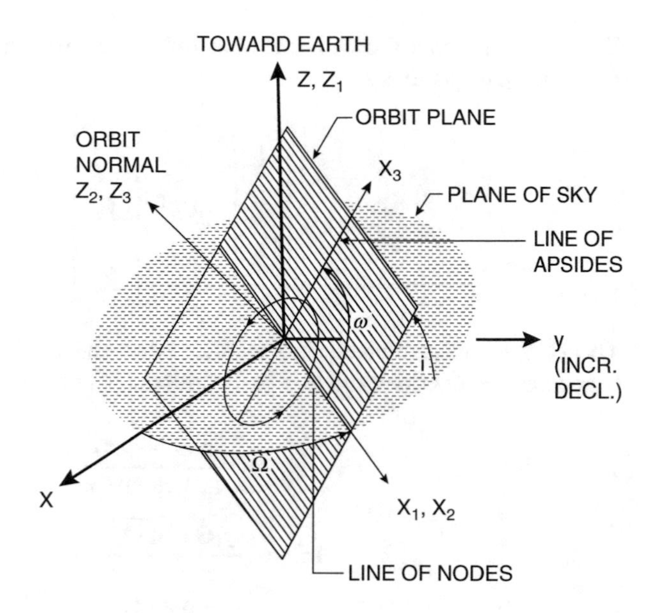

Fig. 6.2 The plane-of-sky coordinate system

$$T = \begin{bmatrix} \cos\omega & \sin\omega & 0 \\ -\sin\omega & \cos\omega & 0 \\ 0 & 0 & 1 \end{bmatrix} \begin{bmatrix} 1 & 0 & 0 \\ 0 & \cos i & \sin i \\ 0 & -\sin i & \cos i \end{bmatrix} \begin{bmatrix} \cos\Omega & \sin\Omega & 0 \\ -\sin\Omega & \cos\Omega & 0 \\ 0 & 0 & 1 \end{bmatrix}$$

The object is to determine an approximate two-body elliptical spacecraft orbit in terms of six elements, the semi-major axis a, the eccentricity e, the time of periapsis t_p, the node Ω, inclination i, and argument of periapsis ω. This set of orbital elements may be readily converted to a Cartesian initial state vector that may be input to orbit determination software for further refinement.

The signature of the spacecraft orbital velocity projected onto the Earth line-of-sight is simply the z component of the spacecraft velocity in the plane-of-sky coordinate system. In the spacecraft orbital plane, the spacecraft position is

$$X_3 = r\cos\eta$$
$$Y_3 = r\sin\eta$$
$$Z_3 = 0.$$

where η is the true anomaly, or periapsis-central body-spacecraft angle, and r is the distance from the central body to the spacecraft. Thus the velocity components are:

$$\dot{X}_3 = \dot{r}\cos\eta - r\dot{\eta}\sin\eta$$
$$\dot{Y}_3 = \dot{r}\sin\eta + r\dot{\eta}\cos\eta$$
$$\dot{Z}_3 = 0.$$

To get velocity coordinates in the plane-of-sky, we transpose the rotations through Ω, i, and ω, respectively.

$$\begin{bmatrix} \dot{X} \\ \dot{Y} \\ \dot{Z} \end{bmatrix} = T^{T} \begin{bmatrix} \dot{X}_3 \\ \dot{Y}_3 \\ \dot{Z}_3 \end{bmatrix}$$

The Doppler measurement $K\dot{\rho}$ is $K\dot{Z}$. After the application of some standard trigonometric identities and two-body orbit element formulas,

$$r = \frac{a\,(1 - e^2)}{1 + e\cos\eta}$$

$$r\,\dot{\eta} = \frac{\sqrt{GM\,a\,(1 - e^2)}}{r}$$

$$\dot{r} = \frac{r\,\dot{\eta}\,e\,\sin\eta}{1 + e\cos\eta}.$$

the following equation for the Doppler signature is obtained.

$$K\,\dot{\rho} = \frac{\sqrt{GM}}{\sqrt{a}}\,\frac{\sin i}{\sqrt{1 - e^2}}\,[\cos(\eta + \omega) + e\cos\omega]\,K \tag{6.25}$$

GM is the gravitational constant of the central body of the spacecraft orbit, and K is a scaling constant which is related to the transmitted frequency. It converts the Doppler observable into Hertz:

$$K = \frac{2\,C_3\,F_t}{c}$$

where F_t is the transmitted frequency, $C_3 = 240/221$ is the spacecraft turnaround ratio, and c is the speed of light. $F_t = 2,112,200,640$ Hz for X-band Doppler.

The representation of the Doppler signature is a function of the time-varying true anomaly η rather than an explicit function of time. The signature may be plotted as a function of time by performing the following computations: for a fixed time t, time of periapsis t_p, and parameters a, e, i, and ω, we may solve for the eccentric anomaly E from

$$t = t_p + 1/2\pi\,(E - e\sin E)\,P \tag{6.26}$$

P is the period, which is related to the semi-major axis a by the standard formula from two-body elliptical motion,

$$P = 2\pi\sqrt{\frac{a^3}{GM}} \tag{6.27}$$

Newton's method with an initial condition $E_o = \pi/2$ is effective for this computation. Given the eccentric anomaly E, the true anomaly η may be obtained from

$$\eta = \tan^{-1}\left(\frac{\sin\eta}{\cos\eta}\right) \tag{6.28}$$

where

$$\sin\eta = (\sin E)\,(a/r)\,(1 - e^2)^{1/2} \quad \text{and} \quad \cos\eta = \frac{\cos E - e}{1 - e\cos E}.$$

6.6.2 Parameterization of Doppler Signature

Figure 6.3 displays the slowly increasing or decreasing amplitude of a simulated Doppler signature, on which six target positions are identified. In the case of an actual mission, the raw Doppler data may be processed by an orbit determination program to remove the known components of the signature due to the relative motions of the Sun, Earth, tracking stations and central body of the spacecraft orbit. The resulting quasi-periodic signature, corresponding to the spacecraft-central body component, may be similarly plotted and the six target positions identified. When the spacecraft passes through the plane-of-sky, the observed signature is at a maximum or minimum. A maximum is attained moving away from the Earth, and a minimum is attained moving toward the Earth. At the zero crossings, the spacecraft range to Earth is at a local minimum or maximum. If the slope is negative at the zero crossing, the spacecraft is on the far side of the central body away from the Earth; and if the slope is positive at the zero crossing, the spacecraft is on the near side of the central body toward the Earth.

The six target positions defined on the signature yield the set of target parameters $\{z_i\}$ below:

$z_1 = t_5 - t_1$	a semi-major axis
$z_2 = t_1$	t_p time of periapsis
$z_3 = h_2 + h_4$	e eccentricity
$z_4 = h_2 - h_4$	i inclination
$z_5 = t_4 - t_2$	ω argument of periapsis
$z_6 = h_6 - h_2$	Ω the node

The six parameters relate to the plane-of-sky orbit elements as indicated. The difference in time between t_5 and t_1 is approximately the orbit period, from which

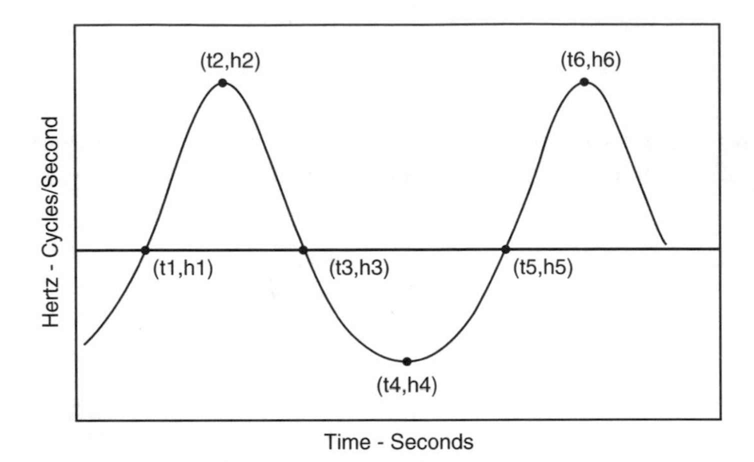

Fig. 6.3 The Doppler signature with target positions

the semi-major axis is obtained from (13). The parameter z_2 is the time of the zero crossing with positive slope, which is related to the time of periapsis, and the third parameter z_3 is related to the eccentricity, because $e \cos \omega$ determines the bias offset, or integrated average, of the graph. The amplitude

$$z_4 = \frac{\sqrt{GM}}{\sqrt{a}} \frac{\sin i}{\sqrt{1 - e^2}}$$

is closely related to the inclination in the plane-of-sky. Also, the argument of periapsis affects the downward slope of the signature, which is related to z_5, but the dependence is not apparent from Eq. (6.25), which is not time-dependent. The nonlinear relationship between ω and time can be observed in Eqs. (6.26)–(6.28). Finally, the node in the plane-of-sky is related to a slow change in amplitude, measured by z_6, that may be attributed to the relative motion of the asteroid with respect to the Earth. The node Ω changes as the planet or asteroid rotates around the Sun, and consequently around the Earth. A change in node indirectly changes the inclination i and argument of periapsis ω, which causes an upward or downward shift in the Doppler signature, measured by the difference in two peaks.

To the first order, the true anomaly is the only time-dependent parameter in the mathematical model of the spacecraft-central body component of the Doppler signature (11). The equation does not explicitly incorporate the node in the plane-of-sky. The relation to Ω may be obtained by solving Kepler's equation. We may obtain a first order equation that includes the node by introducing the rotation of the central body about the Earth as second order inclination and argument of periapsis rates. We have for the rotation of the planet or asteroid about the Earth

$$\boldsymbol{\Omega}_{\mathbf{e}} = \frac{\mathbf{R}_z \times \mathbf{V}_z}{|\mathbf{R}_z|^2}$$

where $\mathbf{R}_z$ and $\mathbf{V}_z$ are the position and velocity of the central body with respect to the Earth. Transforming into the plane-of-sky we have

$$\boldsymbol{\Omega}_{\mathbf{pos}} = T\ \boldsymbol{\Omega}_{\mathbf{e}}$$

The following Euler angle rates may be obtained from the rotation of the plane-of-sky coordinate system.

$$\begin{bmatrix} \frac{di}{dt} \\ 0 \\ 0 \end{bmatrix} = -\begin{bmatrix} \cos\Omega & \sin\Omega & 0 \\ -\sin\Omega & \cos\Omega & 0 \\ 0 & 0 & 1 \end{bmatrix} \boldsymbol{\Omega}_{\mathbf{pos}}$$

$$\begin{bmatrix} 0 \\ 0 \\ \frac{d\omega}{dt} \end{bmatrix} = -\begin{bmatrix} 1 & 0 & 0 \\ 0 & \cos i & \sin i \\ 0 & -\sin i & \cos i \end{bmatrix} \begin{bmatrix} \cos\Omega & \sin\Omega & 0 \\ -\sin\Omega & \cos\Omega & 0 \\ 0 & 0 & 1 \end{bmatrix} \boldsymbol{\Omega}_{\mathbf{pos}}$$

6.6.3 Solution by Newton-Raphson

The determination of the spacecraft orbit convergence problem thus reduces to finding the plane-of-sky orbit elements that generate the observed signature. We estimate the vector of orbital parameters $\mathrm{O} = (\mathrm{a}, \mathrm{t_p}, \mathrm{e}, \mathrm{i}, \omega, \Omega)^{\mathrm{T}}$ by measuring the vector of target parameters $\mathrm{Z} = (\mathrm{z_1}, \mathrm{z_2}, \mathrm{z_3}, \mathrm{z_4}, \mathrm{z_5}, \mathrm{z_6})^{\mathrm{T}}$. Values computed for six points shown in Fig. 6.3 may be compared with the observed values in a Newton-Raphson algorithm to solve for the plane-of-sky orbit elements.

The vector of target parameters $\mathrm{Z_o}$ is computed from the spacecraft-central body component of the Doppler signal. An initial set of orbital elements $\mathrm{O_o}$ is processed by an orbit determination program, which includes a model of the irregular gravity field of the asteroid, or other central body. The computed signature is obtained by integrating the equations of motion of the spacecraft, projecting the spacecraft velocity vector into the line-of-sight relative to Earth, and scaling by the constant K. The target parameters $\mathrm{Z_1}$ are computed from the simulated signal.

The Newton-Raphson method involves the recursive solution of the following equation:

$$O_{n+1} = O_n + \left(\frac{\partial Z}{\partial O}\right)^{-1} (Z_n - Z_0)$$

where Z_n is the vector of target parameters computed from the estimated vector of spacecraft orbital elements O_n. Partial derivatives of the target parameters with respect to the elements of O_j are computed by perturbing each orbital element in O_j from the nominal, generating a signal, measuring the perturbations in the target parameters, and computing

$$\frac{\Delta z_i}{\Delta o_j}, \qquad 1 \leq (i, j) \leq 6.$$

A new set of reference orbital elements O_{n+1} and corresponding set of reference target parameters Z_{n+1} are then computed. The process is continued to convergence. Since the computed and measured Z may be obtained along with the partial derivatives with high precision, the major weakness is the initial guess of the orbit elements. Since a diagonally dominant system will converge from just about any initial guess, it is desirable to define a parameter set that has a matrix of partial derivatives that is as close to diagonal as practical. The above parameter set (z_i) was selected with this purpose in mind. The near diagonal dominance of the matrix of partial derivatives associated with this parameter set ensures convergence over a wide range of initial starting points. Also, the converged orbit elements may be transformed to a high-precision state vector that may be input directly to orbit determination software. The accuracy of the Newton-Raphson solution depends only on the accuracy of the independent parameters (measurements) and their relation to the dependent parameters (the model). The accuracy does not depend on the initial guess or partial derivatives which control the convergence to a solution.

6.6.4 Magellan Example

As an example of the application of the above method, consider the determination of the Magellan spacecraft orbit about Venus. Two orbits of raw Doppler data from DSS 15 at Goldstone, California and DSS 45 at Canberra, Australia were acquired on February 7–8, 1991, and are shown in Fig. 6.4.

The gaps in the data coverage were due to a loss of Doppler data while the spacecraft was being rotated for the purpose of science data acquisition. Similar gaps could be caused by the unavailability of tracking stations or the occultation of the spacecraft by the planet. The target parameters described above require complete data, in the absence of which the parameter set must be modified to make best use of the data available. Because in general complete data is available, the second approach was to generate simulated data to supplement the actual data. Figure 6.5 displays the Magellan spacecraft-Venus component of the Doppler signal with simulated data. From a cursory view of Fig. 6.5, some of the plane-of-sky orbit elements are apparent. The period is observed to be approximately three hours, and the amplitude indicates a high inclination. The eccentricity is less obvious, but the slight negative shift of the curve indicates a small eccentricity. The large difference between the upward and downward slopes indicates ω is near $\pm 90°$.

With complete data the ideal parameters were chosen as described in Fig. 6.3. The resulting convergence ranges for each orbital parameter are displayed in Fig. 6.6. The arrows indicate the true orbital parameters for the Magellan mission at that time. The intervals displayed represent convergence by introducing an erroneous initial condition for one orbital parameter at a time, and using the correct values for the other orbit elements. It was not possible, given time constraints, to completely characterize a six-dimensional convergence region. However, convergence was tested by varying all orbit elements from the true values at once, and results indicated a substantial region of convergence.

A second approach was to use only the real data, choosing alternative target parameters. We select as parameters positions on the signal consisting of the times of two consecutive zero crossings which approximately defines the period and phasing of the spacecraft orbit and four other points at fixed epochs near the maxima and minima of the Doppler signature. The loss of near diagonal dominance resulted in a slower convergence rate than other cases, but even with the sparse data, convergence was observed over a wide range of initial values (Fig. 6.7).

A comparison of the convergence obtained by processing a complete data set including simulated data (Fig. 6.7) with the convergence obtained by processing the actual sparse data set (Fig. 6.6) indicates a substantial range of convergence for

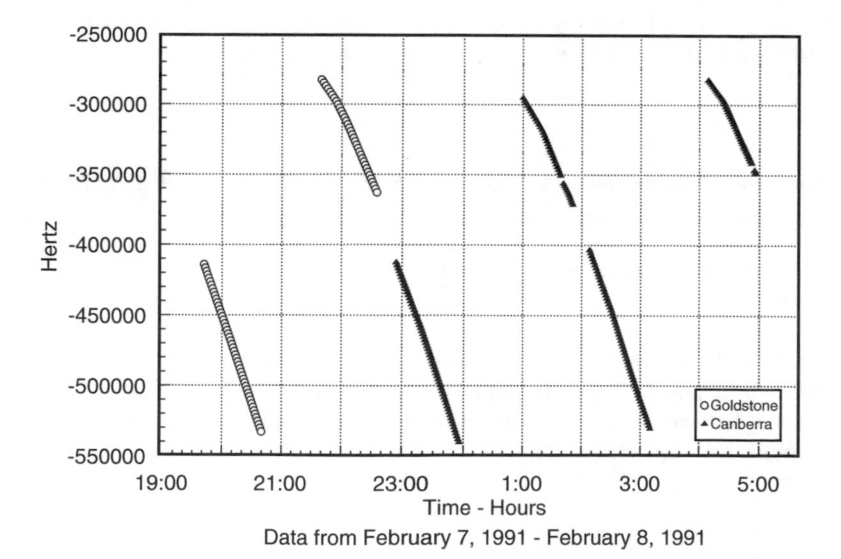

Fig. 6.4 Magellan Doppler signature—raw data

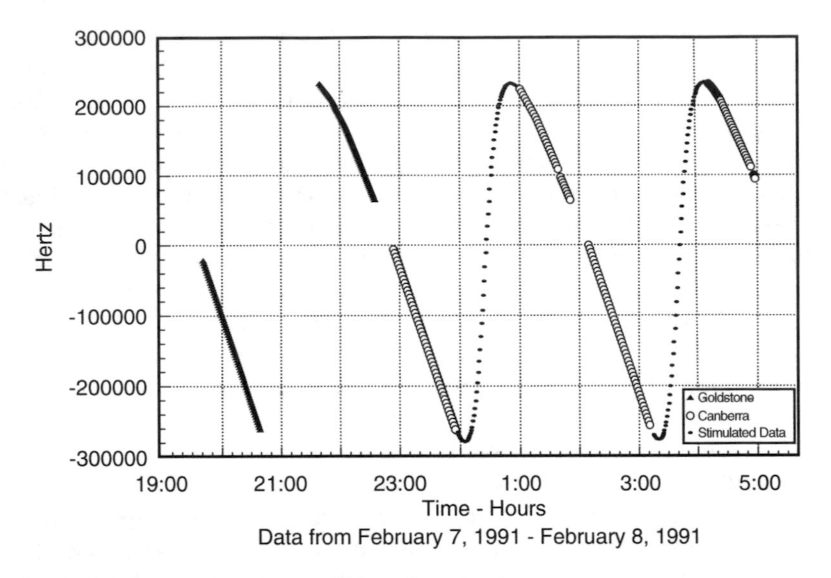

Fig. 6.5 Simulated Magellan spacecraft Venus Doppler signature

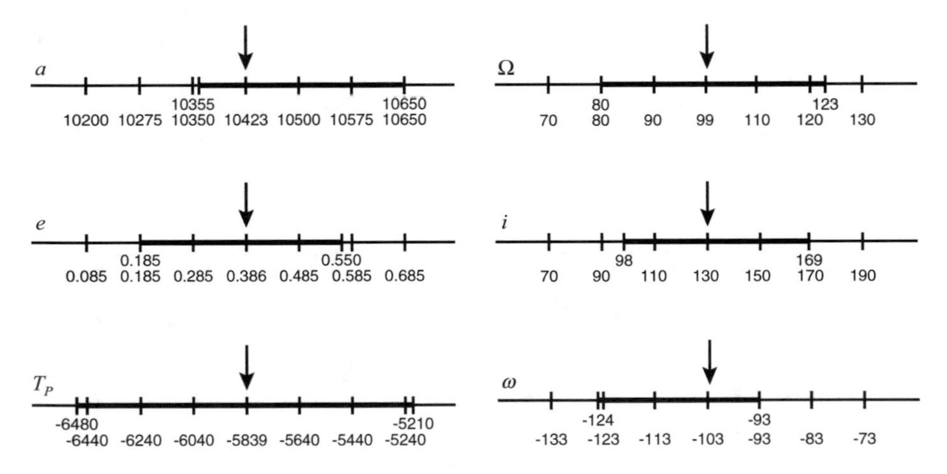

Fig. 6.6 Convergence region for Magellan with simulated data

both. However, the node Ω was better determined by the ideal target parameters used with the simulated data. The node is the most difficult of all the orbital elements to determine, because its effect on the signature is indirect, and because it changes so little over a few orbital periods.

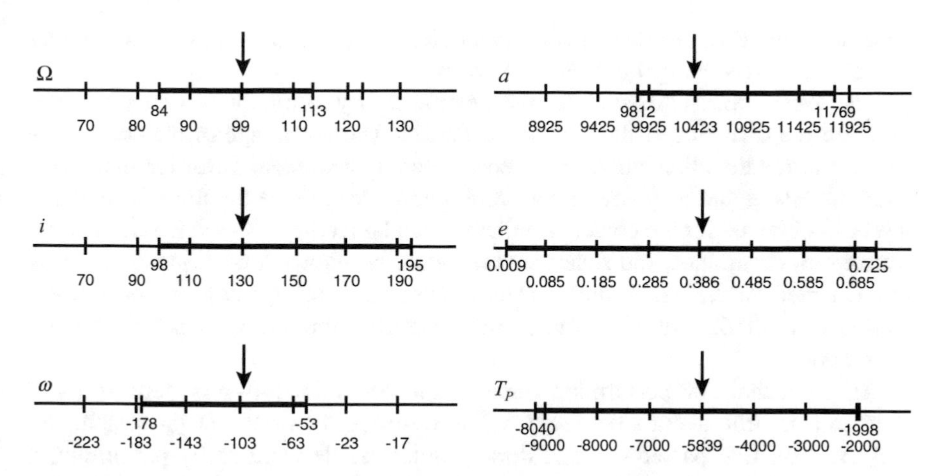

Fig. 6.7 Convergence region for Magellan with real data

6.7 Summary

Orbit determination is performed by searching for a set of parameters that minimize the error in the measurements computed from a model or by direct solution of the orbit from a minimum set of observations ignoring the error in the measurements. Included in the parameters is the initial spacecraft state or orbit elements. The orbit determined during flight operations is almost exclusively obtained by fitting data to a model and direct orbit determination is seldom used. This is contrary to orbits determined before the invention of the computer when orbit determination was by direct observation. The computer algorithm for searching for a solution is called a data filter. A data filter filters out bad data points. There are many types of data filters that operate on the error of the measurements and the covariance of the estimated parameters, the inverse of the covariance, the square root of the covariance, and the square root of the covariance inverse. Covariance filters may be based on weighted least squares or the Kalman filter algorithm. It can be shown that all of the filter options solve the same problem and get the same result and the performance is more dependent on the implementation than the filter algorithm.

Planetary spacecraft navigation uses weighted least squares implemented in a square root information filter. When large number of parameters are estimated, the data filter formulation requires the inversion of a large matrix. The Kalman filter algorithm is a clever algorithm for matrix inversion. An upper triangular square root covariance is easier to invert than a covariance matrix or at least this is the opinion of square root filter advocates. The SRIF algorithm follows directly from work of Gauss and has survived to this day without any significant improvement. During the NEAR mission, 600 parameters were estimated every 3 days for over a year with only minor problems in obtaining solutions for the spacecraft orbit, the Eros ephemeris, the Eros attitude, gravity harmonics, landmark locations, solar pressure model parameters, and propulsive maneuvers.

The discreet form of a data filter may be replaced by continuous differential equations for the elements of the covariance matrix. The advantage of the continuous formulation is the elimination of the need to compute a state transition matrix for data processing and mapping of stochastic parameters. Once the filter formulation has been obtained for the covariance filter, it can be readily adapted to information, square root covariance, and square root information filters. A continuous filter was implemented for the NEAR mission but was not used during navigation operations. Continuous filtering requires more study before committing to actual mission operations.

An algorithm for performing direct orbit determination was derived for a spacecraft in orbit about a body. Six points were spotted on the Doppler signature and the orbit that passed through these points was determined by performing a Newton-Raphson search. Since there are six orbit elements, there is only one orbit solution and this orbit can be determined to a higher precision than one may suspect. Doppler data is very accurate. Direct orbit determination is an interesting backup procedure, but has yet to be implemented during flight operations. In practice, difficulties in determining orbit solutions have been solved by doing more searches.

Exercises

6.1 The Kalman filter algorithm is essentially matrix inversion. Show how a matrix can be inverted by formulating a parameter estimation problem and processing the "data" with the Kalman gain matrix and the weighted least square gain matrix.

6.2 An orbit determination analyst notices a ramp of about 3.5 mHz over 20 min in the Doppler data. This corresponds to an unknown acceleration of 1.906×10^{-10} km/s^2. The spacecraft is near Earth and a planet is suspected. In the early days of the space program, planets were sometimes omitted from the equations of motion. What planet is the best guess? It wasn't Jupiter.

6.3 A single point smoothed estimate requires an update to the computed covariance at the smoothing epoch. Show how this update may be accomplished by processing six artificial measurements at the smoothing epoch.

6.4 An orbit determination estimate X_1 with covariance P_1 is obtained by processing data up to t_1. Some additional data ($Z_{1,2}$) is obtained from t_1 to t_2 and another estimate X_2 is computed with covariance P_2. If the new data is weak, X_2 will be nearly the same as X_1. In a Monte Carlo program, a sample drawn from P_2 must nearly equal a sample drawn from P_1. Thus, we must know the correlation between P_1 and P_2 in order to draw the correct sample at t_2. For an optimum filter without stochastic parameters, determine the cross correlation matrix $P_{1,2} = E(X_1 X_2^T)$.

6.5 The samples in Exercise 6.3 may be drawn by forming the matrix

$$P = \begin{bmatrix} P_1 & P_{1,2} \\ P_{1,2} & P_2 \end{bmatrix}$$

and drawing a sample from P. Drawing a sample from P involves computing the eigenvalues and eigenvectors of P, generating independent normally distributed random variables with sigmas equal to the square root of the eigenvalues and multiplying these random variables by the matrix of eigenvectors. If there are more than two orbit determination estimates in the sequence, the dimension of the P matrix may become too large. Determine a sampling algorithm that involves drawing a sample X_1 from P_1 and computing a conditional sample X_2.

6.6 A spacecraft that is a great distance from Earth is being tracked by the DSN. The tracking station is 5000 km off the Earth's spin axis. The tracking station longitude error is 1.5 m and the spin axis error is 1.0 m. The Earth's radius and rotation rate are 6150 km and 7.26×10^{-5} rad/s, respectively. Three Doppler data points are obtained at station rise, zenith, and station set. Determine the contribution of the station location errors in determining the errors in the right ascension and declination of the spacecraft as a function of declination.

6.7 Given the same data as in Exercise 6.6, determine the error in right ascension and declination of the spacecraft for perfect station location knowledge and a Doppler measurement error of 1.0 mm/s.

6.8 Repeat Exercise 6.7 for a range measurement error of 1 m.

Bibliography

Bierman, G. J., 1977, *Factorization Methods for Discrete Sequential Estimation*, Academic Press, Inc., New York.

Boggs, D.H., "A Partial Step Algorithm for the Nonlinear Estimation Problem", *AIAA Journal*, Vol. 10, No. 5,675–679, May 1972.

Dyer, P., and S. R. McReynolds, "Extension of Square Root Filtering to Include Process Noise," *Journal of Optimization Theory and Applications*, Vol.3, pp.444–458, 1969.

Grewal, M. S., and A. P. Andrews, *Kalman Filtering Theory and Practice*, Prentice Hall, Englewood Cliffs, NJ, 1993.

Kaminski, P. G., A. E. Bryson, Jr., and S. F. Smith, "Discrete Square Root Filtering:A Survey of Current Techniques," *IEEE Transactions on Automatic Control*,Vol. AC-16,pp.727–736,1971.

Masters, W. C., "Design and Implementation of Estimation Methods in ODP", Internal document, 2 December 1991

Masters, W. C., P. J. Breckheimer, "Updating Colored Noise and Spacecraft State from One Batch to the Next", Internal document, 19 September 1991

Miller, J. K., "A Single Point Smoothing Algorithm for Athena with Application to Mariner Temple 2", IOM 314.4–543, 22 September 1986.

Miller, J. K., "An Integrated Current State Square Root Information Filter-Smoother ", Internal document, 27 November 1991.

Miller, J. K. and Belbruno, E. A., “A Method for the Construction of a Lunar Transfer Trajectory using Ballistic Capture,” AAS 91–100, AAS/AIAA Spaceflight Mechanics Meeting, Houston, TX, Feb 11–13, 1991.

Miller, J. K., “A Continuous Square Root Ibformation Filter-Smoother with Discrete Data Update”, AAS 94–169, AAS/AIAA Spaceflight Mechanics Meeting, Cocoa Beach, FL, February 14, 1994.

Potter, J. E., and R. G. Stern, “Statistical Filtering of Space Navigation Measurements”, Proc. 1963 AIAA Guidance and Control Conference, 1963.

Ryne, M.S. and Wang, T.C. “Applications of Singular Value Analysis and Partial-Step Algorithm for Nonlinear Orbit Determination”, AAS/AIAA Spaceflight Mechanics Meeting, AAS 91–190, Houston, Texas, February, 1991.

Scheeres, D. J., “Derivation of a Generalized Form of the SRIF Differential Equation,” Internal document, 13 July 1993.

Weeks, C. J., Rosario, S. and A. Buschelman, “Direct Determination of a Spacecraft Orbit from the Doppler Data Signature”, AAS 95–144, AAS/AIAA Spaceflight Mechanics Meeting, Albuquerque, NM, February 13, 1995.

Chapter 7
Measurements and Calibrations

The measurement system is a collection of instruments on board the spacecraft and on the ground that provide observations of the spacecraft motion with respect to Earth and specific target bodies. Instruments of this kind are the Deep Space Network (DSN), a solid-state imaging (SSI) device, and a laser altimeter. The DSN tracking stations transmit radio frequency signals to the spacecraft and receive signals via the spacecraft transponder and antenna. The received signals constitute observations of range and Doppler data by conventional methods and observations of angles by VLBI methods. An SSI allows optical observations of planets, satellites, comets, and asteroids to be made against the background of the fixed stars and direct observation of landmarks. A laser altimeter bounces laser beams off the surface of a body and measures the round trip light time.

7.1 Radiometric Tracking Data

The major components of the radio tracking system include the telecommunication subsystem on-board the spacecraft, the Deep Space Stations (DSSs) of the Deep Space Network (DSN), and the general purpose computers of the Space Flight Operations Facility (SFOF) located at the Jet Propulsion Laboratory (JPL) in Pasadena or other control centers.

The parts of the telecommunications subsystem used for orbit determination are S-band and X-band receivers and coherently driven S- and X-band transmitters. They provide a coherent two-way communications link for tracking observables, Doppler and range. Also, the spacecraft carrier is modulated with special tones for wideband ΔVLBI. The DSN is a network of tracking stations located around the globe at Goldstone, California; Madrid, Spain; and Canberra, Australia. Each DSS is a complex of a 72-m antenna and several 34-m antennae and special purpose hardware and computers for extracting Doppler, range, and VLBI observables from

J. Miller, *Planetary Spacecraft Navigation*, Space Technology Library 37,
https://doi.org/10.1007/978-3-319-78916-3_7

received spacecraft signals and VLBI from extragalactic radio sources. This data is relayed via high-speed data lines to the control center where software on general purpose computers process the data to obtain estimates of the spacecraft trajectory and compute propulsive maneuvers for trajectory control.

7.1.1 Doppler Data

Doppler data is the work horse of the measurement system. Most missions can be navigated with Doppler data as the only data type input to orbit determination software. Doppler data provides a direct measure of line-of-sight velocity of a spacecraft relative to a tracking antenna. The accuracy of this measurement is about 1 mm/s when the two-way Doppler count is integrated for 1 min. A single Doppler measurement provides no information on position or velocity normal to the line-of-sight. For those phases of the mission where the spacecraft is being accelerated rapidly, such as near a planetary encounter, a series of Doppler measurements permit a quite accurate complete orbit determination by observing the orbit dynamics signature. When the spacecraft is far from a planet, comet, or asteroid, the gravitational accelerations are not sufficient to observe this signature. However, the "velocity parallax" due to the tracking stations rotation with Earth provides a measure of position normal to the line-of-sight. By measuring the amplitude and phase of the tracking stations signature, the right ascension and cosine of declination may be determined to about 0.25 μrad. Thus, at Jupiter distance, the Earth-relative orbit determination error is about 200 km.

The functional definition of Doppler data as line-of sight velocity is useful for analyzing the orbit determination errors that are spacecraft or trajectory dependent but is of little use for analyzing error sources close to the actual measurement such as media or hardware errors. The actual measurement is a count derived from the signal received from the spacecraft and a frequency standard maintained at the tracking station that controls the frequency of the transmitted signal. Thus, a precision model of the Doppler observable includes a model of the signal path as well as hardware elements. In practice, the hardware errors are small compared to media, station location, and spacecraft dynamics errors.

7.1.2 Doppler Measurement Model

A model of the Doppler observable has been developed that idealizes some of the hardware error sources yet precisely models the external environment. This model is sufficiently precise for computation of the observable and is essentially the model contained in orbit determination software. Of particular interest are models that are external to the tracking station hardware yet pertain directly to the signal path. Media and the effect of general relativity on station clocks are examples. Other models, such as station locations and polar motion, though not directly part of the Doppler measurement system, may be treated as measurement calibrations.

The Doppler measurement is simply an electronic count of the number of cycles from a frequency standard (N_c) minus the number of cycles of the spacecraft signal received by the tracking station (N_r) and scaled by the count time interval (ΔT_c). Thus we have

$$Z_m = \frac{(N_c - N_r) + n}{\Delta T_c}$$

where n is the measurement noise which is about 1/10 of a cycle. The received frequency and standard frequency need not be counted individually and differenced but may be added together electronically and the beat frequency counted. This is a detail that is dependent on the hardware implementation. The numerical value of Z_m is the number that is recorded on a tracking data file and used for orbit determination.

In the orbit determination software we need to obtain a computed value for Z_m as a function of parameters that are available. This function can be derived from the equations of motion and a physical model of the system. We start by developing a frequency standard that can be compared with the frequency of the transmitted and received signals. The frequency standard is obtained by scaling a reference oscillator frequency f_q, obtained from an atomic clock, to equal the transmitted frequency (f_t) times the spacecraft turn around ratio (C_3) which would nominally be the received frequency if there were no spacecraft Doppler shift or additional delay. The turn around ratio is necessary so that the downlink will not interfere with the uplink.

$$N_c = C_3 \, f_t \, \Delta T_c$$

where for S band Doppler,

$$C_3 = \frac{240}{221}$$

$$f_t = 96 \, f_q$$

$$\Delta T_c = T_{3_e} - T_{3_s}$$

The count time (ΔT_c) is defined as the difference between the reception time at the start of the count time interval (T_{3_s}) and the reception time at the end of the interval (T_{3_e}). For a schematic representation of these times, see Fig. 7.1. In the above equation, all of the parameters are constant or arbitrarily specified including the reception times. The real information content of the measurement is contained within the count N_r. Thus, in order to obtain a complete equation for the computed measurement, we need an equation for N_r. It is tempting to differentiate and work in the frequency domain; however, the hardware works with phase which makes it convenient to formulate the measurement in terms of phase thus bypassing an explicit equation for the received frequency.

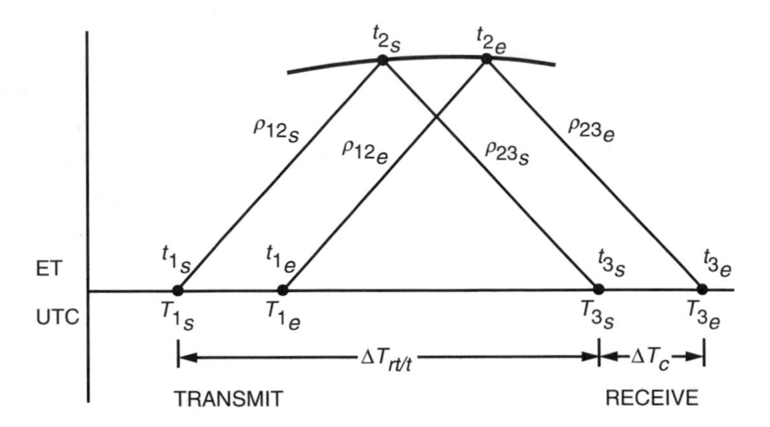

Fig. 7.1 Doppler observable schematic diagram

The equation that relates the measurement to the observable parameters is

$$N_r = \frac{240}{221} N_t$$

where

$$N_t = f_t \,(T_{1_e} - T_{1_s})$$
$$f_t = 96 \, f_q$$

The above equation for N_r states that the number of cycles counted at the receiver is equal to the number of cycles transmitted N_t times the spacecraft turn around ratio. This equation is true because they are effectively the same cycles. Thus, the information content of the measurement is now contained in the transmit times T_{1_e} and T_{1_s}. Since both of these times are unknown, we need some additional equations to tie into the observable quantities. At this point in the development, we have the following equation for the computed measurement.

$$Z_c = (T_{3_e} - T_{3_s} - T_{1_e} + T_{1_s}) \, \frac{C_3 f_t}{\Delta T_c}$$

We need equations for the times in the above equation and these will be developed as functions of ephemeris time t. We have for the atomic clock at the station

$$T = t + F(t, x, y)$$

The station time T is equal to the ephemeris time t modified by a small correction due to general relativity and any other parameter that may affect the running of the clock. The calibration function (F) is a function of t, the state of the solar system (x), and other constant parameters (y). Here, x and y can be thought of as parameter vectors. The relevant times shown in Fig. 7.1 relating to the Doppler measurement are

$$T1_s = t1_s + F(t1_s, x, y)$$
$$T1_e = t1_e + F(t1_e, x, y)$$
$$T3_s = t3_s + F(t3_s, x, y)$$
$$T3_e = t3_e + F(t3_e, x, y)$$

Making the above substitutions, the equation for the computed measurement becomes

$$Z_c = (t3_e - t3_s - t1_e + t1_s)\,\frac{C_3 f_t}{\Delta T_c} + \left[F(t3_e, x, y) - F(t3_s, x, y) - F(t1_e, x, y) + F(t1_s, x, y)\right]\frac{C_3 f_t}{\Delta T_c} \qquad (7.1)$$

Since the speed of light is constant in any reference frame, we may obtain by integrating along the light path

$$t3_e - t1_e = \frac{\rho 12_e + \rho 23_e}{c} + \Delta t^m_{1_e} + \Delta t^m_{3_e}$$
$$t3_s - t1_s = \frac{\rho 12_s + \rho 23_s}{c} + \Delta t^m_{1_s} + \Delta t^m_{3_s}$$

where the ρ terms represent the integrated distance along the light path and the t^m terms represent the additional delay caused by media. The distances along the light path are obtained by integrating the equations of motion.

$$\rho 12_s = \iint_{t1_s}^{t2_s} \ddot{\rho}\; dtdt$$
$$\rho 23_s = \iint_{t2_s}^{t3_s} \ddot{\rho}\; dtdt$$
$$\rho 12_e = \iint_{t1_e}^{t2_e} \ddot{\rho}\; dtdt$$
$$\rho 12_e = \iint_{t2_e}^{t3_e} \ddot{\rho}\; dtdt$$

These equations are referred to as the light time equations and are solved iteratively for the arguments of integration. The media delay is included in the measurement equation by evaluating the calibration function (G) at the appropriate times.

$$\Delta t^m = G(t, x, y)$$
$$\Delta t^m_{1_s} = G(t_{1_s}, x, y)$$
$$\Delta t^m_{1_e} = G(t_{1_e}, x, y)$$
$$\Delta t^m_{3_s} = G(t_{3_s}, x, y)$$
$$\Delta t^m_{3_e} = G(t_{3_e}, x, y)$$

The final equation for the computed measurement includes the observable equations as well as clock and media calibration functions.

$$\begin{aligned} Z_c = & \frac{\rho 12_e + \rho 23_e - \rho 12_s - \rho 23_s}{c} \frac{C_3 f_t}{\Delta T_c} \\ & + \left[F(t_{3_e}, x, y) - F(t_{3_s}, x, y) - F(t_{1_e}, x, y) + F(t_{1_s}, x, y)\right] \frac{C_3 f_t}{\Delta T_c} \\ & + \left[G(t_{3_e}, x, y) - G(t_{3_s}, x, y) + G(t_{1_e}, x, y) - G(t_{1_s}, x, y)\right] \frac{C_3 f_t}{\Delta T_c} \end{aligned} \tag{7.2}$$

7.1.3 Data Noise

Recall that the measurement noise is scaled by the count time (ΔT_c).

$$Z_m = \frac{N_c - N_r}{\Delta T_c} + \frac{n}{\Delta T_c} \tag{7.3}$$

The data noise is approximately 1/10 of the cycle count and is independent of frequency. For a 60 s count time, the data noise is 1.66 mHz. Doppler data is scaled by the count time to make the recorded measurement proportional to velocity. The Doppler measurement sensitivity to line-of-sight velocity is given by

$$Z_{\dot{\rho}} \approx \frac{2\, C_3 f_t}{c}\, \dot{\rho} \tag{7.4}$$

At S-band frequency, typical values for the constants in the above equation are

$$C_3 = \frac{240}{221}$$
$$f_t = 96\, f_q$$
$$f_q = 22 \times 10^6\ \text{Hz}$$
$$c = 299792.458\ \text{km/s}$$

and for $\dot{\rho} = 1$ mm/s the measurement gives the well-known result $Z_{\dot{\rho}} = 15.3$ mHz. In order to obtain this result, the above equation must be entered with a consistent set of units for each of the parameters. At X-band frequency, 1 mm/s velocity corresponds to about 56 mHz.

7.1.4 One-Way Doppler Data

The one-way Doppler data type provides a measure of the change in the line-of-sight range between a DSN station and a spacecraft over some interval of time. The measurement makes use of the Doppler frequency shift of a source when the receiver is moving with respect to the source. When the change in range is divided by the time interval, a measure of the average range rate over the time interval is obtained. The source is a radio signal whose frequency is controlled by an oscillator or clock on the spacecraft. The transmitted radio wave is received by a DSN tracking station and the individual cycles are counted. If the spacecraft is stationary with respect to the DSN antenna or there is no net change in range over the count interval, the measured cycle count divided by the time interval will equal the frequency of the radio signal transmitted by the spacecraft. If the spacecraft is moving with respect to the antenna, the difference between the cycle count transmitted and the cycle count received times the wavelength of the transmitted radio signal is a measure of the range change over the count time interval and is the measurement that may be used to determine the orbit of the spacecraft. A problem with one-way Doppler is the frequency of the transmitted radio signal. The frequency control of a radio signal on a spacecraft is marginal at best. The spacecraft does not have access to high-precision atomic clocks to control the frequency and the measurement is accurate to about 1 cm/s even with ultra-stable oscillators on the spacecraft. An atomic clock on the spacecraft is needed and is under development.

The first step in processing a one-way Doppler data point is to read the first data record from the tracking data file and obtain the time tag, frequency, and the measurement. For the first one-way Doppler point, the count time is subtracted from the recorded time tag to initialize the station time associated with receipt of the Doppler measurement at the start of the Doppler count. For subsequent one-way Doppler points, the start of the Doppler count is exactly the end time of the previous Doppler count. Thus, the first Doppler point requires two solutions of the light-time equation and subsequent points require only one. When the continuity is interrupted, the time tags are not separated by the exact count time and the one-way Doppler count is restarted.

The computation of the observable involves integrating the transmitted and received frequencies over the appropriate time intervals. The transmitted frequency is controlled by the oscillator on the spacecraft which tends to drift. The frequency may be modeled as a polynomial function of time given by

$$f_t = FRQ_0 + FRQ_1(t - t_f) + FRQ_2(t - t_f)^2$$

where FRQ_i are the coefficients of the polynomial and t_f is the ephemeris time associated with FRQ_0. The FRQ_0 coefficient is set equal to the frequency obtained from the tracking data file data record and the linear and quadratic coefficients are generally initially set to zero. The time interval for the integration is from the ephemeris time at the spacecraft obtained from the light time solution at the start of the count (t_{2s}) to the end of the count (t_{2s}). The required integral is given by

$$fq_t = \int_{t_{2s}}^{t_{2e}} f_t \, dt$$

$$fq_t = FRQ_0(t_{2s} - t_{2e}) + \frac{FRQ_1}{2}\left[(t_{2s} - t_f)^2 - (t_{2e} - t_f)^2\right]$$
$$+\frac{FRQ_2}{3}\left[(t_{2s} - t_f)^3 - (t_{2e} - t_f)^3\right]$$

The received frequency (f_r) may be modeled as a polynomial function of received station time by mapping the function for the transmit time to the Earth and converting to station time.

$$f_r = FRQ_0 + FRQ_1(t - t_f) + FRQ_2(t - t_f)^2$$

The received radio signal is integrated over the count time interval defined by the time tags.

$$fq_{dn} = \int_{t_{3s}}^{t_{3e}} f_r \, dt$$

$$fq_{dn} = FRQ_0(t_{3s} - t_{3e}) + \frac{FRQ_1}{2}\left[(t_{3s} - t_f)^2 - (t_{3e} - t_f)^2\right]$$
$$+\frac{FRQ_2}{3}\left[(t_{3s} - t_f)^3 - (t_{3e} - t_f)^3\right]$$

The one-way Doppler observable (Z_{owd}) is given by

$$Z_{owd} = \frac{C3(fq_{dn} - fq_t)}{COUNT}$$

7.1.5 Three-Way Doppler Data

The three-way Doppler data type provides a measure of the change in the total line-of-sight range from a transmitting DSN station to a spacecraft and back to a separate receiving station over some interval of time. The measurement makes use of the Doppler frequency shift of a radio signal source that is moving with respect to a receiver. When the change in range is divided by twice the time interval, an approximate measure of the average range rate over the time interval is obtained. If the transmitting station is widely separated from the receiving antenna, the interpretation of the measurement as range rate is ambiguous. The three-way Doppler measurement is essentially the same as the two-way Doppler measurement. The only significant difference is the inclusion of a second separate tracking station on the downlink. The three-way Doppler measurement has a geometrical advantage over conventional two-way single station Doppler which may be attributed to the baseline between the participating tracking stations. However, the frequency standard must be maintained at two stations resulting in some loss of coherence and accuracy. The data processing for three-way Doppler is the same as for two-way Doppler except that a second station is substituted in computing the downlink.

7.1.6 Range Data

The range data type provides a measure of the range between a DSN station and a spacecraft. The range is inferred from the time it takes a radio signal to travel from the DSN station to the spacecraft and back to the station. The radio signal is transmitted to the spacecraft where it is received and retransmitted back to the tracking station. The round trip light-time is determined by impressing a pattern, referred to as a range code, on the transmitted carrier and detecting this pattern in the received radio signal. The range code provides time markers in the transmitted and received radio signal that may be measured with high precision by an atomic clock. For orbit determination, a computed value of the measurement is obtained from a mathematical model similar to the model used for Doppler data.

Range data has essentially the same information content as Doppler data. Range data provides the integral of Doppler data over some time interval. This integrated Doppler can be determined by differencing two range measurements. The integrated Doppler is more accurate than differenced range. However, the range data provides the constant of integration. Doppler data alone determines range through the orbit dynamics. For this reason, an orbit determination strategy has evolved to process a single loosely weighted range point for each station pass to initialize the Doppler. Processing range and Doppler together at the same weight can result in aliasing. Both data types determine the right ascension and declination independently and they may disagree.

The first step in computing the two-way range observable is to assemble all the input data required by the orbit determination software. These are essentially the same models as used for Doppler data with some minor exceptions. Models of the transmission media, station locations and the affect of General relativity are virtually the same as used for computing the Doppler observable. One exception is the sign of delays associated with charged particles in the ionosphere and solar plasma. For Doppler data, the charged particles speed up the velocity of the carrier and the delay is subtracted. For range data, the velocity of the carrier is slowed down and the delay is added as for all the other media delays. The magnitude of the velocity increase associated with charged particles for Doppler data is equal to the velocity decrease or delay for range data.

A range data point is read from the tracking data file to obtain the time tag ($TIMTAG$), frequency ($FRQCY$), lowest ranging component ($NLOW$), and the measurement ($ROBS$). The light-time equation is solved for the transmit and spacecraft times t_1 and t_2. The station receive time (t_3) is equal to $TIMTAG$.

The next step is to integrate the ramp tables for the uplink and downlink. The ramp tables keep the received signals in the center of the carrier bandwidth. Since the range code is modulated on the carrier, the range traveled by the radio signal equals the sum of the wavelengths associated with all the cycles between the spacecraft and the DSN antenna and is equal to the cycle count times the speed of light after correcting for media. The cycle count (f_q) is obtained by integrating the uplink ramp table from t_1 to t_3. The range observable (Z_r) is computed from the output of the ramp table integration which is scaled by an integer ratio corresponding to the frequency dividers used in the actual hardware implementation to give the measurement in range units (R_u). For S-band frequency, the conversion to range units is

$$R_u = \frac{1}{2} fq$$

For X-band frequency, the conversion for the 34-m Az-EL high efficiency antenna (HEF) is

$$R_u = \frac{11}{75} fq$$

and for 34-m Block 5 Receivers (BVR)

$$R_u = \frac{221}{749 \times 2} fq$$

The range code is a pattern consisting of square waves whose frequency decreases by powers of two. Thus, the range code is repeated at a rate determined by the lowest frequency square wave or range component. This results in an ambiguity in the determination of range that must be resolved by introducing information from other sources, most notably the Doppler measurement which has no ambiguity. The range ambiguity manifests itself as a roll over to zero in the range unit counter. Thus, if the computed number of range units is greater than the ambiguity, the ambiguity is repeatedly subtracted from the computed measurement until it is in the proper range. The range ambiguity is computed from $NLOW$, which is obtained from the tracking data file, and is given by

$$AMBIG = 2^{(NLOW+6)}$$

The number of roll overs of the range unit counter is the integer part of

$$N_a = Integer\left(\frac{R_u}{AMBIG}\right)$$

The adjusted value for the range unit count is then

$$R_{ua} = R_u \; - \; N_a \; AMBIG$$

At the time the computed observable rolls over, it cannot be determined from the computed range alone whether the actual observable has just rolled over or is about to roll over. This ambiguity may be resolved by inspecting the measurement residual. The measurement residual is simply

$$RESID = ROBS - R_{ua}$$

where $ROBS$ is obtained from the tracking data file. If the absolute value of $ROBS$ is greater than $1.5 \times AMBIG$, the ambiguity resolution is skipped. Otherwise, the following adjustment is made to the computed observable.

$$If(RESID \; > \; 0.5 \times AMBIG) \; Z_r = R_{ua} + AMBIG$$

$$If(RESID \; < \; 0.5 \times AMBIG) \; Z_r = R_{ua} - AMBIG$$

Otherwise, $Z_r = R_{ua}$. The residual (RESID) is then recomputed with the new value for Z_r.

$$RESID = ROBS - Z_r$$

This correction to the range measurement can be dangerous. If the ambiguity is set too low and the computed measurement is not known within the ambiguity, the range measurement will be in error and the resultant orbit solution can be off by several hundred kilometers. The range and Doppler residuals could be flat making it difficult to detect this error.

7.1.7 Very Long Baseline Interferometry

Plane radio waves from the spacecraft or an extragalactic radio source (EGRS) are received at two tracking stations separated by an intercontinental baseline as shown in Fig. 7.2. As originally implemented, the received signals are clipped, digitized, and recorded on video recorders at each tracking station. Special tones are impressed on the spacecraft transmitted signal to obtain the required bandwidth. The EGRS signal is passed through filters to obtain the proper spectral bandwidth. Since the spacecraft signal is considerably stronger than that obtained from an EGRS, the tracking strategy consists of recording an EGRS for about 20 min, slewing the

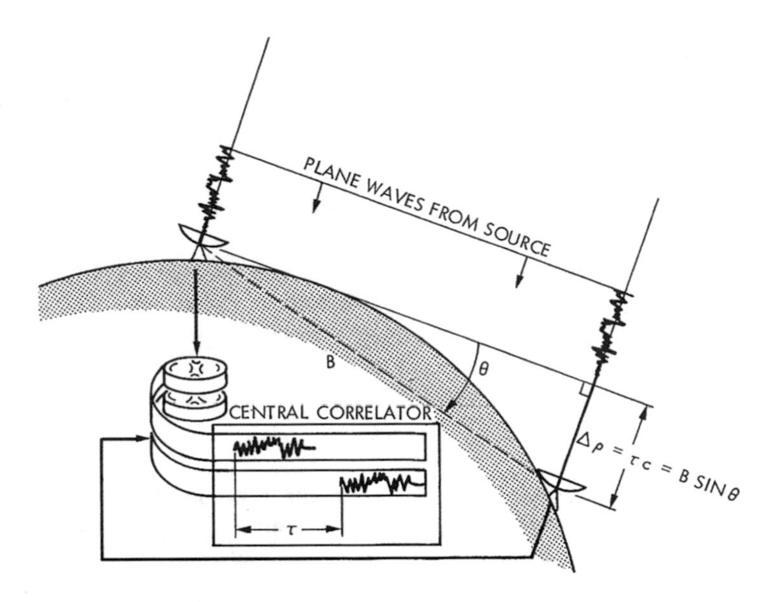

Fig. 7.2 VLBI functional implementation

antennas to the spacecraft and recording for 5 min, and if necessary slewing back to the EGRS and recording for another 20 min.

After recording the radio signals, it is necessary to bring the two recordings together at a central correlator. The tapes may be physically transported to the correlator or played back across high-speed data lines. The function of the correlator is to match the two recordings and determine the delay. Correlation involves shifting, multiplying, and integrating the bit streams together until a maximum is found. The delay is a measure of the angle between the baseline and the direction of the plane radio wave. Differencing the spacecraft and EGRS delay effectively provides a measure of their angular separation. Combining observations on an East-West baseline (Goldstone/ Madrid) with a North-South baseline (Goldstone/ Canberra) gives a precise measure of spacecraft right ascension and declination accurate to about 5 nrad relative to the EGRS. As currently implemented, the configuration shown in Fig. 7.2 has been updated and replaced by more sophisticated software and hardware, but the function and accuracy remains about the same as originally implemented.

7.1.8 Differential Wide Band VLBI

Differential Very Long Baseline Interferometry (ΔVLBI) consists of near simultaneous interferometric tracking of a spacecraft and an angularly nearby EGRS. The accuracy of ΔVLBI is dependent on the angular separation of the spacecraft and EGRS. Separation angles less than about ten degrees are needed in order to achieve cancelation of errors due to media effects and station location uncertainties. An extensive survey of the sky has developed a radio source catalog. To be useful for VLBI observations, the source flux must be greater than 0.20 Jy and the source structure should be smaller than about 5 nrad in diameter. Another consideration is the variability of source strength. Once identified, precise VLBI measurements of source position are made for inclusion in the source catalog. In the search for radio sources, emphasis is given to the portion of the sky near the ecliptic plane because missions to the planets place spacecraft on trajectories that are within a few degrees of the ecliptic plane. For orbit determination, the difference between the EGRS VLBI measurement and the spacecraft VLBI measurement is processed as a data type.

7.1.9 Differential Narrow Band VLBI

The narrow band VLBI observable provides a measure of the change in the angle between the wave front of a radio signal from a radio source and the baseline between two tracking stations in the plane of the tracking stations and radio source over an interval of time. Narrow band VLBI differs from wide band VLBI by the manner of detection of the received radio signals. For a spacecraft, narrow band VLBI tracks the carrier and determines a count of the number of cycles received over an interval of time referred to as the count time. It is directly analogous to Doppler. A problem with narrow band VLBI is the existence of a singularity in the measurement at zero declination relative to the Earth's equator. Therefore, it is generally only used when wide band VLBI cannot be obtained. Low signal level or the absence of tones on the carrier would preclude wide band VLBI.

For quasar VLBI, the same algorithm for the light time solution is used as for Doppler and spacecraft VLBI. The same approach is used for the quasar as for the spacecraft only the equations for ρ_{12}, ρ_{23} and their derivatives with respect to time must be modified. Since the quasar is effectively located at infinity, an invariant plane is defined as illustrated in Fig. 7.3 to provide a reference for defining t_2. The invariant plane is defined, for the purpose of illustration, to be 7000 km from the center of the Earth and perpendicular to the direction of the quasar being observed. The actual distance of the invariant plane from the center of the Earth is arbitrary. Plane radio waves from the quasar will cross this invariant plane at the same time which is taken to be the t_2 time. The modifications necessary for the downlink portion of the light time algorithm relate to the computation of ρ_{23} and its derivatives. First, the spin vector of the Earth in inertial space is computed. For a quasar, some additional signal processing is required. Since the quasar radiates essentially white noise, it is necessary to condition the signal to obtain the monochromatic tone required by narrow band VLBI. This is accomplished by passing the quasar signal through a pass band filter before recording on the video recorder. The two recordings obtained from the participating stations are correlated

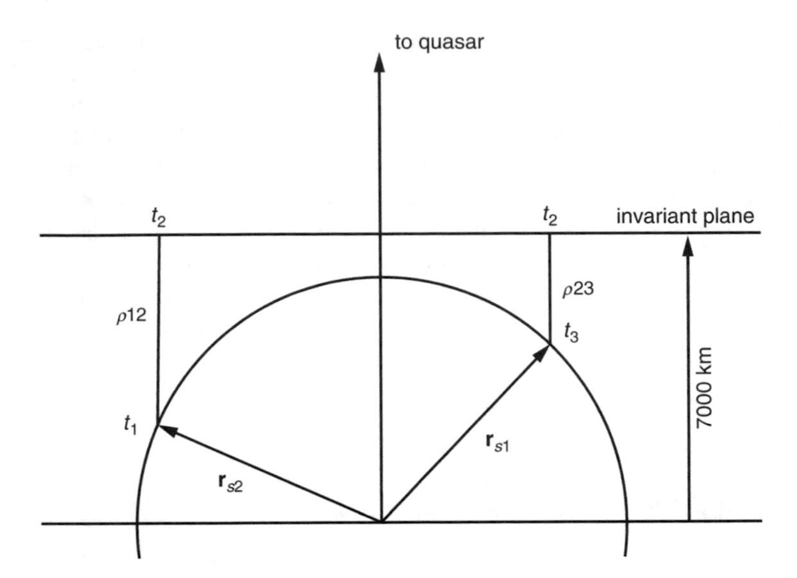

Fig. 7.3 VLBI quasar geometry

which produces a difference in the cycle count over the count time interval. This cycle count provides a measure of the change in delay which is a measure of the change in the angle (θ) between the baseline (B) and the direction of the plane radio wave front as illustrated in Fig. 7.2. For orbit determination, a computed value of the measurement is obtained from a mathematical model similar to the model used for Doppler data.

7.2 Radiometric Data Calibrations

The radiometric data observables require high-precision models of the hardware and path length from the DSN tracking stations to the spacecraft as well as the location of the tracking stations relative to the center of Earth. Some of the contributors to the path length are small enough to be ignored or require only approximate models. The solar radiation pressure on the planets and the momentum transfer from charged particles impacting the spacecraft can be ignored. Other physical perturbations to the path length are included as calibrations, but are not included in the equations of motion of the spacecraft. Examples of calibrations that are included are troposphere and ionosphere and the effect of General Relativity on the curvature of space and time. The tectonic plate motion of the continents and solid Earth tides enter as calibrations. For very high-precision VLBI, the Lorentz contraction of the Earth resulting from its velocity relative to the Sun's barycenter contributes a few centimeters.

7.2.1 *Clock Calibration*

According to the special theory of relativity, a clock running in a frame of reference that is moving with respect to an observer's frame of reference will appear to run slower by the observer. According to the general theory of relativity, a clock running in a gravitational potential field will run slower than a clock removed from the field. Therefore, an observer that is stationary with respect to the solar system will see the atomic clocks at the tracking stations running slower than his hypothetical clock. The observer is placed stationary with respect to the barycenter of the solar system because the equations of motion are written with respect to this center and placed far away to escape the effect of the gravitational acceleration of the Sun and planets. The coordinate time thus defined is called post-Newtonian time (PNT).

The relationship between PNT and the proper time measured by an atomic clock is given by the metric. For a particle moving in an orbit around the Sun, the metric in isotopic Schwarzschild coordinates is given by

$$ds^2 = \frac{\left(1 - \frac{U}{2c^2}\right)^2}{\left(1 + \frac{U}{2c^2}\right)^2} c^2 dt^2 - \left(1 + \frac{U}{2c^2}\right)^4 \left(dx^2 + dy^2 + dz^2\right)$$

Retaining terms to order c^2, the metric may be approximated by

$$ds^2 = \left(1 - \frac{2U}{c^2}\right) c^2 dt^2 - v^2\, dt^2$$

where

$$v^2 = \left(\frac{dx}{dt}\right)^2 + \left(\frac{dy}{dt}\right)^2 + \left(\frac{dz}{dt}\right)^2$$

Solving for proper time ($ds^2 = c^2\, d\tau^2$) we obtain

$$\frac{d\tau}{dt} = \sqrt{1 - \frac{2U}{c^2} - \left(\frac{v}{c}\right)^2}$$

which may be further approximated by

$$\frac{d\tau}{dt} = 1 - \frac{\mu_s}{c^2 r} - \frac{1}{2}\frac{v^2}{c^2} - \frac{\mu_e}{c^2 r_e}$$

where the Earth's gravitational potential is separated from the Sun's. The atomic clock time (τ) is obtained as a function of t by integrating the metric in conjunction with the equations of motion.

$$\tau = \int_{t_0}^{t} (1 - L)\, dt$$

where

$$L = \frac{\mu}{c^2 r} + \frac{1}{2}\frac{v^2}{c^2} + \frac{\mu_e}{c^2\, r_e}$$

The function L can be separated into a constant term (L_0), secular terms that grow with time (L_s), and periodic terms (L_p). Thus we have

$$L = L_0 + L_s + L_p$$

The constant term (L_0) is obtained by averaging L over all time and can be represented by

$$L_0 = \frac{1}{c^2}\left(\frac{\mu}{r_0} + \frac{1}{2}\, v_0^2\right) + \frac{\mu_e}{c^2\, r_e}$$

where r_0 and v_0 are constants that give the correct average value for L_0. For the Earth's orbit about the Sun, r_0 is approximately the semi-major axis of the orbit and v_0 approximately the mean orbital velocity. Since the orbit is nearly an ellipse,

$$\frac{\mu}{a} = \frac{2\mu}{r} - v^2$$

and for $r = a$,

$$L_0 \approx \frac{3\mu}{2c^2\, a} + \frac{\mu_e}{c^2\, r_e}$$

The secular terms L_s are assumed to be zero because of conservation of energy and momentum. This leaves the periodic terms and these are given by

$$L_p = \frac{1}{c^2}\left(\frac{\mu}{r} - \frac{\mu}{r_0} + \frac{1}{2}\, v^2 - \frac{1}{2}\, v_0^2\right)$$

and

$$\tau = t + \int_{t_0}^{t} -L_0 - \frac{1}{c^2}\left(\frac{\mu}{r} - \frac{\mu}{r_0} + \frac{1}{2}\, v^2 - \frac{1}{2}\, v_0^2\right) dt \tag{7.5}$$

An approximate analytic formula for the periodic terms, derived by Brooks Thomas, is given by

$$\tau \approx t - L_0\,(t - t_0) - \frac{2}{c^2}(\dot{\mathbf{r}}_b^s \cdot \mathbf{r}_b^s) - \frac{1}{c^2}(\dot{\mathbf{r}}_b^c \cdot \mathbf{r}_e^b) - \frac{1}{c^2}(\dot{\mathbf{r}}_e^c \cdot \mathbf{r}_b^e) - \frac{1}{c^2}(\dot{\mathbf{r}}_s^c \cdot \mathbf{r}_b^s)$$
$$- \frac{\mu_j}{c^2(\mu_j + \mu_s)}(\dot{\mathbf{r}}_j^s \cdot \mathbf{r}_j^s) - \frac{\mu_{sa}}{c^2(\mu_{sa} + \mu_s)}(\dot{\mathbf{r}}_{sa}^s \cdot \mathbf{r}_{sa}^s) \tag{7.6}$$

In the notation used above, the position of the body identified by the subscript is with respect to the body identified by the superscript, where c = the solar system barycenter, s = the Sun, b = the Earth-Moon barycenter, e = the Earth, j = Jupiter, and sa = Saturn.

7.2.2 *Troposphere Calibration*

A radio signal passing through the Earth's troposphere will be delayed depending on the dielectric constant of the media and path length.

$$\Delta t^t = G_t(t, x, y)$$

The troposphere delay has been conveniently separated into wet and dry components that are functions of delay at zenith (z) and elevation angle (γ).

$$G_t(t, x, y) = R_d + R_w$$

The first term in the above equation represents the nonlinearity of the dry troposphere mapping function and the second term represents the variation in the dry troposphere z height due to local weather. The next two terms are the same quantities for the wet troposphere. The troposphere wet and dry mapping functions are tabulated as delay as a function of spacecraft elevation angle. Empirical formulas for these mapping functions are given by

$$R_d = \frac{z_d}{\sin\gamma + \dfrac{A_d}{B_d + \tan\gamma}}$$

$$R_w = \frac{z_w}{\sin\gamma + \dfrac{A_w}{B_w + \tan\gamma}}$$

where

$$\sin\gamma = \cos\delta\cos\lambda\cos\phi + \sin\lambda\cos\phi\ + \sin\phi\sin\delta$$

$$\lambda = \omega_e t + \lambda_s - \alpha$$

The dry component of the troposphere (R_d) is a function of the delay at zenith (z_d), the elevation angle (γ), and constants A_d and B_d that are provided to model the bending at low elevation angles. The wet component (R_w) is similarly defined. The elevation angle (γ) is computed as a function of the latitude of the tracking station (ϕ), the declination of the spacecraft (δ), and the local hour angle with respect to the spacecraft (λ). The local hour angle is zero when the spacecraft is at zenith and is a function of the Greenwich hour angle ($\omega_e t$), the station longitude (λ_s), and the right ascension of the spacecraft (α).

The troposphere dry component is assumed to be stable and most of the variability is associated with the wet component. The variation in the wet component may be modeled as a periodic variation in the z height (z_w). The hourly variation in the wet component of the troposphere appears as a random walk that would require a high order Fourier series to represent analytically. The variation may be modeled as a simple sinusoid with amplitude and frequency selected to be representative of the short-term variation.

$$z_w = z_{w0} + z_{w1} \sin(\omega_{w_1} t)$$

7.2.3 *Ionosphere Calibration*

A radio signal passing through the ionosphere experiences a reduction in group velocity and an equal increase in phase velocity that is a function of the frequency and the number of charged particles along the signal path. The Doppler measurement is dependent on the phase velocity and the advance of the signal is functionally defined by

$$\Delta t^i = G_i(t, x, y)$$

An empirical formula for the effect of the ionosphere on the Doppler measurement is given by

$$G_i = \frac{-1}{c} \sum_{j=0}^{n} k\, C_j\, X^j$$

$$X = 2\left(\frac{t - t_a}{t_b - t_a}\right) - 1$$

where the C_j,s are coefficients of a polynomial in time (t) from t_a to t_b normalized over the interval of -1 to $+1$ and k is a proportionality factor.

7.2.4 Earth Platform

The accuracy of radio metric data is strongly dependent on the calibration of the Earth as an observational platform. It is essential to know the location of each tracking station on the Earth's crust to within several centimeters and the location of the pole and prime meridian in inertial space to the same accuracy. The locations of the DSN stations are computed in a geocentric cylindrical coordinate system. Before VLBI data was available, solutions for the coordinates were obtained from two-way Doppler tracking of spacecraft encounters with the planets. Included were the Mariner and Viking class spacecraft encounters with the inner planets and Voyager and Pioneer class spacecraft encounters with Jupiter and Saturn. As each spacecraft encounters one of the planets, the Doppler tracking provides a means of precisely estimating the spacecraft orbit relative to the encountered planet and also the coordinates of the tracking stations. For each planetary encounter data arc, a strong solution for the tracking stations spin radii and longitudes are obtained. A station location database was developed. Since there is no information content in the Doppler tracking on the height of the station above the Earth's equator, the station location database was augmented with survey data. At the current time, station locations are obtained from VLBI observations of quasars and the accuracy is less than a meter in all coordinates.

7.2.5 Polar Motion

The Earth is slowing down irregularly. In addition, the Earth's principal axis wobbles about its spin axis with an amplitude slowly varying between 0 and 10 m. Timing and polar motion corrections are determined astronomically and provided to the orbit determination software by the Earth Orientation Parameters file.

7.2.6 Continental Drift

The continents are slowly drifting on the Earth's magma at a rate of about 3 cm per year. Over a period of 120 million years South America drifted away from Africa and formed the South Atlantic ocean. The continents are still drifting at about the same rate and carrying along the tracking stations. The rates are tabulated below for each tracking station complex in Earth body fixed Cartesian coordinates from the initial epoch of January 1, 2003. These rates are used to adjust the tracking station coordinates.

Continental drift

DSN station	x-cm/year	y-cm/year	z-cm/year
Goldstone	−1.80	0.65	−0.38
Madrid	−3.35	−0.41	3.92
Australia	−1.00	2.42	1.56

7.2.7 Solid Earth Tide

As the Earth rotates on its axis the tidal forces from the Sun and Moon raise and lower the oceans resulting in the tides everyone is familiar with. The Earth's crust and magma are also subjected to the same tidal forces which may be computed by taking the gradient of Eq. (1.9). Every rock or mountain on the planet is stretched a small amount by these forces. The result is a raising and lowering of the tracking stations every 12 h. The amplitude of the displacement of tracking stations from there nominal locations can be as large as 30 cm. A model of solid Earth tides is used to adjust the station location coordinates.

7.2.8 Plane Wave Propagation Through Ionized Gas

The propagation of a plane wave through an ionized gas such as the Sun's corona is described by Maxwell's equations, specifically the laws of Faraday and Ampere in vector form.

$$\nabla \times \mathbf{E} = -\boldsymbol{\mu_0} \frac{\partial \mathbf{H}}{\partial \mathbf{t}}$$

$$\nabla \times \mathbf{H} = \boldsymbol{\epsilon_0} \frac{\partial \mathbf{E}}{\partial \mathbf{t}} + \mathbf{J}$$

If we assume a plane transverse wave in the z direction with associated electrical field **E**, magnetic field **H**, and current density **J**, Maxwell's equations reduce to the one dimensional wave equation in E.

$$\frac{\partial^2 E}{\partial z^2} = \mu_0 \epsilon_0 \frac{\partial^2 E}{\partial t^2} + \mu_0 \frac{\partial J}{\partial t} \tag{7.7}$$

Equations describing the motion of free electrons in a time varying electric field are also needed. The force (**F**) on an electron in an electric field (**E**) is proportional to the charge (e), and equal to the mass of the electron times its acceleration in the direction of (**F**):

$$\mathbf{F} = e\mathbf{E} = m \frac{d^2\mathbf{r}}{dt^2}$$

The current density is simply the electron charge (e) times the electron flux, or the number of electrons that pass through a given area per unit time. The electron flux is the product of the electron density (N) and the velocity of the electrons.

$$\mathbf{J} = Ne\frac{d\mathbf{r}}{dt}$$

For a plane transverse wave these equations for the electron reduce to

$$m\frac{d^2r}{dt^2} = eE$$

$$J = Ne\frac{dr}{dt}$$

which, when substituted into the wave equation (7.7) results in the following.

$$\frac{\partial^2 E}{\partial z^2} = \mu_0\epsilon_0\left(\frac{\partial^2 E}{\partial t^2} + \frac{Ne^2}{m\epsilon_0}E\right) \tag{7.8}$$

A solution to this one-dimensional wave equation for an ionized atmosphere is

$$E = E_0 \sin\left[\omega t - kz\right] \tag{7.9}$$

from which we obtain

$$\omega_p^2 = \frac{Ne^2}{m\epsilon_0}$$

$$k^2 = \omega^2\mu_0\epsilon_0\left\{1 - \frac{Ne^2}{\omega^2 m\epsilon_0}\right\} = \omega^2\mu_0\epsilon_0\left\{1 - \frac{\omega_p^2}{\omega^2}\right\} \tag{7.10}$$

The phase velocity of the wave is defined by the locus of points along z where E_z is constant. Thus we have

$$\omega t - kz = \text{constant}$$

which implies

$$v = \frac{dz}{dt} = \frac{\omega}{k}.$$

Substituting k from Eq. (7.10), the phase velocity of the wave is

$$v = \frac{c}{\left(1 - \frac{\omega_p^2}{\omega^2}\right)^{\frac{1}{2}}} \tag{7.11}$$

making use of the well-known relation $c = 1/\sqrt{\mu_0\epsilon_0}$.

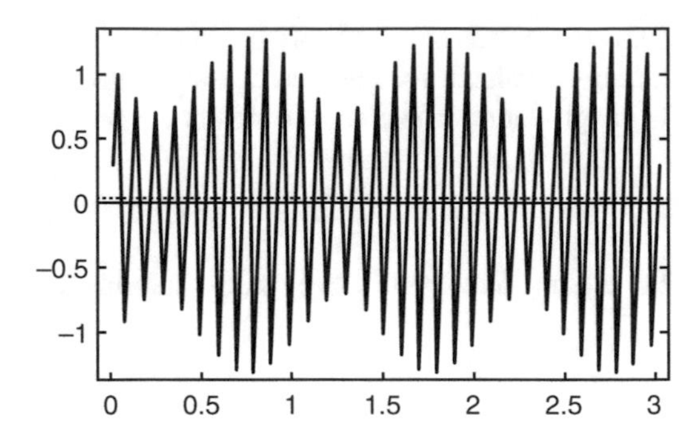

Fig. 7.4 Wave packets

We next consider the delay in the group velocity, which is associated with the range data. The concept of group velocity arises when we have electromagnetic waves that are nearly the same frequency traveling in the same direction through the same medium. Because linearity holds for electromagnetic waves, any electromagnetic wave may be regarded as the sum of its individual frequency components. Consider the case of two electromagnetic waves that differ in frequency and wave number by an infinitesimal amount $\delta\omega$ and δk, respectively. When added together we obtain the wave packets illustrated in Fig. 7.4.

The resultant wave is the higher frequency carrier that moves with phase velocity v as described above and the wave packets formed by the beating of the two nearly equal in frequency waves that move at a different group velocity (u). Doppler tracking data is associated with the phase velocity and range data is associated with the group velocity. In a vacuum, the phase and group velocities are equal to the speed of light. The two electromagnetic waves alluded to above are given by

$$\psi_1 = \sin(\omega t - kz)$$
$$\psi_2 = \sin\left[(\omega + \delta\omega)t - (k + \delta k)z\right]$$

We must perturb both the frequency and wave number in order to get the correct velocity which is controlled by the medium. The resultant wave is obtained by adding. After some trigonometric substitutions we have

$$\psi = \psi_1 + \psi_2 = 2\cos\left(\frac{\delta\omega t - \delta k z}{2}\right)\sin\left[(\omega + \frac{\delta\omega}{2})t - (k + \frac{\delta k}{2})z\right]$$

The carrier is given by the sine term and the modulation of the carrier is given by the cosine term. In a dispersive medium, the carrier wave moves at a velocity greater than the speed of light as shown above and the wave packet described by the cosine term moves at a velocity slower than the speed of light. The velocity of the wave packet is obtained in the same manner as described above for the carrier. The locus of points along z where the amplitude of the wave packet is constant is given by

$$\frac{\delta \omega t - \delta k z}{2} = \text{constant}$$

The group velocity is thus

$$u = \frac{dz}{dt} = \frac{\delta \omega}{\delta k} = \frac{d\omega}{dk}$$

and we get the desired group velocity.

$$u = \frac{k}{\omega} c^2 = c^2 \left(1 - \frac{\omega_p^2}{\omega^2}\right)^{\frac{1}{2}}$$

Because the group and phase velocities are close to the speed of light, ω_p is small and we may make the following approximations.

$$v \approx c\,(1 + \frac{\omega_p^2}{2\omega^2})$$

$$u \approx c\,(1 - \frac{\omega_p^2}{2\omega^2})$$

7.2.9 Solar Plasma Time Delay

The phase velocity in a dispersive medium is always greater than the speed of light. This apparent contradiction of special relativity is possible because the radio signal phase velocity does not describe the actual velocity of mass or energy, but rather the velocity of a pattern, or mathematical entity. Since it is critical to our analysis that ω_p/ω be less than 1.0, it is helpful to estimate it at this time. For this it is necessary to estimate N, the electron density in the plasma, which depends on the distance of the signal path from the Sun. Because a spacecraft is occasionally occulted by the Sun, the closest approach distances of the signal path goes to zero, but at less than 18 Solar radii, the signal is often degraded beyond usability. At 18 Solar radii, previous estimates have placed N at on the order of 10^3 electrons per cubic centimeter, which yields an ω_p of less than 1.0 MHz. For an X-band signal, $\omega = 2\pi f$ where f is approximately 8.9 GHz. Thus (ω_p^2/ω^2) is small, on the order of 10^{-8}. Since N decreases with increasing distance from the Sun, this is an upper bound.

The range delay associated with a plane wave passing through the Sun's corona is obtained by integrating the group velocity of propagation along the path length.

$$\int_{t_1}^{t_2} dt = \int_{z_1}^{z_2} \frac{dz}{u_z} = \int_{z_1}^{z_2} \frac{1}{c}\left[1 + \frac{e^2}{2m\epsilon_0\omega^2} N(z)\right] dz$$

The electron density varies approximately as the inverse square of the distance from the Sun.

$$N(z) = \frac{N_0 r_s^2}{r^2} = \frac{N_0 r_s^2}{R^2 + z^2}$$

R is the perpendicular distance from the center of the Sun to the light path or the distance of closest approach to the Sun. The constant N_0 is the effective electron density at the surface of the Sun and r_s is the radius of the Sun. Carrying out the integration, we have for the delay,

$$t_2 - t_1 = \frac{z_2 - z_1}{c} + \frac{e^2}{2\,\mathrm{cm}\epsilon_0 \omega^2} \frac{N_0 r_s^2}{R} \left\{ \tan^{-1}\left(\frac{-z_1}{R}\right) + \tan^{-1}\left(\frac{z_2}{R}\right) \right\}$$

The time advance of the Doppler signal, which is associated with the phase velocity, is the same equation as above except with a minus sign.

7.3 Optical Data

Optical imaging is a powerful data type for determining the position of a spacecraft relative to a nearby central body. This method involves imaging the target body on a star background. For this purpose, the Solid State Imager (SSI) science instrument on board the spacecraft is well suited as a precision optical measurement instrument for navigation. The use of science imaging instruments for navigation was developed during the Mariner 6, 7, and 9 missions and provided prime navigation measurements for the Viking, Voyager and many other missions. The SSI instrument was developed specifically for Galileo, replacing standard vidicon instruments used on previous missions. The principal attributes of the SSI instrument affecting navigation are its low image distortion and high sensitivity. The low distortion virtually eliminates the need for special calibration and the high sensitivity minimizes the exposure time required for imaging dim stars. For optical navigation, the imaged satellite lit limb may be fit and the center determined to better than one pixel. The star background may also be determined to less than one pixel even though the star images may spread over several pixels. From the focal length and pixel spacing, the angular accuracy may be computed and is about 10 μrad. For navigation analysis there are some systematic errors associated with SSI imaging that must be accounted for. These include shape and albedo variations that cause the center of brightness to not coincide with the center of mass. Image distortion also may impair satellite center determination. For conservatism, an optical center finding error of 1% of the satellite radius is assumed.

Optical data provides a measure of the direction of a vector from a spacecraft to a point on a target body. The target body may be a planet, asteroid, comet or satellite of one of these bodies. When combined with a data type that provides a measure of distance, such as range, a complete three-dimensional fix of the spacecraft may be inferred from the data providing a determination of the spacecraft orbit and physical parameters describing the central body gravity and inertial properties.

There are many variations of optical measurement systems that are used for orbit determination. The most accurate optical measurement systems focus on a specific point on the target body. This point may be the center-of-mass or a point associated with a feature on the surface of the body. The feature may be a crater or the intersection of fracture lines. The point of interest for a crater is the geometric center of the rim and is referred to as a landmark. Tracking the center of mass of primary body satellites, as was done for the Mariner, Viking, Voyager, Galileo and Cassini missions, provided sufficient accuracy for these missions but the error associated with determining the center-of-mass from limb data limits the accuracy obtainable. Landmark tracking accuracy is limited only by the resolution of the camera. For this reason landmark tracking was used for the NEAR mission where high accuracy optical and radio metric navigation was required.

7.3.1 Optical Data Processing

The raw data required for computing the optical observable consists of a sequence of images of the target body, the shutter time of each image, the spacecraft and camera attitude at the shutter time and ancillary data such as camera parameters and a star catalog. The raw data is processed to extract certain geometric parameters that are written to a picture sequence file. The parameters on the picture sequence file define an interface between navigation and measurement data preparation. The detailed data processing required to produce the picture sequence file is analogous to the data processing required for Doppler calibration or VLBI correlation.

The picture sequence file contains a header with the camera focal length, pixel and line spacing, the focal plane alignment matrix, camera distortion parameters and the boresight offset from the camera axis. The header parameters are obtained from preflight and inflight test images. A sequence of optical image data records are written for each image. Each optical image data record contains the image number, the image shutter time or time tag, the spacecraft attitude, filter setting, exposure time and a sequence of records for each landmark identified on the image. The landmark records contain a unique landmark number and the measurement which is the pixel and line location of the landmark in the image. The detection, identification, and numbering of landmarks may be performed visually by an optical navigation analyst or by a computer algorithm without human intervention. In addition to the picture sequence file, a separate landmark location file is generated containing *a priori* landmark locations for each landmark. The line and pixel coordinates of landmarks that appear near the limb in an image are not observed very well. The elevation angle of the spacecraft above the horizon when viewed from the landmark may be used as a test to reject data points.

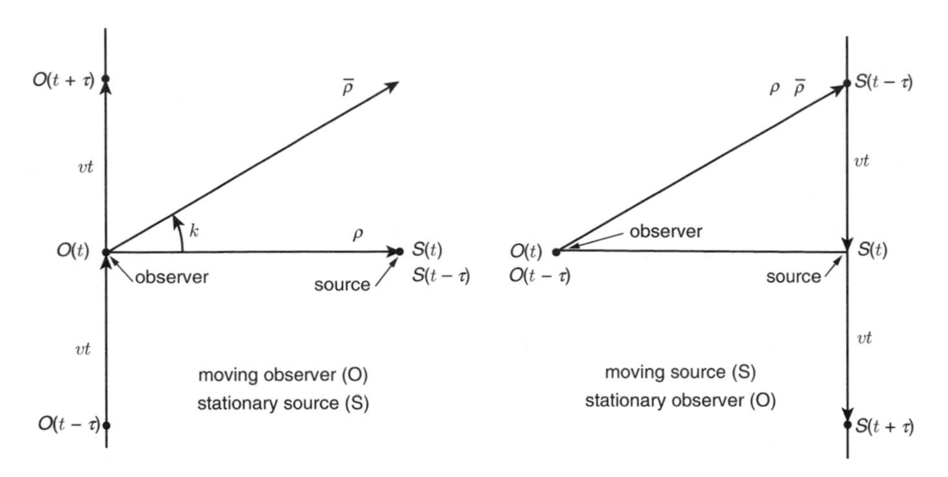

Fig. 7.5 Observer and source relative motion

7.3.2 Planetary and Stellar Aberration

The observed direction of light from a distance source differs from the actual direction obtained by solution of the light time equation due to the velocity parallax of the observer with respect to the photons or incoming wave front. This velocity parallax is referred to as aberration by astronomers and is aptly named. The first definition of aberration in the Webster's second edition dictionary is deviation from what is right, natural or normal. Light is red shifted or blue shifted in frequency depending whether the source is moving away from or toward the observer. The Doppler frequency shift is determined by the relative velocity. The observed direction of the light wave front is also affected by the motion of the source with respect to the observer but only the source velocity contributes to aberration. This apparent contradiction of Special Relativity may be resolved by examining the light time solution in conjunction with aberration in an inertial frame. Consider the case of an observer moving with respect to a stationary source as shown on the left side of Fig. 7.5.

Assume that closest approach occurs at time t. The observer at time t, identified by $O(t)$, will receive an incoming light wave from the direction $\boldsymbol{\rho}$ which is the solution of the light time equation. The observer was at $O(t - \tau)$ when the photons were emitted by the source so aberration should not be confused with the solution of the light time equation. Because of aberration, due to the relative motion of the source with respect to the photons, the observer will see the source in the direction defined by the vector $\bar{\boldsymbol{\rho}}$. The geometry is analogous to rain drops falling straight down. If the person starts to move, the rain drops appear slanted with respect to the local vertical.

Now suppose the observer is stationary and the relative motion of the source with respect to the observer is the same. In this case, shown on the right side of Fig. 7.5, the source appears to be moving in the opposite direction with velocity v.

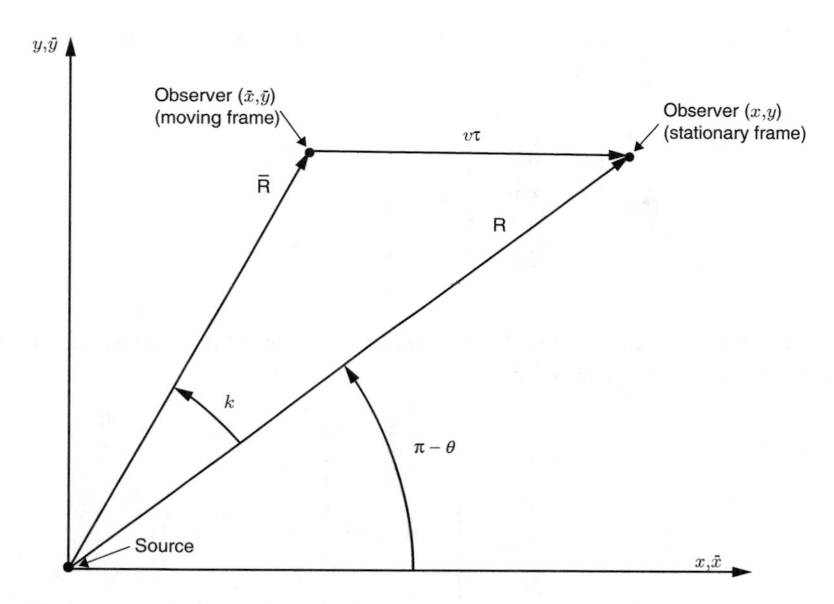

Fig. 7.6 Observer velocity relative to source

The observer at time t sees a plane wave emanating from the location of the source at $t - \tau$. There is no aberration. This is consistent with Special Relativity since the observer sees the source in the same direction for both cases. Since the Doppler shift is dependent on the relative velocity, the observer sees the same Doppler shift for both sides of Fig. 7.5. Furthermore, the speed of light will be the same for both observations due to time dilation and Lorentz contraction.

The geometry associated with an observer moving with respect to a source is illustrated in Fig. 7.6. In the stationary inertial frame, the source emits a photon at time $t = 0$ from the origin and this photon arrives at the observer at time $t = \tau$ with coordinates (x, y). The light time solution vector is $\mathbf{R}$ and the travel time is given by

$$\tau = \frac{R}{c}$$

If a coordinate system is defined that is moving with the observer $(\bar{x}, \bar{y})$ in the $+x$ direction with velocity v and the origins coincide at time $t = 0$, the observed direction of the photon is given by $\bar{\mathbf{R}}$. The angle between $\mathbf{R}$ and $\bar{\mathbf{R}}$ is the aberration angle κ and

$$\sin(\kappa) = \frac{|\mathbf{R} \times \bar{\mathbf{R}}|}{R\,\bar{R}} \tag{7.12}$$

The angle θ is between the velocity vector and the vector from the observer to the source $(-\mathbf{R})$.

If classical Galilean motion is assumed, where the speed of light is not constant, the observed vector is given by

$$\bar{\mathbf{P}} = \begin{bmatrix} 1 & 0\,0\,0 \\ -v & 1\,0\,0 \\ 0 & 0\,1\,0 \\ 0 & 0\,0\,1 \end{bmatrix} \mathbf{P} = \begin{bmatrix} \frac{R}{c} \\ -\frac{Rv}{c} - R\cos\theta \\ R\sin\theta \\ 0 \end{bmatrix} \tag{7.13}$$

where time is artificially carried along as the first component of **P** and **R** is contained in the last three components of **P**.

$$\mathbf{P} = \begin{bmatrix} \frac{R}{c} \\ -R\cos\theta \\ R\sin\theta \\ 0 \end{bmatrix}$$

The angle κ is obtained by substituting the position vectors obtained from the second through third components of **P** and $\bar{\mathbf{P}}$ into Eq. (7.1).

$$\sin\kappa = \frac{v\sin\theta}{c}\left\{\frac{1}{\sqrt{1+\frac{v^2}{c^2}+\frac{2v\cos\theta}{c}}}\right\}$$

Making use of the approximation

$$\frac{1}{\sqrt{1+\frac{v^2}{c^2}+\frac{2v\cos\theta}{c}}} \approx 1 - \frac{v\cos\theta}{c} - \frac{v^2}{2c^2}$$

and

$$\sin\theta\cos\theta = \frac{1}{2}\sin(2\theta)$$

the aberration angle may be approximated to second order by

$$\sin\kappa \approx \frac{v}{c}\sin\theta - \frac{1}{2}\frac{v^2}{c^2}\sin(2\theta) + \ldots$$

The classical result assumes that the speed of light in the moving frame is different from the speed of light in the inertial frame at rest. The Lorentz transformation from Special Relativity is used to get the correct result.

$$\bar{\mathbf{P}} = \begin{bmatrix} \dfrac{1}{\sqrt{1-\dfrac{v^2}{c^2}}} & \dfrac{-v}{c^2\sqrt{1-\dfrac{v^2}{c^2}}} & 0 & 0 \\ \dfrac{-v}{\sqrt{1-\dfrac{v^2}{c^2}}} & \dfrac{1}{\sqrt{1-\dfrac{v^2}{c^2}}} & 0 & 0 \\ 0 & 0 & 1 & 0 \\ 0 & 0 & 0 & 1 \end{bmatrix} \mathbf{P}$$

The Lorentz transformation is given in conventional engineering coordinates where time has the units of time and the existence of c is explicitly acknowledged. Since c is a constant, a system of space-time coordinates can be defined with $c = 1$ and time given the dimension of length. For these coordinates, favored by relativists, the Lorentz transformation matrix is symmetrical. The advantage of the conventional coordinates used here is that it is immediately obvious that the approximation to first order reduces to the Galilean transformation given by Eq. (7.2) in the limit as c approaches infinity. Since the "at rest" coordinate system is arbitrary, the inverse of the Lorentz transformation matrix can be obtained by changing the sign of v. The Galilean transformation also has this property. Another property of the Lorentz transformation is that the Minkowski metric must be preserved.

$$ds^2 = c^2dt^2 - dx^2 - dy^2 - dz^2 \tag{7.14}$$

Since ds^2 is null for a photon ($ds^2 = 0$) then $d\bar{s}^2$ must also be null. The observation vector in the moving frame is given by

$$\bar{\mathbf{P}} = \begin{bmatrix} \dfrac{R}{c\sqrt{1-\frac{v^2}{c^2}}} + \dfrac{Rv\cos\theta}{c^2\sqrt{1-\frac{v^2}{c^2}}} \\ -\dfrac{Rv}{c\sqrt{1-\dfrac{v^2}{c^2}}} - \dfrac{R\cos\theta}{\sqrt{1-\dfrac{v^2}{c^2}}} \\ R\sin\theta \\ 0 \end{bmatrix}$$

When $\bar{P}$ is substituted into the Minkowski metric (Eq. 7.9) it is demonstrated that $d\bar{s}^2$ is null. Proceeding as for the classical solution, the angle κ corrected for Special Relativity is obtained.

$$\kappa = \arcsin\left(\frac{\frac{v \sin\theta}{c} + \left(1 - \sqrt{1 - \frac{v^2}{c^2}}\right) \sin\theta\cos\theta}{1 + \frac{v}{c}\cos\theta}\right)$$

Making use of the approximations

$$\sqrt{1 - \frac{v^2}{c^2}} \approx 1 - \frac{1}{2}\frac{v^2}{c^2}$$

$$\frac{1}{1 + \frac{v}{c}\cos\theta} \approx 1 - \frac{v}{c}\cos\theta$$

the first two terms of the series expansion for $\sin\kappa$ are

$$\sin\kappa \approx \frac{v}{c}\sin\theta - \frac{1}{4}\frac{v^2}{c^2}\sin(2\theta) + \ldots$$

The aberration corrected vector is in the same plane as the source velocity vector and light time solution vector. The calculation of this vector from the aberration angle κ and the angle between the velocity vector and the vector from the observer to the source (θ) is illustrated in Fig. 7.7. The vector $\boldsymbol{\rho}_{in}$ is the light time solution from the observer to the source. The vector $\bar{\boldsymbol{\rho}}$ is the direction that the source is observed and the direction that one would point a telescope. The angle κ between these vectors is the aberration angle as defined above. From the geometry, the vector formula for planetary aberration is simply

$$\bar{\boldsymbol{\rho}} = \left\{\boldsymbol{\rho}_{in} + \left[\frac{\rho_{in}\sin\kappa}{v\sin\theta}\right]\mathbf{v} - \left[\frac{\sin\kappa}{\tan\theta}\right]\boldsymbol{\rho}_{in}\right\}$$

If the source is a star or remote object, the magnitude of the light time solution vector approaches infinity. The formula for stellar aberration ($\bar{\boldsymbol{\rho}}_s$) may be obtained by taking the limit and

$$\bar{\boldsymbol{\rho}}_s = \left\{\hat{\boldsymbol{\rho}}_{in} + \left[\frac{\sin\kappa}{v\sin\theta}\right]\mathbf{v} - \left[\frac{\sin\kappa}{\tan\theta}\right]\hat{\boldsymbol{\rho}}_{in}\right\}$$

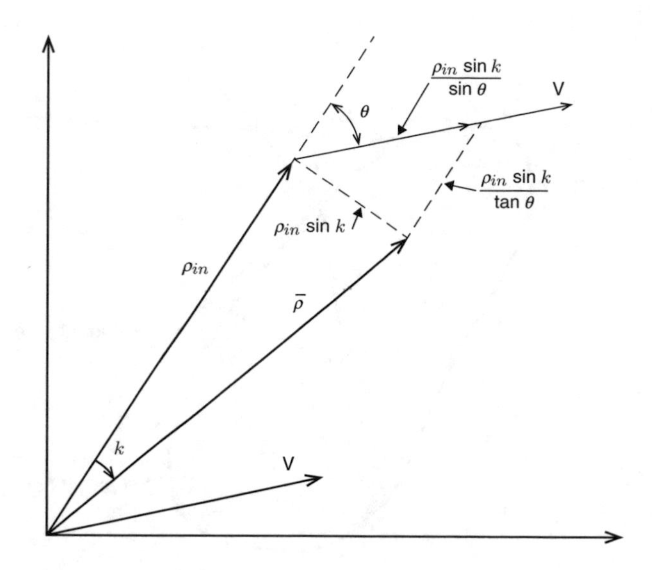

Fig. 7.7 Aberration correction

7.4 Altimetry

Altimetry provides a measure of the magnitude of a vector from a spacecraft to a target body that may be a planet, asteroid, comet, or a satellite of one of these bodies. The altimetry measurement is the distance or slant range from the spacecraft to a point on the surface of the target body which is inferred from the time that it takes an electromagnetic wave to traverse the distance. The electromagnetic wave is transmitted and the reflected signal received by the altimeter instrument and the slant range is determined by multiplying the signal delay time by the speed of light. The altimeter may transmit and receive a radar signal or laser beam. Radar based altimeters are limited by range and have only been used for landing spacecraft on a target body such as the Surveyor spacecraft on the Moon or the Viking spacecraft on Mars. Laser altimeters can operate out to several hundred kilometers and are thus useful for measurements in orbit about a target body. When combined with optical imaging of landmarks, altimetry provides a complete three-dimensional fix of the spacecraft orbit. However, since the accuracy is limited by the error in determining the surface of the target body, laser altimetry is only marginally useful for ground-based spacecraft orbit determination. In the future, laser altimetry may be used for medium accuracy autonomous navigation since the measurement may be readily obtained on board the spacecraft. An important application of laser altimetry is in determining the shape of a target body given the orbit of the spacecraft. This application was used on the NEAR mission to determine a high precision shape model of Eros that was used to support landing operations.

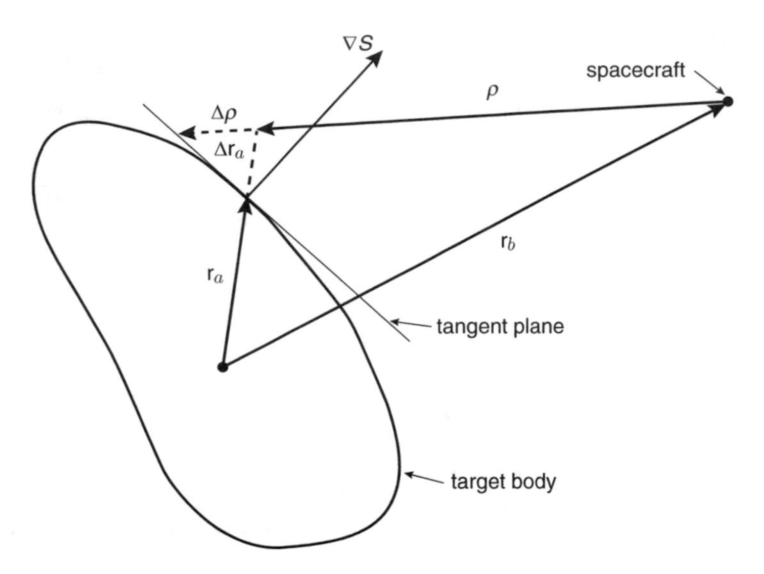

Fig. 7.8 Altimetry observable iteration geometry

7.4.1 Altimetry Data Measurement Model

The observed vector from the spacecraft to the target body surface is referenced to a coordinate system that is fixed to the instrument platform that is movable if the instruments are mounted on a scan platform or is fixed to the spacecraft body. The z axis is in the direction of the nominal boresight of the instruments, y is to the right and x is down. The definition of down is arbitrary, but is generally taken to be in the direction of decreasing declination on the star background. The transformation matrix (TC) describes the pointing direction which is generally taken to be the camera boresight.

The altimeter may be mounted on the spacecraft at a location that may be as far away as a meter from the spacecraft center-of-gravity. The altimeter position must be corrected for this offset and

$$\Delta\mathbf{r}_{cg} = [TC]^T\ \mathbf{r}_{cg}$$

where $\mathbf{r}_{cg}$ is the location of the altimeter focal plane relative to the spacecraft center-of-gravity in instrument platform coordinates.

The altimeter boresight is offset slightly from the instrument axes as defined by two angles ψ_a and χ_a. The vector that defines the altimeter boresight is given by

$$\hat{\mathbf{B}}_a = [\cos\chi_a \sin\psi_a,\ \sin\chi_a,\ \cos\chi_a \cos\psi_a]^T$$

The vector from the spacecraft to the surface is computed as illustrated in Fig. 7.8.

The vector from the center-of-mass of the target body to the altimeter in target body fixed coordinates is given by

$$\mathbf{r}_b = [T(\alpha, \delta, W)](\mathbf{r} + \Delta\mathbf{r}_{cg})$$

where $[T(\alpha, \delta, W)]$ is the transformation matrix from J2000 coordinates to the pole and prime meridian of the target body as defined by Fig. 1.3. The unit vector $\hat{\boldsymbol{\rho}}$ is the altimeter boresight direction in target body fixed coordinates given by

$$\hat{\boldsymbol{\rho}} = [T(\alpha, \delta, W)][TC]^T\, \hat{\mathbf{B}}_b$$

An initial guess is needed for the magnitude of $\boldsymbol{\rho}$ and this is taken to be

$$\rho_i = r - r_{ao}$$

where r_{ao} is the average radii of the target body. The first step of the iteration for the observation vector $\boldsymbol{\rho}$ is to compute the vector from the center-of-mass of the target body to the surface.

$$\mathbf{r}_{ai} = \mathbf{r}_b + \rho_i \hat{\boldsymbol{\rho}}$$

A test is performed to determine if $\mathbf{r}_{ai}$ is on the surface of the target body. The surface of the target body is obtained from a harmonic expansion of Legendre polynomials and associated functions as a function of latitude and longitude.

$$r_a = \sum_{n=0}^{\infty} \sum_{m=0}^{n} P_n^m(\sin\phi_a)\{A_{nm}\ \cos m\lambda_a + B_{nm}\ \sin m\lambda_a\}$$

where λ_a and ϕ_a are the longitude and latitude at the solution point and A_{nm} and B_{nm} are the harmonic coefficients. The longitude and latitude of the surface point are given by

$$\lambda_a = \tan^{-1}\left(\frac{r_{aiy}}{r_{aix}}\right)$$

$$\phi_a = \sin^{-1}\left(\frac{r_{aiz}}{r_{ai}}\right)$$

respectively and the harmonic coefficients (A_{nm} and B_{nm}) are input constant parameters. The error in computing the surface radius vector is

$$\Delta r_a = r_{ai} - r_a$$

If $|\Delta r_a|$ is less than an input tolerance, nominally 10^{-7} km, convergence is obtained and the observable is computed from ρ_i. If convergence is not obtained, then $\boldsymbol{\rho}$ is lengthened or shortened by $\Delta\rho$ as shown in Fig. 7.8. In order to compute $\Delta\rho$, a model of the target body surface is needed. This model consists of a vector to a point on the surface and the local tangent plane and $\mathbf{r}_a$ is the vector to the point on the surface given by

$$\mathbf{r}_a = [r_a \cos\phi_a \cos\lambda_a,\ r_a \cos\phi_a \sin\lambda_a,\ r_a \sin\phi_a]^T$$

The surface normal vector may be computed by taking the gradient of the surface defined by

$$S = r_a - \sum_{n=0}^{\infty}\sum_{m=0}^{n} P_n^m(\sin\phi_a)\{A_{nm}\ \cos m\lambda_a + B_{nm}\ \sin m\lambda_a\}$$

where the surface of interest corresponds to $S = 0$. The normal vector is then given by

$$\mathbf{S} = \left[\frac{\partial S}{\partial r_a}\ \frac{\partial S}{\partial \phi_a}\ \frac{\partial S}{\partial \lambda_a}\right]\begin{bmatrix} \cos\phi_a \cos\lambda_a & \cos\phi_a \sin\lambda_a & \sin\phi_a \\ -\dfrac{\sin\phi_a \cos\lambda_a}{r_a} & -\dfrac{\sin\phi_a \sin\lambda_a}{r_a} & \dfrac{\cos\phi_a}{r_a} \\ -\dfrac{\sin\lambda_a}{r_a \cos\phi_a} & \dfrac{\cos\lambda_a}{r_a \cos\phi_a} & 0 \end{bmatrix}$$

and

$$\frac{\partial S}{\partial r_a} = 1$$

$$\frac{\partial S}{\partial \phi_a} = -\sum_{n=0}^{\infty}\sum_{m=0}^{n} \frac{\partial P_n^m(\sin\phi_a)}{\partial \phi_a}\{A_{nm}\ \cos m\lambda_a + B_{nm}\ \sin m\lambda_a\}$$

$$\frac{\partial S}{\partial \lambda_a} = -\sum_{n=0}^{\infty}\sum_{m=0}^{n} P_n^m(\sin\phi_a)\{-A_{nm}\ m \sin m\lambda_a + B_{nm}\ m \cos m\lambda_a\}$$

The change in the slant range ($\Delta\rho$) may be computed by application of the law of sines to the small triangle shown in Fig. 7.8 whose sides are the extension of the $\mathbf{r}_a$ vector (Δr_a), the extension of the $\boldsymbol{\rho}$ vector ($\Delta\rho$), and the intersection of the local tangent plane with the plane containing $\mathbf{r}_a$, $\boldsymbol{\rho}$ and $\mathbf{r}_b$. The intersection of the local tangent plane ($\mathbf{t}_i$) with the plane of Fig. 7.8 is given by

$$\mathbf{t}_i = (\mathbf{r}_b \times \mathbf{r}_a) \times \mathbf{S}$$

The angle opposite the $\Delta\rho$ side of the triangle is

$$\alpha_\rho = \arccos\left(\frac{\mathbf{t}_i \cdot \mathbf{r}_a}{t_i\ r_a}\right)$$

and the angle opposite the Δr_a side of the triangle is

$$\alpha_{r_a} = \arccos\left(\frac{\mathbf{t}_i \cdot \boldsymbol{\rho}}{t_i\ \rho}\right)$$

The lengthening or shortening of the $\boldsymbol{\rho}$ vector is then obtained from the law of sines and

$$\Delta\rho = \frac{\sin\alpha_\rho\ \Delta r_a}{\sin\alpha_{r_a}}$$

$$\rho_{i+1} = \rho_i + \Delta\rho$$

Using ρ_i, a new ρ_{i+1} is computed and repeated until convergence is obtained. If convergence is not obtained after several iterations, the data point is rejected. Since two solutions are possible, a check is made to determine if the surface intersection point is in view from the spacecraft. The elevation angle (E_ℓ) of the spacecraft above the horizon when viewed from the surface may be used as a test.

$$E_\ell = \sin^{-1}\left(-\frac{\boldsymbol{\rho} \cdot \mathbf{S}}{\rho\ \nabla S}\right)$$

If $E_\ell < 0$, the solution is rejected and the range is shortened by

$$\rho_{i+1} = \rho_i - 2r_a \sin E_\ell$$

After several more iterations, the data point is rejected if a valid solution has not been found. Once a valid solution has been found, the observable is computed.

7.4.2 Altimetry Variational Partial Derivatives

The partial derivatives of the altimetry observable with respect to state and constant parameters are given by

$$\frac{\partial Z_a}{\partial(\mathbf{r}_0, \dot{\mathbf{r}}_0, q)} = \frac{\partial Z_a}{\partial\rho}\left[\frac{\partial\rho}{\partial\mathbf{r}_b}\frac{\partial\mathbf{r}_b}{\partial\mathbf{r}}\frac{\partial\mathbf{r}}{\partial(\mathbf{r}_0, \dot{\mathbf{r}}_0, q)} + \frac{\partial\rho}{\partial q}\right]$$

where

$$\frac{\partial Z_a}{\partial\rho} = 1$$

$$\frac{\partial\mathbf{r}_b}{\partial\mathbf{r}} = [T(\alpha, \delta, W)]$$

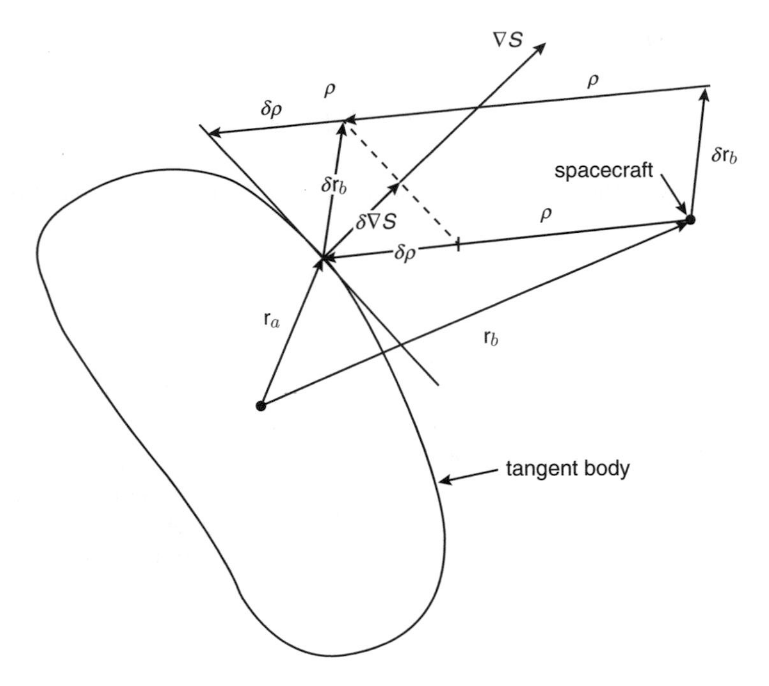

Fig. 7.9 Altimetry observable variation geometry

and the partial derivative of ρ with respect to the target body fixed spacecraft state ($\mathbf{r}_b$) may be determined by geometrical construction. The observational geometry is shown in Fig. 7.9 where the parallelogram formed by the vector $\boldsymbol{\rho}$ and the spacecraft position variation $\delta\mathbf{r}_b$ and the vector constructions are all in the plane of Fig. 7.9. The normal vector $\mathbf{S}$ and target body center are not necessarily in the plane of the figure. The spacecraft position variation vector is projected on to the normal vector and

$$\delta\nabla S = \frac{\mathbf{S} \cdot \delta\mathbf{r}_b}{\nabla S}$$

The extension of the observation vector ($\delta\rho$), projected on to the normal vector, has the same magnitude as $\delta\nabla S$.

$$\delta\nabla S = \delta\rho\,\frac{-\boldsymbol{\rho}\cdot\mathbf{S}}{\rho\ \nabla S}$$

Solving for $\delta\rho$ gives

$$\delta\rho = -\rho\,\frac{\mathbf{S} \cdot \delta\mathbf{r}_b}{\boldsymbol{\rho}\cdot\mathbf{S}}$$

The required partial derivative may be obtained by taking the limit as $\delta \mathbf{r}_b$ approaches zero. In this rudimentary application of differential geometry, the parallelograms and triangles with sides defined by variational symbols shrink to a point at the end of the $\mathbf{r}_a$ vector and

$$\frac{\partial \rho}{\partial \mathbf{r}_b} = -\rho \, \frac{\mathbf{S}}{\boldsymbol{\rho} \cdot \mathbf{S}}$$

7.5 Summary

The measurements that are used for orbit determination may be separated into several broad categories. These include radiometric tracking data from the DSN and optical imaging and altimetry from the spacecraft. Radiometric measurements include Doppler, range, and VLBI. Orbit determination involves a three-dimensional determination of the spacecraft position and velocity. Radiometric data is one dimensional and the process of extracting a measurement from instrumentation is cleanly separated from navigation. Navigation instrumentation can be designed in a laboratory with little knowledge of its ultimate use. Therefore, navigators are primarily concerned with the physical quantity being measured and a model of the measurement that can be programmed into orbit determination software and need not be concerned with the details of instrument design. Electrical engineers design the instrumentation and navigators navigate. The external environment has an effect on measurements. If the motion of the spacecraft is not perturbed, these environmental disturbances are modeled and a correction to the measurement is computed by the orbit determination software. Examples are the troposphere, ionosphere, solar plasma and polar motion for radiometric data, aberration for optical data and a shape model of the target body for altimetry.

Bibliography

Chao, C. C., "New Tropospheric Range Corrections with Seasonal Adjustment," Deep Space Network Progress Report 92-1526, vol. 6, Jet Propulsion Laboratory, Pasadena, California, pp. 67–73, December 15, 1971.

Chao, C. C., "A New Method to Predict Wet Zenith Range Correction From. Surface Measurements," Deep Space Network Progress Report 92ů1526, vol. 14, Jet Propulsion Laboratory, Pasadena, California, pp. 33–41, April 15, 1973.

Curkendall, D, W., "Radio Metric Technology for Deep Space Naviga- tion: A Development Overview," Paper 78-1395, presented at the AAS/AIAA Astrodynamics Conference, Palo Alto, California, August, 1978.

Davis, R. R., "Interplanetary Optical Navigation using Charge- Coupled Devices," Paper 80-1652, presented at the AIAA/ASS Astro- dynamics Conference, Danvers, Massachusetts, August,1980.

Ellis, J. 1980. Large Scale State Estimation Algorithms for DSN Tracking Station Location Determination. J. Astronaut. Sci. 28, 15–30.

Hamilton, T. W. and Melbourne, W. G., "Information Content of a Single Pass of Doppler Data from a Distant Spacecraft," *The Deep Space Network Space Programs Summary 37–39*, Vol. III, Jet Propulsion Laboratory, Pasadena, California, 31 May 1966, pp. 18–23.

Hawkins, S. E., et al., 1997. Multi-Spectral Imager on the Near Earth Asteroid Rendezvous Mission. Space Science Review 82, 31–100.

Landau, L. D. and Lifshitz, E. M., *The Classical Theory of Fields*. Butterworth-Heinemann; 4th edition (October 1997).

Miller, J. K. and K. H. Rourke, "The Application of Differential VLBI to Planetary Approach Orbit Determination", *The Deep Space Network Progress Report 42-40*, May-June 1977, pp 84–90.

Miller, J. K., "The Effect of Clock, Media and Station Location Errors on Doppler Measurement Accuracy", *TDA Progress Report 42-113*, Jet Propulsion Laboratory, May 15, 1993.

Moyer, T. D., *Formulation for Observed and Computed Values of Deep Space Network Data Types for Navigation*, JPL Publication 00-7, October 2000.

Royden, H. N., D. W. Green, and G. R. Walson, "Use of Faraday-Rotation Data from Beacon Satellites to Determine Ionosophere Corrections for Interplanetary Spacecraft Navigation," Proc. Satellite Beacon Symposium, Warszawa, Poland, May 19–23, 1980, pp. 345–355, 1981.

Stratton, J. A., "Electromagnetic Theory", McGraw-Hill, New York, 1941.

Chapter 8
Navigation Operations

Navigation operations is the process of identifying a destination, finding the path to the destination, and performing the necessary tasks to transport a vehicle to the destination along with the passengers which may be people or science instruments. Navigation operations are synonymous with exploration and it is performed by engineers and explorers or ordinary people when they get in their car, turn on the GPS, and drive to the mall. The GPS navigation system requires a GPS receiver and a map. GPS alone does not navigate. Spacecraft navigation to the bodies in the solar system is a bit more complicated than driving to the mall. The closest analogy to planetary spacecraft navigation is the navigation performed on eighteenth-century and earlier sailing ships. The navigator knows his home port and has a vague idea of the location of his destination. Once the ship sails beyond the horizon, its location is not known very well. The major problem is determining time. In order to determine longitude, the navigator must know the time in his home port where 12:00 p.m. is high noon. If high noon occurs at 1:00 p.m. he knows he is about 700 miles West of his home port depending on his latitude that can be determined from the elevation of stars above the horizon. Magellan carried 18 h glasses for his voyage around the world. In determining a route, he must have some knowledge of wind and ocean currents. For planetary navigation, the navigator knows his home port, namely the Earth. He has a vague idea of his destination. If his destination is a planet, he can go out at night and look up and see his destination. The problem of determining where the spacecraft is located and plotting a route is shared with sailors of antiquity. Determination of time with atomic clocks enables the navigator to determine range to the spacecraft and is the key to planetary orbit determination. The major advantage of planetary spacecraft navigators over sailors of antiquity is that they perform navigation in a comfortable flight operations facility and do not die if they make a mistake.

Planetary navigation operations consist mainly of collecting data from a number of sources, inputing this data to navigation software, determining the orbit, computing trajectory correction maneuvers, and transmitting maneuver commands to

J. Miller, *Planetary Spacecraft Navigation*, Space Technology Library 37,
https://doi.org/10.1007/978-3-319-78916-3_8

the spacecraft. This process does not require a deep understanding of mathematics or navigation and can be performed by anyone with a rudimentary knowledge of navigation and computers. However, the navigation operations team is comprised of individuals who are highly educated and dedicated. Their job is to analyze the navigation system and prepare solutions for problems that may arise as well as perform the routine navigation operations. The job can be frustrating because the interesting problems arise quickly and often have no solution. Most of the time is spent doing routine operations.

8.1 Navigation System

The navigation system is a collection of hardware, instrumentation, and computer software that enable navigation. For planetary spacecraft navigation, the navigation system is comprised of a spacecraft, the Deep Space Network (DSN), and procedures that are encoded in software that resides on the spacecraft or on the ground at a space flight operations facility. The spacecraft hardware includes transponders, imagers, an attitude control system, and a propulsion system. The DSN is comprised of tracking stations located at Goldstone California, Madrid Spain, and Canberra Australia and the Space Flight Operations Facility (SFOF) located at the Jet Propulsion Laboratory (JPL) and at other locations depending on the mission. The SFOF houses the software required to extract and format the data and the navigation team required to operate the navigation software. For some missions, the navigation team and software may be located elsewhere.

8.1.1 Deep Space Network

The tracking stations that comprise the DSN are in a complex containing a 72-m antenna, several 34-m antennae, and special- purpose hardware end computers for extracting Doppler, range, and VLBI observables from received spacecraft signals and VLBI from extragalactic radio sources. This data is relayed via high-speed data lines to the SFOF. Propulsive maneuver commands are sent to a DSS via the same high-speed data lines and then transmitted to the spacecraft.

8.1.2 Spacecraft

The spacecraft is designed to enable acquisition of data by the DSN and provide propulsive maneuvers to enable the spacecraft to arrive at the target body. The hardware systems onboard the spacecraft include transmitters and receivers with a Doppler transponder, imaging and altimetry instruments, a telecommunications system for downloading images and receiving maneuver commands, an attitude control system for pointing cameras and propulsive motors in the desired direction, and a propulsion system for changing the flight path of the spacecraft.

8.2 Orbit Determination

The launch vehicle injects the spacecraft on a trajectory that goes from the launch site to the target body. The initial problem for planetary spacecraft navigation is to find the spacecraft after launch. The DSN searches the sky in the direction of the predicted launch trajectory. This is a three-dimensional search in tracking station pointing angles and carrier frequency. Near the Earth the search is difficult; however, as the spacecraft departs the Earth's gravity the search is easier. The spacecraft location in the sky and carrier frequency becomes well defined, provided the spacecraft is headed toward the target planet. Once the spacecraft is located, orbit determination software determines the trajectory. A block diagram of the orbit determination software is shown in Fig. 8.1. In the example used here, the Near Earth Asteroid Rendezvous software configuration is described. The rectangular

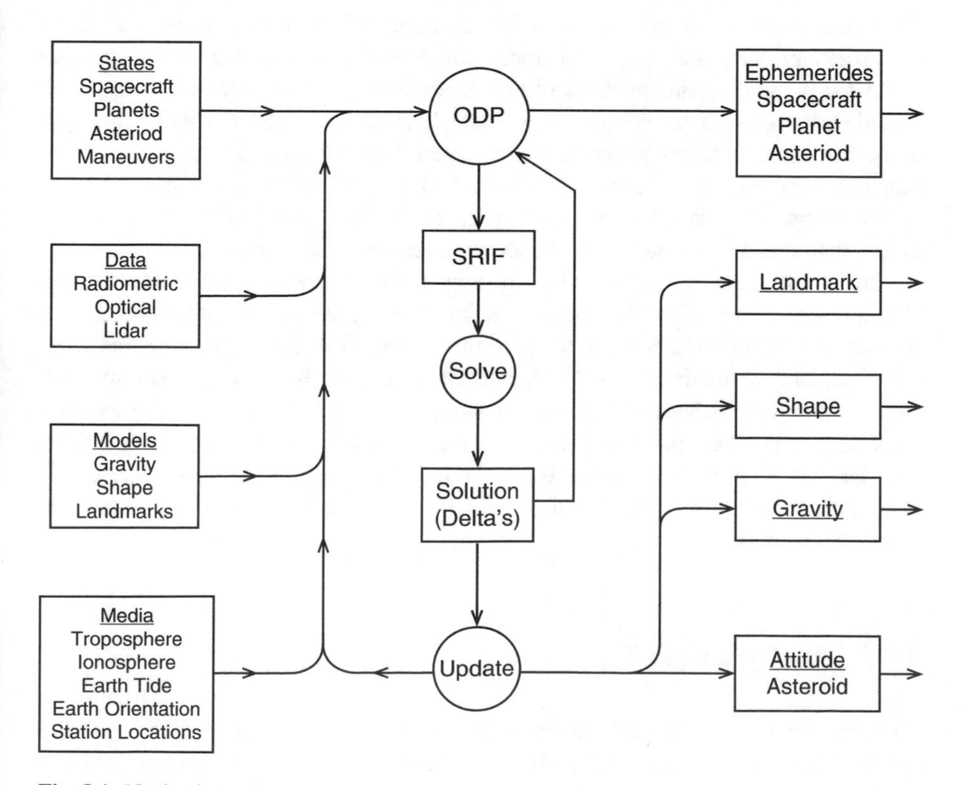

Fig. 8.1 Navigation software

blocks are files that need to be created. The circles are the programs that create the files. Prior to launch, state vectors for the spacecraft and planets are obtained and included in a file called STATES. The file MODELS contains the Earth's gravity model, propulsive maneuvers, station locations, Earth orientation, solid Earth tides, tectonic plate motion, and landmark locations. The troposphere, ionosphere, and solar plasma are contained in the MEDIA file.

The first file to arrive at the SFOF after launch is the tracking data file (DATA) containing Doppler and range data. The orbit determination program outputs a SRIF matrix and files of spacecraft ephemeris, central body ephemeris, and central body attitude consisting of segments of Chebyshev polynomials. The program SOLVE inverts the SRIF along with the measurement residuals to obtain corrections for all the estimated parameters. The corrected estimated parameters are fed back to ODP, and another solution is attempted. When convergence is achieved, the program UPDATE is executed to provide files that may be exported to the science team and others outside of navigation.

8.2.1 Orbit Determination Strategy

Orbit determination strategy involves selecting the estimated parameters, acquiring the necessary data, assigning error uncertainties to the measurements and *a priori* estimated parameters and analysis of the residual errors. The estimated parameters include spacecraft state, central body state, central body Euler angels and spin rates, gravity coefficients, propulsive maneuvers, landmark locations, solar pressure model parameters, and other parameters that are needed for trajectory propagation.

The amount of data to be processed by the orbit determination filter, or the length of the data arc, must also be specified. A given orbit determination run presents the analyst with many options. One strategy is to start with a short data arc and obtain convergence. Then the data arc is lengthened, admitting more data, until a satisfactory solution is obtained. If the data arc is too long, the filter may choke on too much data. Unmodeled errors will eventually destroy the validity of the solution and computed statistics. Introduction of stochastic noise in the form of a random walk may help. The stochastic parameters cause the filter to deweight earlier data, and thus the filter tends to forget and rely on more recent data. A problem with stochastic parameters is they tend to falsely smooth out the residual errors and introduce instability.

8.2.2 Multiple Data Types

The problem of orbit determination is exacerbated when multiple data types are processed. For a single data type, which would generally be Doppler only data, the solution is not dependent on the data weight provided all the measurements are given equal weight and there are no stochastic parameters. When multiple data types are processed, the solution becomes dependent on relative weight. For example, when range and optical data are processed with Doppler data the solution will depend on how the optical data is weighted compared to the Doppler data. For planetary approach, the optical and Doppler data determine components of the spacecraft state that are orthogonal and a natural separation is obtained. The optical data determines the coordinates in the B plane, and Doppler data determines flight time and the approach velocity vector. Range data is redundant to Doppler data but provides the constant of integration. In orbit, the separation is less apparent. For the NEAR mission, optical data provided the orientation of the orbit in space and Doppler data provided a measure of the distance traveled by the spacecraft. These measurements were complimentary and provided about one meter in orbit measurement accuracy. This was important for orbit prediction. It was necessary to predict the orbit ten days into the future to an accuracy of about 100 m which required orbit determination accuracy of about 1 m. This accuracy could be achieved relative to the center of mass of Eros even though the location of landmarks on the surface was known only to about 50 m. In computing the optical observable, the offset of the camera from the center of mass of the spacecraft was included.

8.2.3 Simulated Data

Another technique for detecting problems with orbit convergence is analysis of the signatures in the data residuals. For example, a timing error during a planetary encounter has a distinct signature that looks like the tangent function near 90°. Analysis of residual errors requires considerable experience. The first time a new error appears in the data, the signature is generally not recognized. Experience in recognizing these data signatures is best obtained by processing simulated data prior to conducting navigation operations.

Simulated data is an important tool for analysis of orbit determination performance. It proved to be valuable during the NEAR mission which involved introduction of new data types. A simulated tracking data file was prepared that consisted essentially of time tags of the Doppler and range data points. The simulated tracking data file also included header data such as station number necessary for computing the observable. In assigning time tags, care had to be taken to assure that the spacecraft was above the horizon. Simulated images of the asteroid were prepared by ray tracing asteroid brightness on the surface of the asteroid to each pixel in the camera and a picture sequence file prepared with image time and camera pointing angles. A simulated LIDAR data file was generated that contained the time of each measurement and instrument pointing angles. Media and clock calibration files were prepared and were assumed to be the same for each station.

The simulated data files were input to the orbit determination software. As each data point was processed, the computed measurement was modified by adding noise obtained from a random number generator to generate a measurement. New simulated data files were output by the orbit determination software with the simulated measurement. The spacecraft state, Eros ephemeris, Eros attitude, and all the input model parameters were modified and given to a second party along with the data files and simulated images. The modification of the input *a priori* data used the same process as adding noise to the measurement; however, for some inputs the error was considerably larger than the *a priori* error would indicate. The second party did not know the real assumed value of the estimated parameters or the assumed value of the measurements. The second party extracted line and pixel locations from the simulated images, created a landmark tracking file, and ran the orbit determination software to determine the simulated orbit. If the second party failed to determine the orbit, the process would be repeated until he was successful. There was more than one second party and they all got it right the first time, but we repeated the exercise to be sure. On the Viking mission, we repeated simulated navigation operations many times until we got it right. During mission operations, the NEAR navigation team determined many orbits and computed many maneuvers and made no significant mistakes, which proves that practice makes perfect.

8.3 Maneuver Targeting

Once the orbit has been determined, it may be necessary to perform a propulsive maneuver to steer the spacecraft back on course. Since both the position and velocity are in error, requiring the correction of six components of position and velocity, two propulsive maneuvers may be required. A propulsive maneuver can correct only three components.

8.3.1 Interplanetary Maneuvers

When the spacecraft is far from the target body, the position and velocity errors are generally relatively small and can easily be corrected to put the spacecraft back on the nominal trajectory. However, this strategy is not optimum with regard to fuel usage. The fuel optimum strategy is to propagate the trajectory and determine the position error with respect to the target body. This position error is determined in B plane coordinates of $B{\cdot}R$, $B{\cdot}T$, and time of arrival. A single maneuver is targeted to correct the position error at the target body. The arrival velocity is permitted to float. Correcting position at the target body will also tend to correct velocity. If this is not true, then two maneuvers will need to be performed. The magnitude of the velocity correction at the time of the planned maneuver may be very small. The magnitude of the velocity correction maneuver increases inversely proportional to time of flight to the target body. If encounter time is 2 years in the future, then delaying the propulsive maneuver by 1 year will double the required velocity change. This may be a small penalty to pay for minimizing the number of maneuvers. The propulsive trajectory correction maneuver may be delayed and combined with a deterministic maneuver.

8.3.2 In Orbit Maneuvers

If the mission involves inserting the spacecraft into orbit about the target body, the maneuver strategy for interplanetary maneuvers must be modified a little. One exception is orbits about Jupiter where the approach to the Galilean satellites behaves more like an interplanetary trajectory. The target parameter set becomes three orbit elements. The use of orbit elements in this context should not be confused with the use of orbit elements for patched conic trajectory propagation. The osculating orbit elements computed at a specific time in an orbit provide a measure of energy and angular momentum that is as accurate as the input state vector (see Sect. 3.1.1). Certain distances, angular orientation, and period are close to their actual values, if computed at the right place in the orbit. The input state vector is obtained from a high-precision integrated trajectory. In orbit, the principle that a single maneuver can correct the orbit is tested. Two maneuvers are often required, but, since there are generally many more deterministic maneuvers, the second maneuver can be combined.

8.3.3 K Matrix

The K matrix relates propulsive maneuver velocity change to the change in position at the target body. When the spacecraft is far from the target body, correcting position errors at the target body also tends to correct velocity errors. If the spacecraft trajectory goes from the Earth to the target planet, conservation of energy will result in the spacecraft arriving with the designed velocity. Since velocity errors are small, analysis of interplanetary trajectory errors can be performed by mapping the K matrix to the target and the inverse of the K matrix back to the position of the spacecraft where a propulsive maneuver is executed. The K matrix is a covariant tensor and its inverse is contravariant.

For the K matrix, the position at the target is defined by B plane parameters $(B \cdot R, B \cdot T, t_l)$. These parameters may be directly related to the hyperbolic orbit elements defined in Sect. 3.1.2.

$$B \cdot R = b \cos(\theta)$$

$$B \cdot T = b \sin(\theta)$$

$$t_l = t_p - \sqrt{\frac{a^3}{GM}} \ln(e)$$

The parameter t_l is called linearized flight time. As a spacecraft approaches a planet, it is accelerated by the gravity of the planet and the time of periapsis is a function of B, the magnitude of the **B** vector. In order to remove this dependency, t_l is defined as the time of arrival if the planet had no mass. The modified B plane parameters are thus

$$\mathbf{O}_h = [B \cdot R, B \cdot T, t_l, V_\infty, \alpha_\infty, \delta_\infty]$$

The state transition matrix defines the mapping from a maneuver to the target body in Cartesian coordinates. The state transition matrix may be determined by finite difference of conic propagation of the trajectory or directly by integration of the translational variational equations. The Cartesian state at the target body is transformed to B plane parameters by multiplying by the local Jacobian or partial derivative matrix of B plane parameters with respect to Cartesian state.

$$B = \frac{\partial \mathbf{O}_h}{\partial \mathbf{X}(t)} \frac{\partial \mathbf{X}(t)}{\partial \mathbf{X}(t_0)} = \begin{bmatrix} \frac{\partial(B \cdot R, B \cdot T, t_l)}{\partial(x_0, y_0, z_0)} & \frac{\partial(B \cdot R, B \cdot T, t_l)}{\partial(\dot{x}_0, \dot{y}_0, \dot{z}_0)} \\ \frac{\partial(V_\infty, \alpha_\infty, \delta_\infty)}{\partial(x_0, y_0, z_0)} & \frac{\partial(V_\infty, \alpha_\infty, \delta_\infty)}{\partial(\dot{x}_0, \dot{y}_0, \dot{z}_0)} \end{bmatrix}$$

The K matrix is the upper right 3 by 3 partition of the 6 by 6 transformed state transition matrix.

$$K = \frac{\partial(B \cdot R, B \cdot T, t_l)}{\partial(\dot{x}_0, \dot{y}_0, \dot{z}_0)}$$

For the first 15 years of the space program, interplanetary maneuver analysis involved B planes and K matrices almost exclusively. For example to compute $\mathbf{\Delta}$V, we obtain the miss in B plane parameters ($\mathbf{\Delta}\,\mathbf{O}_k$) and multiply by K inverse,

$$\mathbf{\Delta}\,\mathbf{V} = K^{-1}\,\mathbf{\Delta}\,\mathbf{O}_k$$

where $\mathbf{\Delta}\,\mathbf{O}_k = (\Delta B \cdot R, \Delta B \cdot T, \Delta t_l)$

K matrices were also used to map orbit determination errors to the B plane. Mission design and science objectives could also be mapped to the B plane. The capture radius of a planet and the region in the B plane where occultation occurs can be plotted along with the orbit determination error ellipse, and the probability of impact or occultation can be computed. Another useful application of K matrices is analysis of singularities. The state transition matrix can never be singular, but K matrices are singular for 180–360° transfers. A Hohmann transfer, which is 180°, will pass through the line of nodes connecting the Earth and the target body. The mapping of orbit determination errors to the line of nodes on the far side of the body will be positive semi-definite. For 360° transfers, such as the MESSENGER mission returning to the Earth, the spacecraft will return to the same point in its orbit about the Sun.

As the space program moved on to orbiting and landing on various bodies, interest in B planes waned. Everyone grew tired of looking at B planes. However, the concept was adapted for analysis of orbits about planets and asteroids. The local Jacobian of B-plane elements was replaced by a local Jacobian of elliptical orbit elements, and the same methods that were useful for interplanetary maneuver analysis were adapted to in orbit and landing analysis.

8.4 Summary

Navigation operations are conducted by a navigation team in a spaceflight operations facility. Data is received from the spacecraft at a DSN tracking station, formatted into files, and transmitted to the navigation team. The navigation team receives the tracking data and telemetry from the DSN and also receives files containing calibration data from a variety of sources. These files are processed in navigation software to obtain solutions for the spacecraft trajectory and other model parameters. Examples of model parameters are gravity harmonic coefficients, propulsive maneuver thrust or velocity change, and solar pressure acceleration coefficients. The orbit determination solutions are written to files and distributed to the science team, spacecraft team, and DSN. The trajectory is propagated to the target, and propulsive maneuver components are computed to correct the trajectory. The propulsive maneuvers are forwarded to the spacecraft team, and maneuver commands are formatted and transmitted to the spacecraft via the DSN.

During flight operations, the DSN, spacecraft team, science team, and navigation team work independently and are not generally colocated. Prior to flight operations, the format and content of the files that are communicated are agreed upon. It is important to get the file interface correct because of the high degree of compartmentalization of the participants. During the navigation design, it is necessary for the participating teams to work together. However, the compartmentalization often carries over to the design of the navigation system and the resulting design is often compromised due to lack of communication.

Chapter 9
Navigation Analysis

Navigation analysis is performed to aid mission design and to verify the veracity of the navigation system. The latter can be separated into three time ordered phases being pre-flight, during mission operations, and post-flight. Pre-flight analysis is probably the most important. It involves imagining problems that the navigation system could have during flight and determining the performance of the proposed navigation system. These problems are resolved and the spacecraft and navigation system design can be modified as appropriate. After launch, it is too late to alter the spacecraft design. During mission operations the spacecraft is generally in the cruise mode for long periods of time. Navigation analysis continues as before launch with an emphasis on navigation operations. There is not much interest in discovering during cruise that the spacecraft design cannot achieve mission success. Another source of problems arise from unexpected performance of the navigation system. These problems can be mission catastrophic. However, sometimes there is enough time to solve the problem and salvage the mission. This generally only happens if the problem is simple to solve. During mission operations, a computer solution that takes a week to implement will probably be too late. Post-flight analysis could be useful for problems that may arise during future mission operations. However, funds are generally not available if the mission was determined to be a success.

Often the navigation analyses described below were performed before the spacecraft was launched and the actual spacecraft and mission were not the same as analyzed preflight. These differences are usually minor when the analyses are for the original mission design. All of the navigation analyses described below were for missions that were actually flown. In the sections that follow, problems that occurred during navigation design or navigation operations are defined and analyzed in detail. Some of these problems are of interest to navigators and some were useful for the design and implementation of the navigation system. Most of navigation analysis is concerned with problems associated with a potential failure of the navigation system. Since the probability of failure is small, most of the failure modes analyzed have never occurred. However, these low probability failure modes are often the

J. Miller, *Planetary Spacecraft Navigation*, Space Technology Library 37,
https://doi.org/10.1007/978-3-319-78916-3_9

most interesting from an analysis point of view. The analyses described below are a small subset of all the analyses performed. The objective is to cover a wide range of the type of problems encountered with emphasis on the first time the problem was encountered.

9.1 Viking

Two Viking spacecraft were launched on separate Titan/Centaur rockets on August 20,1975 and September 9, 1975 from Cape Canaveral on a mission to explore Mars and determine the possible existence of life on Mars. The spacecraft consisted of an orbiter and sterilized lander capsule. Viking 1 was inserted into Mar's orbit on June 19,1976 and the lander touched down on July 20,1976 at the Chryse Planitia. Viking 2 followed and was inserted into Mars orbit on August 7, 1976 and the lander touched down on September 3, 1979 at the Utopia Planitia. For several years the Viking orbiters and landers mapped the Mars topography, searched for life, and analyzed the geology.

9.1.1 Planetary Quarantine

Since the primary purpose of the mission was to determine the existence of life on Mars, a major concern was the possible introduction of life from Earth and contamination of the planet. The purpose of planetary quarantine analysis was to guarantee that the probability of contamination was below an agreed upon probability. The agreements were international and were taken very seriously. The lander was sterilized at considerable cost to the project. Navigation was concerned with the probability that large objects like the orbiter, Centaur and various space junk, like the shroud, would introduce microbes from Earth to the Mars surface. The probability analysis consisted of sub allocating probabilities to the various sources of contamination. The probability equation is similar to Drake's equation for determining the possible existence of life in the universe. Navigation was allocated a probability of 3.027×10^{-5}. This navigation suballocation was further suballocated to the mission phases. In addition, a fuel allocation that was enough to change the velocity of the orbiter and lander by a total of 41 m/s was allocated. The suballocation of the probability and fuel navigation allocation to the various mission phases is given in Table 9.1.

When the spacecraft is injected into its trans-Mars trajectory, the Centaur boost vehicle and associated hardware follow along and also make the trip to Mars. These objects must be aimed far away from the desired target in order to avoid impacting Mars. The injection aim point is about 400,000 km away from Mars or about as far away as the Moon is from Earth. Figure 9.1 shows the geometry drawn to scale for Viking 2. At this scale Mars appears as a small circle as would the Earth and

Table 9.1 Planetary quarantine large impactables allocation

Trajectory phase	Probability sub allocation	ΔV suballocation
Injection	0.227×10^{-5}	7.5 m/s
Midcourse	0.200×10^{-5}	5.0 m/s
MOI	2.500×10^{-5}	28.5 m/s
Orbit trims	0.100×10^{-5}	Nill

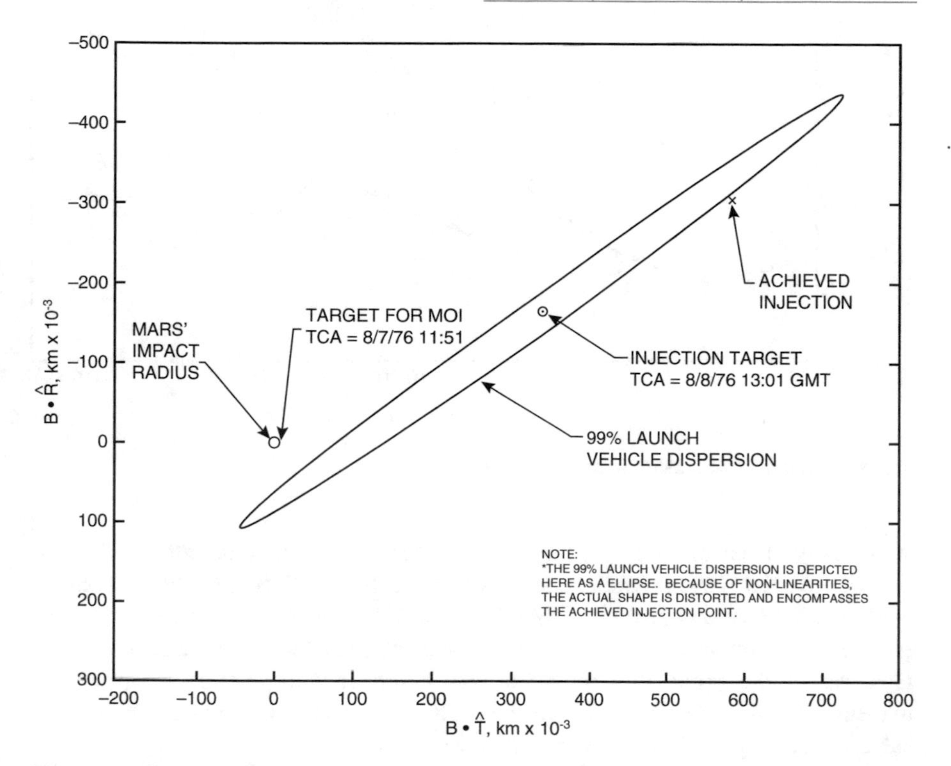

Fig. 9.1 Viking 2 injection dispersion

Moon drawn to the same scale. The ellipse, which is actually a cross section of a triaxial ellipsoid, defines the region where 99% of Earth injections would arrive or the probability is 99% that the spacecraft is inside the ellipse. Observe that the achieved injection is outside the 99% ellipse. This would imply that we had a bad injection. However, careful analysis revealed that, due to nonlinearity, the ellipse shown on the figure is actually bent like a banana. The achieved injection was actually inside the real 99% contour.

The first midcourse maneuver, which is not performed midcourse but near the Earth, delivers the spacecraft to the point in the B-plane shown in Fig. 9.2. Modern terminology refers to maneuvers performed during the interplanetary phase of the mission as Trajectory Correction Maneuvers (TCMs). The achieved trajectory is

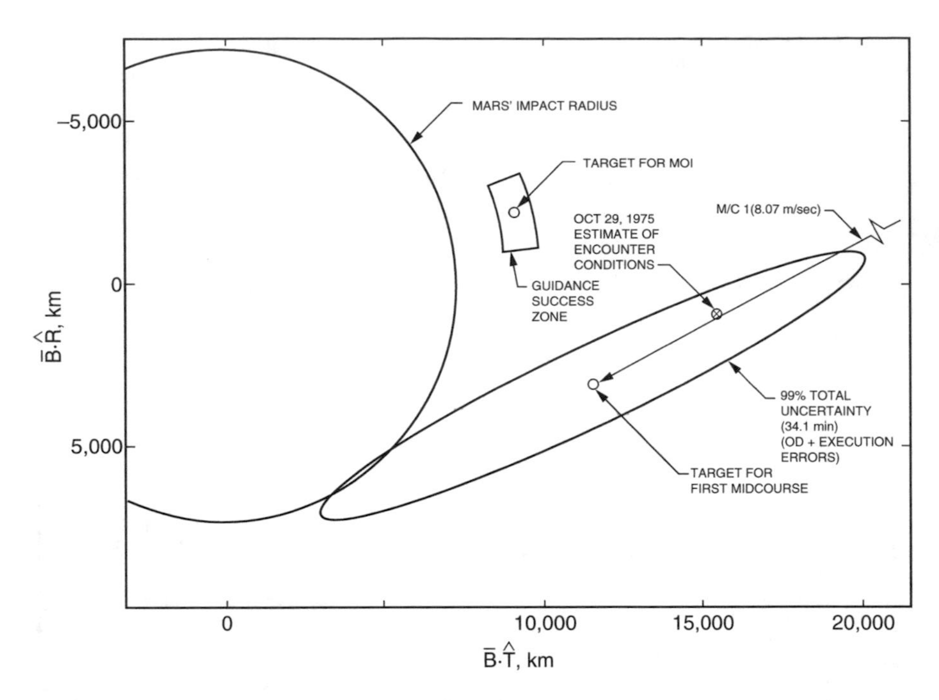

Fig. 9.2 Viking 2 first midcourse maneuver dispersion

statistically closer to the target than at injection and this may be attributed to the luck of the draw. The first midcourse was also biased to satisfy planetary quarantine per the suballocations given in Table 9.1. It would appear from the figure that the probability of impacting Mars is greater than the planetary quarantine allocation. However, if the probability of the orbiter being able to perform a subsequent maneuver to move the spacecraft off the capture circle is included in the probability calculation, the probability is within the allocation.

Near Mars, the mapping of maneuver execution errors is much smaller and planetary quarantine constraints are easier to satisfy. Once in orbit, the orbit trim maneuvers are too small to result in impact. Orbit lifetime may result in eventual orbit decay and impact. This suballocation is considered separately and not included in the navigation suballocation. The strategy for the planetary quarantine bias required for MOI is included in the MOI maneuver design discussed below.

9.1.2 Orbit Insertion Maneuver Design

During approach to Mars, the Viking orbiter and lander was maneuvered into a trajectory that provided the optimum initial condition for a large propulsion motor burn that inserted the orbiter and lander into an orbit about Mars. The design of the approach trajectory and orbit insertion burn involved finding the ignition time, magnitude, and direction of two propulsion maneuvers. The constraints on the trajectory design included the period of the inserted orbit, periapsis altitude, latitude of the landing site and sun elevation angle at the landing site. Since the number of control parameters related to the propulsive motor burns exceeded the number of mission constraints, a solution could be found that satisfied the constraints and minimized a performance criterion. The performance criterion was the total amount of fuel consumed. A numerical solution to a classical constrained parameter optimization problem was needed. A constrained parameter optimization algorithm was devised by the author for this purpose and is referred to as the method of explicit functions described earlier in Chap. 4.

The first problem was to optimize the final approach maneuver in conjunction with the orbit insertion burn. The control parameters are given by the propulsive maneuver components $\Delta\mathbf{V}_1$, $\Delta\mathbf{V}_2$ and the ignition time. The ΔV maneuver components are the integral of the thrust over the finite burn time of about 40 min. The candidate constraint parameters are period of the post MOI orbit (P_0), the periapsis altitude (h_p), the latitude of a point in the orbit, referred to as the PER point,ϕ_{PER}, the longitude of the sub PER point at the time of touchdown (θ_{PER}) and the sun elevation angle of the landing site at touchdown (SEA). The PER point is at a fixed true anomaly on the separation orbit and is directly over the lander at touchdown. The performance criterion is

$$J = |\Delta\mathbf{V}_1| + |\Delta\mathbf{V}_2|$$

The constraint on periapsis altitude is necessary to prevent the optimization algorithm from collapsing onto the planet surface. The optimum solution involves targeting to as low a periapsis as possible and then raising periapsis altitude by doing a maneuver at apoapsis. The periapsis altitude was constrained to be 1500 km to avoid hitting Mars. At that time, radiometric orbit determination was several hundred kilometers. During the actual approach to Mars, optical data was able to determine the orbit to within 25 km however, optical data was not accepted as a primary data source and was regarded as a backup.

After the final approach maneuver was executed, the spacecraft continued about 10 days to encounter with Mars. During this time additional radiometric tracking revealed that the spacecraft was not on course. The orbit insertion burn was thus adjusted to satisfy the important mission constraints. For this optimization, the periapsis altitude was not a target constraint. The desired 1500 km altitude could not be achieved or would require too much fuel. The MOI control parameters, constraint parameters, and performance criterion were reduced to

$$U = (t_2,\ \Delta\mathbf{V}_2)$$

$$\Psi_c = (P_0,\ \phi_{PER})$$

$$J = |\Delta\mathbf{V}_2|$$

and the other parameters were permitted to float. Of course, periapsis altitude could not be permitted to float too far from the nominal value of 1500 km. If the tracking data revealed that the spacecraft was getting too close to Mars or too far away, an emergency propulsive maneuver would be executed to restore the target periapsis altitude.

Another approach to partially restore h_p to its target is to adjust the performance criterion to enable some propellant to be expended to raise or lower h_p. Consider the following modification.

$$J = |\Delta \mathbf{V_2}| + G\ |h_p - 1500.|$$

The gain G would achieve this purpose. A value for G may be obtained from the trajectory optimization performed for the last maneuver targeting. The Lagrange multiplier associated with h_p may be obtained as a by-product of the optimization.

$$\lambda_{hp} = -\frac{\partial J}{\partial h_p}$$

The Lagrange multiplier provides a measure of the cost of maintaining h_p at 1500 km. Recall that the optimum solution would allow h_p to be much lower. If we set $G = -\lambda_{hp}$, the optimization algorithm will partially restore h_p to 1500 km at a cost that has already been committed.

9.2 Galileo

The Galileo spacecraft was launched to Jupiter on October 18, 1989. An atmospheric entry probe was released 150 days prior to Jupiter encounter. Following probe release, the orbiter portion of the spacecraft was deflected for a close flyby of the Galilean satellite Io. The orbiter was then configured for recording and relay to Earth of probe entry data. Immediately following relay of probe data, the orbiter was reconfigured for a Jupiter Orbit Insertion (JOI) motor burn. The orbiter was then inserted into a highly eccentric 200-day orbit about Jupiter. At apojove, a large motor burn raised the perijove radius to a less severe radiation region. The orbiter then began a series of close encounters of the Galilean satellites Europa, Ganymede, and Callisto.

9.2.1 Probe Delivery to Jupiter

The first major orbit determination activity during the Jupiter approach phase is determination of the spacecraft trajectory so that Trajectory Correction Maneuvers (TCMs) can place the probe on the proper Jupiter atmosphere entry trajectory. This process combines trajectory state estimates with the nominal probe-orbiter separation velocity to determine estimates of six Jupiter-relative entry parameters.

The six entry parameters are latitude, longitude, speed, heading angle, time of entry, and entry flight path angle. Entry is at a defined altitude of 450 km above the reference ellipsoid with an equatorial radius of 71,398 km and flattening of 0.065. This ellipsoidal surface is assumed to represent the l-bar pressure level in the Jovian atmosphere. The probe was not expected to encounter perceptible atmosphere until about 100 km below the reference 450 km point, but use of a fixed reference is convenient because of uncertainties in the actual entry point.

During the 2-year interplanetary cruise from Earth to Jupiter, the spacecraft is tracked to monitor the effect of nongravitational accelerations such as solar pressure, gas leaks, and attitude control thruster imbalance. As the spacecraft approaches Jupiter's sphere of influence, the orbit determination error was predicted to be about 500 km. This total error represents a statistical combination of the Jupiter ephemeris error and the 0.25 μrad tracking error characteristic of Doppler tracking.

The probe is released from the orbiter 150 days prior to entry. Control of the six entry parameters is dependent on the initial position error and the mapping of separation velocity errors to entry. Errors in solar pressure and probe outgassing modeling have little effect on the predicted probe trajectory.

9.2.2 *Gravity Focusing*

The mapping of probe position errors from probe separation to entry is best described in an orthogonal rotating frame with the X axis parallel to the velocity vector (the downtrack direction), the Z axis normal to the plane-of-motion (the out-of-plane direction), and the Y axis normal to the velocity vector and in the plane-of-motion (the crosstrack direction). The mapping of a spherical position error in downtrack and crosstrack from far out on the approach asymptote to entry is shown schematically in Fig. 9.3. Observe that the downtrack error increases as the probe

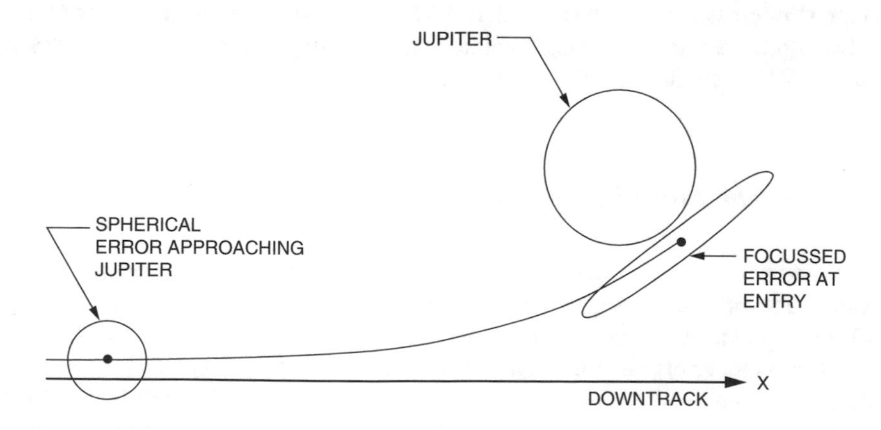

Fig. 9.3 Gravity focusing

approaches Jupiter (defocuses) whereas the crosstrack error decreases (focuses). The out-of-plane error focusing is similar to the crosstrack error focusing.

The exact relationships describing gravity focusing of the orbiter state (X, Y, Z) with respect to the initial state at minus infinity $(X_\infty, Y_\infty, Z_\infty)$ have been derived by Kent Russell and are given by

$$\frac{\partial X}{\partial X_\infty} = \left[1 - \frac{2}{e\ \cosh(F) - 1}\right]^{\frac{1}{2}}$$

$$\frac{\partial X}{\partial Y_\infty} = \left(\frac{-2(e^2-1)^{\frac{1}{2}}}{e}\right)\left[\frac{\exp(F)}{(e^2\cosh^2(F)-1)^{\frac{1}{2}}}\right]$$

$$\frac{\partial Y}{\partial Y_\infty} = \left(1 + \frac{\exp(F)}{e}\right)\left[\frac{e\cosh(F)-1}{e\cosh(F)+1}\right]^{\frac{1}{2}}$$

$$\frac{\partial Z}{\partial z_\infty} = \left(1 - \frac{\exp(F)}{e}\right)$$

where the hyperbolic eccentric anomaly (F) is defined by its relation to the mean anomaly (M) as

$$M = \frac{V_\infty^3}{GM}(t - t_{ca}) = e\sinh F - F$$

where V_∞ is the hyperbolic excess velocity, e is the orbiter eccentricity, t_{ca} is the time of closest approach, and GM is Jupiter's gravity. For the Galileo probe, the approach hyperbola has the orbit elements $V_\infty = 5.86$ km/s, $e = 1.0193$, and entry occurs 276 s before periapsis. From Russell's equations, the crosstrack focusing is 0.19, the out-of-plane focusing is 0.04, and the downtrack defocusing is 10.0. The coupling of crosstrack into downtrack is 1.91. For a 500 km spherical delivery error on the approach asymptote, the downtrack error at entry is 5080 km, the crosstrack error is 98 km, and the out-of-plane error is 20 km.

9.2.3 *Probe Entry Dispersions*

A detailed covariance analysis was performed to determine the dispersions of Jupiter relative atmosphere entry parameters. Computer simulations of data scheduling, trajectory propagation, data filtering, and mapping resulted in the probe delivery entry parameter errors given in Table 9.2. The effect of gravity focusing is somewhat obscured by Jupiter's rotation rate. Since the probe enters near Jupiter's equator at a heading angle that is about due East, the latitude dispersion is predominantly caused by the out-of-plane trajectory error. Similarly, the flight path angle error is related

Table 9.2 Predicted probe delivery entry parameter errors

Entry parameter	Error (99%)
Latitude	0.05°
Longitude	2.7°
Time-of-entry	256 s
Speed	0.035 km/s
Heading angle	0.16°
Flight path angle	0.94°

to the crosstrack error. The downtrack error maps into the time-or-entry error. The longitude error is related to the time-of-entry error by Jupiter's rotation rate.

9.2.4 Probe Entry Flight Path Angle

Probe entry flight path angle, the angle between the relative velocity vector and local horizontal plane, is the most critical entry parameter. It was required to deliver the probe to an entry angle corridor with 99% probability between $-7.2°$ and $-10.0°$. The upper limit is related to skip out of the Jovian atmosphere and the lower limit is related to structural limitations. For the purpose of analysis, it is convenient to relate entry parameters to orbit determination errors in the B-plane coordinate system shown in Fig. 3.4 which has its T-axis parallel to the Jupiter equatorial plane. Since the Jupiter equatorial plane, the ecliptic plane, and the trajectory plane are nearly coplanar, $\mathrm{B} \cdot T$ is essentially in the trajectory plane and $\mathrm{B} \cdot \mathrm{R}$ is essentially perpendicular to the trajectory plane. On the approach asymptote, $\mathrm{B} \cdot T$ is nearly in the crosstrack direction defined above and $\mathrm{B} \cdot \mathrm{R}$ is nearly in the out-of-plane direction. The magnitude of the B-vector is the hyperbolic impact parameter. The S-vector shown in Fig. 3.4 is the unit vector in the direction of the approach asymptote. It is in the direction of the hyperbolic excess velocity (V_∞) and is in the downtrack direction on the approach asymptote.

The entry flight path angle is most strongly dependent on the encounter impact parameter (B) which is a measure of the "miss" of the approach hyperbola asymptote with respect to the center of the planet. The relationship between the impact parameter (B) and the inertial entry angle γ_I is given by

$$\cos \gamma_I = \frac{B}{\sqrt{r_e \left(\frac{2GM}{V_\infty^2} + r_e \right)}}$$

An approximate relationship for the relative flight path angle, which assumes 90° heading, is given by

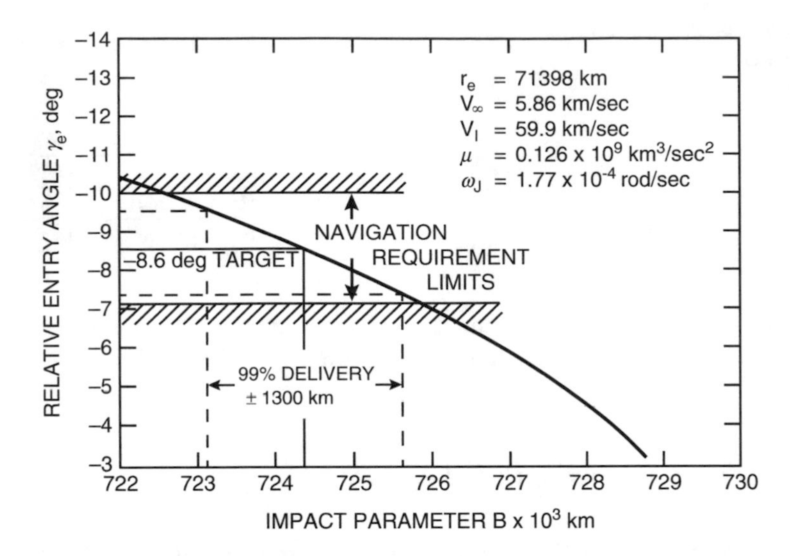

Fig. 9.4 Probe entry angle delivery

$$\tan \gamma_e = \frac{\sin \gamma_I}{\cos \gamma_I - \frac{\omega_J r_e}{V_I}}$$

where ω_J is Jupiter's rotation rate, V_I is the inertial entry velocity, and r_e is the reference entry radius.

In terms of B, expected delivery error to Jupiter was approximately 1300 km (99%). This delivery error is dominated by the Jupiter ephemeris error. The effect of this error on entry angle delivery is shown in Fig. 9.4. The relative entry angle is plotted as a function of the hyperbolic impact parameter. Superimposed on the abscissa is the targeted B and the 99% orbit determination delivery error. Projected onto the ordinate is the 99% relative entry angle dispersion. The entry angle delivery error is shown to be about 1.1°. The margin indicated by these results is small but adequate.

9.2.5 *Probe Entry Angle-of-Attack*

Another aspect of probe delivery is control of the entry angle-of- attack which is the angle between the relative velocity vector at entry and the probe spin axis. Since the probe has no active attitude control, it is necessary to deploy the probe in an attitude such that it will enter the perceptible Jovian atmosphere at near zero angle-of- attack. Since the plane of the orbit is near Jupiter's equatorial plane, the angle-of-attack dispersion is the difference between the relative entry angle and true anomaly or inertial longitude dispersion at entry. The 99% angle-of-attack delivery caused by trajectory errors is about 0.6°. This is well within the Galileo project requirement of 4.5°.

9.2.6 *Trajectory Bending*

As the orbiter approaches Jupiter, the gravitational acceleration causes the orbiter to deviate from the approach asymptote. The resultant bending of the trajectory can be measured with $\Delta VLBI$ and range data and the position of the orbiter relative to Jupiter may be inferred. The trajectory of the orbiter may be obtained from solution of

$$\mathbf{r} = \mathbf{r}_0 + \mathbf{V}_\infty(t - t_0) + \iint \frac{GM}{r^3}\mathbf{r}\,dtdt$$

For orbit determination, the partial derivatives of the observable with respect to the estimated parameters are needed. In the simplified analysis presented here, the observable is orbiter position (X,Y,Z) and the Jupiter position (X_p, Y_p, Z_p) is estimated. When the orbiter is far out on the approach asymptote, these partial derivatives may be approximated as follows:

$$\frac{\partial X}{\partial X_p} = \iint \frac{-2GM}{r^3}\,dtdt$$

$$\frac{\partial Y}{\partial Y_p} = \iint \frac{GM}{r^3}\,dtdt$$

$$\frac{\partial Z}{\partial Z_p} = \iint \frac{GM}{r^3}\,dtdt$$

where

$$r = X_p - X$$

In the limit, the above integrals may be evaluated by making the assumption that velocity (V_∞) remains constant on the approach asymptote and by performing a change of variable from time to r.

$$\lim_{r\to-\infty} \frac{\partial X}{\partial X_p} = \frac{-2GM}{V_\infty^2}\frac{1}{2r} = \frac{-2\exp(F)}{e}$$

$$\lim_{r\to-\infty} \frac{\partial Y}{\partial Y_p} = \frac{GM}{V_\infty^2}\frac{1}{2r} = \frac{\exp(F)}{e}$$

$$\lim_{r\to-\infty} \frac{\partial Z}{\partial Z_p} = \frac{GM}{V_\infty^2}\frac{1}{2r} = \frac{\exp(F)}{e}$$

where

$$r = a(e\cosh(F) - 1)$$

and

$$\lim_{F \to -\infty} \cosh(F) = \frac{\exp(-F)}{2}$$

The above result may also be derived from Russell's equations by taking the limit as F approaches minus infinity. An interesting observation is that on the approach asymptote, the crosstrack and out-of-plane partial derivatives are equal and the downtrack partial derivative is greater by a factor of two. Recall that the crosstrack, out-of-plane, and downtrack directions are nearly coincident with the B·T, B·R, and time-of-flight or S directions, respectively. This implies that the Jupiter approach orbit determination B · R and B · T errors is equal and the S orbit determination error is one half of the B · R and B · T errors. The orbit determination error is proportional to the reciprocal of the measurement sensitivity. Of course these idealized results provide only an approximate insight into the actual errors.

9.2.7 Jupiter Approach Orbit Determination

The Jupiter approach orbit determination error as a function of time from encounter is shown in Fig. 9.5. The orbit determination error is mapped to the Jupiter B-plane. Prior to Encounter (E) minus 40 days, the B-plane errors are dominated by Jupiter's ephemeris error. Starting at about E-40 days, the ΔVLBI measurements begin to sense the gravitational acceleration of Jupiter. The B · R and B · T errors are approximately equal as predicted by the above analysis. The S error is predicted to be less than B · R and B · T errors by a factor of two. The time-of-flight error (TL), which is equivalent to the S error divided by V_∞, is greater than predicted by the above simplified analysis until about E-5 days. However, the qualitative agreement with the results of detailed computer simulations that include optical and Doppler data is good. At about E-5 days, the B · R error levels off at around 25 km. This may be attributed to a systematic optical center finding error of 1% of Io's radius. Near Jupiter periapsis, the orbit determination error decreases rapidly because of the strong orbit dynamics signature in the Doppler data.

The same orbit determination error is shown on the right side of Fig. 9.5 mapped to Io closest approach. The in-plane B · T and TL components tend to follow the Jupiter relative TL component shown on the left side of Fig. 9.5. The ratios of Io relative B.T and TL errors to the Jupiter relative TL error may be computed from the orbiter and Io velocities. The orbiter trajectory crosses Io's orbit at an angle of 36°. From the geometry, the ratios of Io-relative B · T and TL errors to the Jupiter relative TL error are 16 and 1.25, respectively. This result may also be obtained from

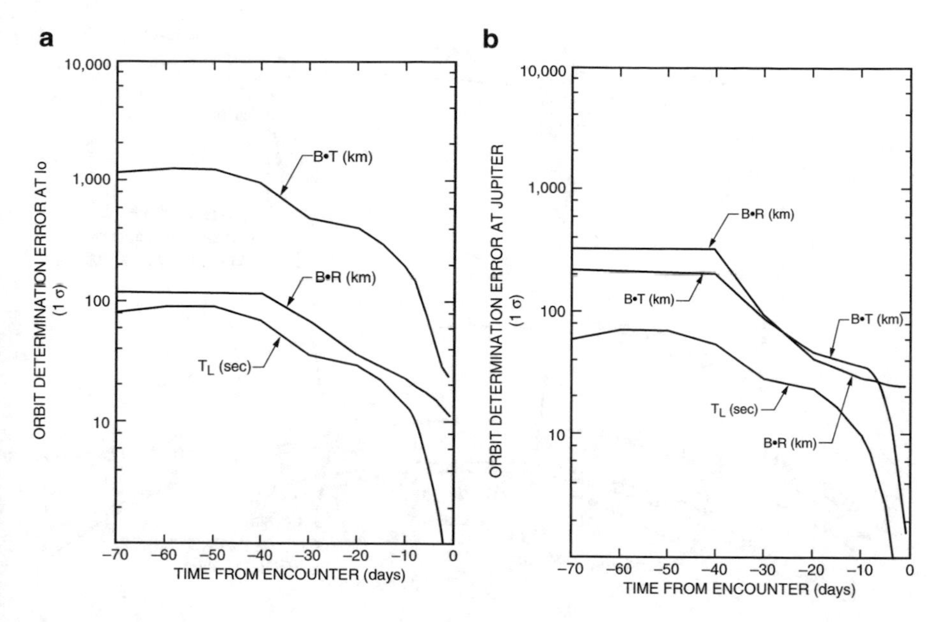

Fig. 9.5 Jupiter approach orbit determination errors

the gravity focusing formulas. The Io-relative B.R component shown in Fig. 9.5 is much better determined because of the effect of Jupiter gravity focusing. The orbit estimate at E-5 days is used for the TCM at E-3 days that delivers the orbiter to Io.

9.2.8 Relay Link

The Galileo mission probe-to-orbiter communications relay link is illustrated schematically in Fig. 9.6. The link begins with transmission of coded science and engineering data from a transmitter within the probe descent module through a relatively broad-beam antenna fixed to the aft end of the probe so that normally its axis is oriented along the local vertical. The signal is received by the relatively narrow-beam Relay Radio Antenna (RRA) on-board the orbiter and retransmitted in real time to the DSN through the orbiter high gain antenna. A few days after probe separation, the RRA is deployed to a position such that the antenna axis will be pointing in the direction of the predicted probe location approximately 20 min after probe entry. This optimizes relay link performance during the first 30 min of the relay. During the latter half of the relay, the RRA is repointed several times so as to minimize the RRA aspect angle. Ten days before probe entry, the RRA pointing is updated in accordance with the improved orbit determination.

The dynamic behavior of the link geometry as a function of time past entry is also illustrated in Fig. 9.6. Note that at signal acquisition, the signal enters the RRA

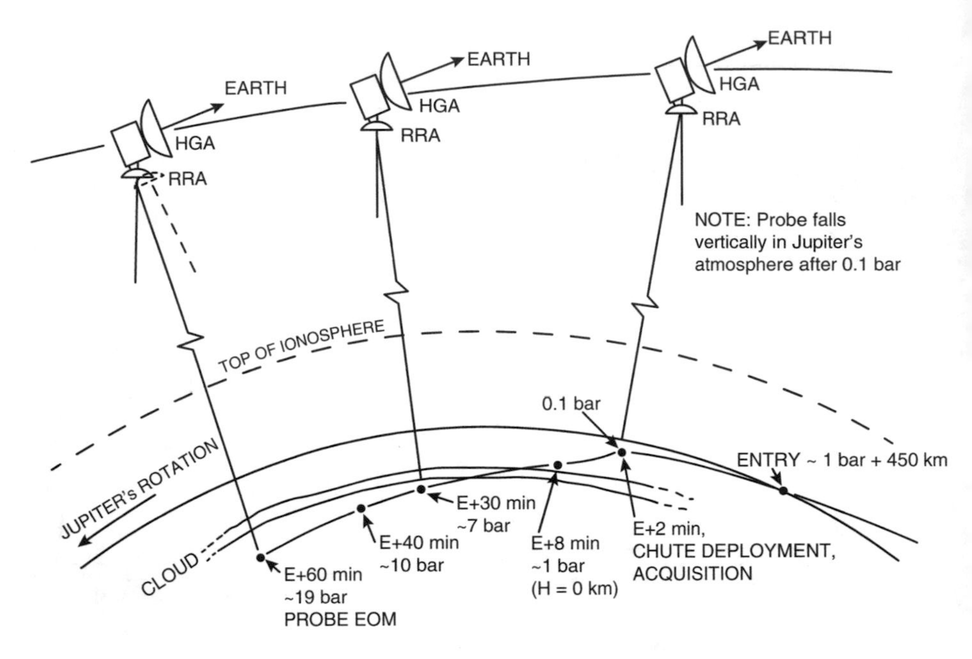

Fig. 9.6 Relay link geometry

to the left of the RRA axis, while 30 min later, it enters to the right of the axis. If the RRA remains inertially pointed, the signal arrives far off the RRA axis at the end of probe mission. The dashed line represents a repointing of the RRA to improve the relay link performance in the later portion of the mission.

Four pertinent antenna angles are defined in Fig. 9.7. The reference direction for the probe aspect angle is the local vertical. The orbiter aspect angle is referenced to the +Z-axis of the orbiter. The direction of the axis of the RRA is also referenced to th orbiter Z-axis. Then, the operating point in the RRA pattern is defined by the difference between these two angles.

These angles may all be calculated at any time during the probe mission, based on reference trajectories for the orbiter and the probe, allowing the RRA and probe antenna gains to be calculated. However, the values for probe and orbiter antenna aspect angles at any given time will be perturbed by deviations in the trajectories of the two spacecraft and by errors and changes in the attitudes of the two spacecraft. It should be noted that for probe and orbiter trajectories, the relative motion of the probe in the orbiter frame of reference is almost exclusively in the cone-angle direction. The RRA may be articulated in this direction by rotation of the boom on which it is mounted. RRA pointing in the clock or cross-cone direction is controlled by orientation of the de-spun portion of the dual-spin orbiter spacecraft.

Delivery orbit determination errors of the relay link parameters for a data arc ending at the time of probe separation and knowledge orbit determination errors at 10 days before encounter are given in the Table 9.3.

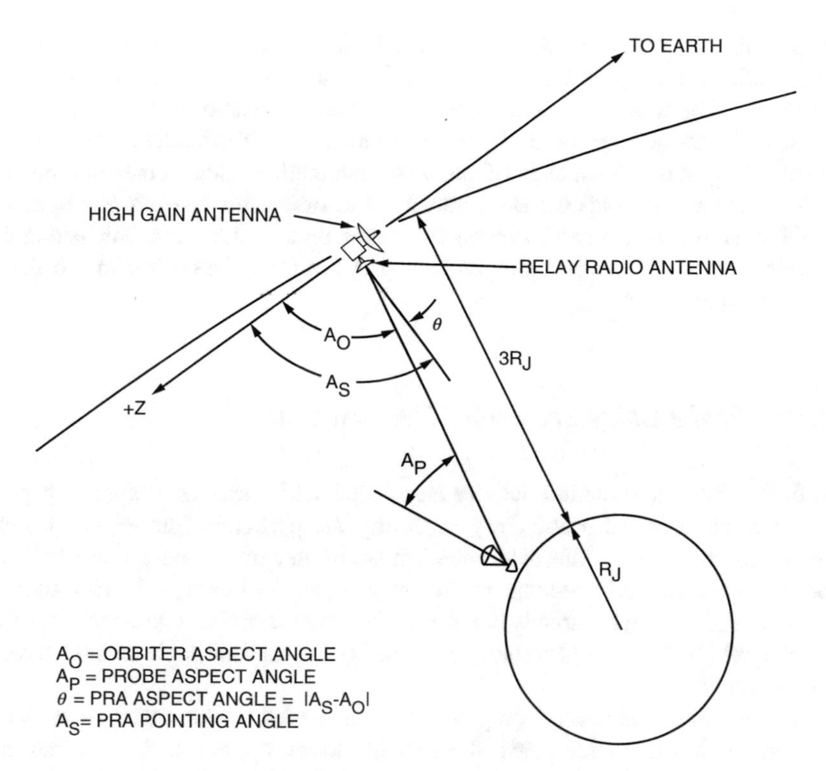

Fig. 9.7 Aspect angle definition

Table 9.3 Predicted probe delivery entry parameter errors

Relay link delivery errors		
Parameter	Requirement error (99%)	Capability error (99%)
Orbiter aspect angle	2.0°	1.8°
Probe aspect angle	3.0°	2.4°
Time-of-entry	480 s	256 s

9.2.9 Jupiter Orbit Insertion

The final event of the Jupiter approach phase is insertion of the orbiter into orbit about Jupiter. Immediately following the relay of probe data, the orbiter is placed into a 200-day period orbit around Jupiter by the Jupiter Orbit Insertion (JOI) maneuver. The initiation of the JOI maneuver is time critical because the relay link ends after Jupiter periapsis and delay results in expenditure of additional propellant to get into orbit. The amount of propellant required for JOI and subsequent orbit trims is also dependent on how accurately the orbiter is delivered to Io and prior knowledge of Io delivery that would permit a late update of the JOI motor burn. A final Jupiter approach TCM is performed 3 days prior to Io encounter. This TCM is based on data taken up to 5 days prior to Io encounter. The Io relative orbit determination error is shown in Fig. 9.5 as a function of time from encounter. At E-5 days the B · T orbit determination error is about 60 km. The post JOI orbit period is most sensitive to the in-plane B · T delivery to Io. The B · T orbit determination error decreases to around 25 km at E-1.5 days. A systematic optical center finding error of 1% of Io's radius holds the B · T and B · R errors at this level. A late update of the JOI maneuver is planned based on data taken up to E-1.5 days. This late update results in a savings of several kilograms of propellant for the orbit trim subsequent to the JOI motor burn.

9.2.10 Probe Entry Trajectory Reconstruction

The final orbit determination activity associated with the Jupiter approach phase is reconstruction of the probe entry trajectory. Of particular interest is the probe entry flight path angle. The determination of Jupiter atmosphere scale height is dependent on accurate knowledge of the entry angle. On-board probe measurement of deceleration cannot accurately determine whether the probe is descending rapidly through a relatively thin atmosphere or entering on a shallow angle through a dense atmosphere.

The entry angle reconstruction procedure consists of tracking the orbiter during the entire Jupiter approach phase through Io closest approach. A smoothed best estimate of spacecraft trajectory state relative to Jupiter is determined at the time of separation. The separation spring impulse is added to the probe velocity and the probe trajectory state is mapped ahead to the reference entry altitude. This state vector and its covariance are provided to the Jupiter atmosphere scientists.

Data types that are used for reconstruction include Doppler, range, $\Delta VLBI$, and optical imaging of Io. Orbiter accelerometer data of the separation impulse is also included. The major error sources that affect entry angle reconstruction are Jupiter's ephemeris, execution errors associated with TCMs, and velocity perturbations of the probe and orbiter that occur during the separation sequence. Radio metric tracking data through Jupiter closest approach provides a powerful solution for Jupiter's ephemeris. When $\Delta VLBI$ data is included, the TCMs both before and after separation including the orbit deflection maneuver may be determined with precision.

The major error sources that affect entry angle reconstruction are velocity perturbations of the probe and orbiter that occur during the separation sequence. During this sequence, the orbiter turns to the attitude required for zero angle-of-attack at entry. A spin-up maneuver is performed to give the probe the required angular momentum to maintain a stable attitude during the 150-day ballistic transit to the Jovian atmosphere. The probe is separated from the orbiter by springs and the orbiter then performs a spin-down and returns to normal cruise attitude.

A feature of the separation sequence is that two-way lock is lost when the orbiter turns off Earth-line. An accurate measurement of separation velocity is obtained from Doppler data. The total orbiter velocity change associated with the turn, spin-up, separation springs, spindown, and return to cruise attitude may be determined to 1 mm/s when the orbiter returns to Earth-line and two-way lock is reestablished. However, the velocity given to the probe cannot be completely separated from the orbiter velocity changes associated with the spin up, spin down, and turns. The probe velocity must be inferred from the observed orbiter velocity change.

Analysis has shown that it is difficult to meet the entry angle reconstruction error requirement of 0.15° (99%) with the sequence defined above. There are several methods being studied to reduce the entry angle reconstruction error. They include use of orbiter accelerometer data, performing an Earth-line separation and entering the Jovian atmosphere at some small angle-of-attack, special calibrations of the attitude control system, and performing a more accurate calibration of the separation springs. Assuming perfect determination of the velocities associated with separation, the probe entry angle could be reconstructed to an accuracy of 0.05° (99%). With worst case estimates of separation velocity errors and no calibrations, the reconstruction error could be as bad as 0.45° (99%). The current best estimate of entry flight path angle reconstruction error is 0.2° (99%). With implementation of some or all of the above methods, the goal is to meet the atmosphere science requirement on entry angle reconstruction.

9.3 Pioneer

In July of 1992, the Pioneer Venus Orbiter (PVO) spacecraft began a series of orbits that entered the Venus atmosphere. At first the orbit periapsis just grazed the upper atmosphere but, because of the perturbing effect of the sun, the periapsis altitude was pushed deeper into the atmosphere. With the limited propellant available, a series of propulsive maneuvers were executed to raise the periapsis altitude and extend the life of the spacecraft.

During the entry phase, navigation is required for support of propulsive maneuvers and entry science. Of particular interest to science is the determination of the velocity change imparted to the spacecraft while in the Venus atmosphere. The velocity change may be directly related to the drag experienced by the spacecraft and hence the atmospheric density. Determination of the spacecraft orbit was extremely hampered by the lack of tracking data. Continued deterioration of solar cells resulted in a critical power shortage that limited the time that the spacecraft transmitter was operated. As a result, tracking coverage was limited to a total of 2 h for each orbit.

A strategy was planned for determining the orbit with the limited data available. The strategy involved determining the velocity change directly and then relating this quantity to the atmospheric drag.

9.3.1 Orbit Determination Strategy

The orbit determination strategy defines the data types and data acquisition required for orbit determination. For PVO, the data type is two way coherent Doppler obtained by tracking the spacecraft from the Deep Space Network. During times of high spacecraft activity, such as during maneuvers or when the spacecraft is perturbed by external forces, it is desired to have continuous tracking coverage. With the solar array degradation being experienced by PVO, the total tracking coverage during the atmospheric entry phase is about 2 h per day. Science playback and commanding for maneuvers require about 30 min of coverage at periapsis and apoapsis. With this as a baseline tracking coverage, a study was undertaken to determine the minimum coverage needed to meet science and mission accuracy requirements. Tracking data was simulated and the strategy for placement of tracking passes varied to determine the optimum tracking coverage. As a general rule, it was discovered that the length of individual tracking passes spaced around the orbit had little effect on orbit determination accuracy. Orbit determination accuracy is primarily determined by the number and geometric placement of the passes. For example, it was found that six tracking passes each 15 min in length and geometrically spaced evenly around the orbit would determine the orbit nearly as well as continuous tracking coverage. However, an hour of tracking at periapsis and another hour in the vicinity of apoapsis does not do nearly as well.

Further investigation of the placement of tracking passes revealed that the number of passes required per revolution could be substantially reduced by processing several consecutive orbits of data. With this strategy, three consecutive orbits containing three periapsis passages would require four tracking passes per revolution to obtain orbit determination accuracy comparable to continuous coverage. The four tracking passes would be 15–30 min in length and placed at periapsis, apoapsis, and roughly geometrically in between at plus and minus 4 h from periapsis.

9.3.2 Orbit Determination Results from 1980

In order to quantify and verify these observations, a series of atmospheric entry passes from 1980 were selected and the data processed in a manner similar to that described above. First a baseline case was run processing all the data available. These were orbits number 503 through 506 extending from April 20 to April 23, 1980. The estimated parameters were spacecraft state, the maneuver velocity components at apoapsis 505, and the magnitude of the velocity change that occurred during each periapsis passage. The atmospheric drag was modeled as a small retro propulsive maneuver in a direction opposite to the nominal direction of the velocity vector at periapsis. The length of the burn was taken to be about 300 s which is approximately the length of time that the spacecraft was in the atmosphere. It will be shown later that the detailed modeling of the atmospheric drag is not critical since the orbit determination filter responds primarily to the total impulse. The filter model also included stochastic acceleration components around the entire orbit and solar radiation pressure.

The results of processing the baseline case for orbit numbers 503 through 506 are shown in Fig. 9.8. The abscissa is time measured in days from periapsis passage 503 which is assigned the day number 503. The ordinate is milli Hertz (mHz) of Doppler residual. Except for the data near periapsis, the fit is good to within 15 mHz which corresponds to a velocity error of 1 mm/s. Analysis of these residuals indicates that the data near periapsis is corrupted by errors in modeling the Venus gravity field. A better fit may be obtained by estimating gravity harmonics or deleting the data near periapsis as has been demonstrated on previous missions.

Here, we have elected to fit through the periapsis data and the results are tabulated below in Table 9.4. Shown is (ΔV) that was estimated at each of the periapsis passages indicated under the heading "Revolution." Atmospheric drag was modeled as a finite motor burn directed opposite to the nominal velocity vector and centered around periapsis with a nominal burn time of 300 s.
The associated orbit period change (ΔP_{base}) was computed from the partial derivative of period with respect to velocity and given by

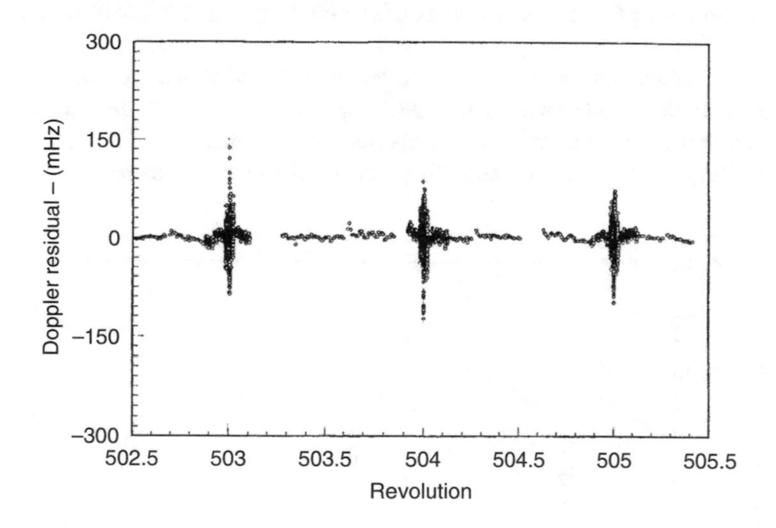

Fig. 9.8 Pioneer orbits 503–505 data residuals—continuous tracking

Table 9.4 Pioneer 1980 orbit determination baseline

Revolution	ΔV (mm/s)	ΔP_{base} (s)	ΔP_{80} (s)
503	1.863	0.573	0.574
504	0.413	0.128	na
505	60.07	18.463	18.522

$$\Delta P_{base} = \frac{\partial P}{\partial v_p} \Delta V$$

$$\frac{\partial P}{\partial v_p} = 6\pi \frac{v_p a^{\frac{5}{2}}}{GM^{\frac{3}{2}}}$$

$$a = \frac{-GM}{h}$$

$$h = v_p^2 - \frac{2GM}{r_p}$$

where v_p is the velocity at periapsis, r_p is the radius of periapsis, and GM is the Venus gravitational constant. For comparison, the period change obtained in 1980 at the time of the actual atmospheric entry is also shown (ΔP_{80}). The period changes agree within 0.06 s even though obtained by somewhat different methods. Subsequent analysis of additional orbit cases has shown general agreement to within 10 ms. The differences may be attributed to small perturbations from the sun and nongravitational accelerations. During the 1992 atmospheric entry phase much less tracking data was available as discussed above. As a test, the sparse data tracking strategy can be applied to the 1980 data. The tracking data residuals are shown in Fig. 9.9.

Since the spacecraft is normally occulted by the planet near periapsis this data is deleted. Included are two 30 min tracking data arcs just before and after the simulated occultation and several 15 min tracking data arcs spaced around the orbit. The resulting $\Delta V's$ and corresponding period changes are shown in Table 9.5.

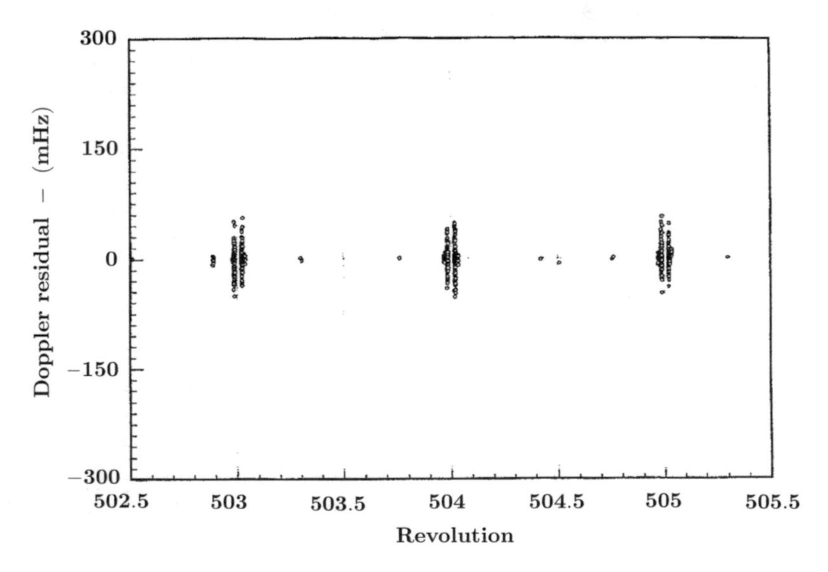

Fig. 9.9 Pioneer orbits 503–505 data residuals—sparse tracking

Table 9.5 Pioneer 1980 orbit determination sparse data

Revolution	ΔV (mm/s)	ΔP_{sparse} (s)	ΔP_{base} (s)
503	1.851	0.569	0.573
504	0.409	0.126	0.128
505	60.05	18.487	18.463

The sparse data solution compares quite favorably with the solution obtained by processing all the available data.

9.3.3 Estimation of Drag

The estimation of drag from orbit tracking data involves separating the velocity imparted to the spacecraft associated with atmospheric drag from the velocity imparted by all other sources including both gravitational and nongravitational accelerations. The significant other sources include the Venus gravity harmonics, solar tide, solar radiation pressure, and any spacecraft propulsive or attitude maneuvers. Over a relatively short data arc of a few hours, the solar tide and nongravitational accelerations are predictable and do not contribute significantly to the drag estimation error. The Venus gravity harmonics, on the other hand, result in a large perturbation of the orbit that reaches a maximum near periapsis just where the drag acceleration attains a maximum. Thus, the main orbit determination problem is separating the gravity harmonic perturbation from the drag perturbation.

In order to gain some insight into the problem of drag estimation it is useful to examine the response of the tracking data to errors in the values of key parameters involved in the estimation process. The tracking data response to estimated parameters is most useful when displayed as perturbations on the data residual which is referred to as the parameter signature. Figure 9.10 shows data residual signatures of the two key parameter sets describing gravity harmonics and drag. The drag is described by the atmospheric density and is the curve consisting of the o's, the lower curve. The gravity harmonic truncation error is approximated by the sum of the perturbations caused by the gravity harmonics of degree 21 and is the curve described by the x's, the upper curve. The representation of the gravity field truncation error by the highest degree harmonics available is somewhat arbitrary. The actual error in the truncation of the gravity field is some weighted average of all the harmonics that have been omitted from the solution.

The data residual signatures of atmospheric density and gravity harmonics shown in Fig. 9.10 were obtained by the following procedure. First, a single orbit of Doppler data was fit and residuals similar to those shown in Fig. 9.8 were obtained. These residuals are simply a plot of the difference between the actual measurement obtained at the tracking station and the measurement computed from a model of the system after the parameters of the model have been adjusted by least squares to minimize the residual error. At this point we intervene and generate a simulated

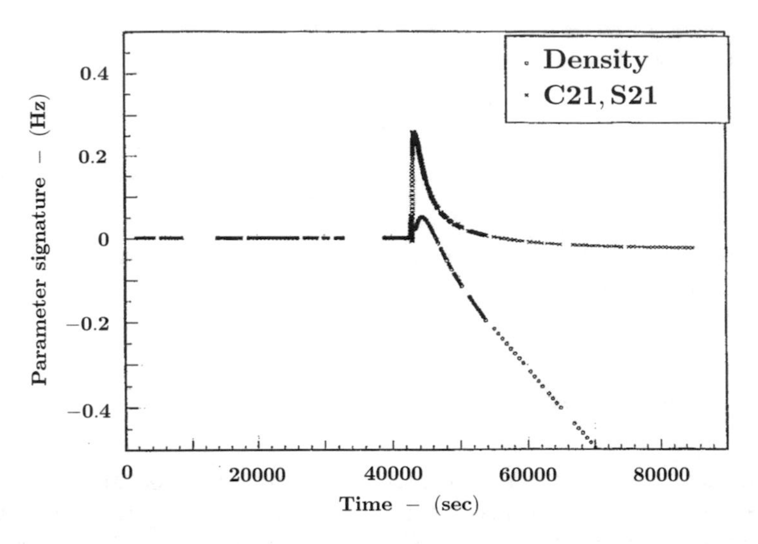

Fig. 9.10 Pioneer drag and gravity harmonic signatures

data set of measurements that are equal to the computed measurements. When we process this new data set, the residuals are exactly zero for each data point. Next, we adjust some parameter in the model that we are interested in and process the simulated data thus exposing the sensitivity of the adjusted parameter. Shown in Fig. 9.10 is the result of performing this procedure for atmospheric density and the degree 21 gravity harmonics. The orbit determination filter effectively looks at these two curves and separates one parameter from another based on the characteristic response or signature of the parameter. If we restrict the orbit determination solution to data near periapsis, say from 42,000 to 46,000 s on Fig. 9.10, the filter will not be able to determine the drag since the perturbation of the spacecraft is dominated by gravity harmonics in this region. However, the gravity harmonics tend to conserve energy around a closed orbit and their signature is periodic whereas the atmospheric drag reduces the energy resulting in a signature that grows with time. An hour or so of tracking data after periapsis reveals a secular growth in the atmospheric drag signature that may be easily separated from the gravity harmonics signature by the filter. Comparison of the atmospheric density residual with the gravity harmonic residual indicates that the orbit period change associated with the drag may be estimated to an accuracy of about 15 ms.

9.3.4 Relating Drag ΔV *to Period Change*

The orbit determination software provides an estimate of the ΔV attributable to drag as the spacecraft flies through the Venus atmosphere. The problem faced by science is to relate this ΔV to the parameters of a detailed spacecraft aerodynamic and planet atmosphere model. The procedure that has been used in the past is to map the spacecraft state to some reference epoch and compute the difference between the osculating orbit periods with and without the effect of atmospheric drag. The atmospheric scientists may then adjust the parameters of their model until they obtain the same period difference at the same reference epoch. The theory of small perturbations then applies provided the velocity change associated with the drag acceleration is small compared to the spacecraft velocity at periapsis. This is certainly the case since the spacecraft is moving at approximately 10 km/s relative to the planet and the velocity change attributable to drag is on the order of 1 m/s.

Consider a simplified model consisting of flat plate drag and a simple exponential atmosphere. We thus have for the force model,

$$A_d = -\frac{C_d\, A}{m}\, q$$

$$q = \frac{1}{2}\, \rho\, v^2$$

$$\rho = \rho_o\, e^{\frac{-(r - r_o)}{h_o}}$$

where ρ_o, r_o, and h_o are the atmosphere base density, reference altitude, and scale height, respectively and C_d, A, m, and v are the spacecraft drag coefficient, reference area, mass, and speed, respectively. The atmospheric density (ρ), drag acceleration (A_d), and dynamic pressure (q) are functions of these quantities.

The above drag equations in conjunction with the two body equations of motion are solved for ΔV_R.

$$\Delta V_R = \frac{C_d\, A}{m} \rho_{r_p} \sqrt{\frac{\pi G M h_o}{2e}}\, (1 + e)$$

$$\rho_{r_p} = \rho_o\, e^{\frac{-(r_p - r_o)}{h_o}}$$

The actual Venus atmosphere has considerable more structure than is represented by the simple exponential atmosphere and furthermore it is not convenient to solve for atmospheric model parameters in orbit determination software. Since the filter primarily responds to the total impulse and is insensitive to the time variation of the acceleration in the atmosphere, a much simpler model of the drag should suffice. After some experimentation, a model was selected where the drag is a constant acceleration over a fixed interval of time. The direction is opposite to the velocity vector and therefore the spacecraft executes a gravity turn trajectory during the time that the acceleration is active and is ballistic when the acceleration is inactive. The actual acceleration describes a bell shaped curve as a function of time centered at periapsis and this is replaced by a constant acceleration over a fixed time interval centered at periapsis. The width of the time interval is selected such that the modeled acceleration profile is close to the actual acceleration profile. The total integrated drag acceleration is thus given by

$$\Delta V_c = A_c \, \Delta T_c$$

We desire a value for the time interval (ΔT_c) such that the constant acceleration (A_c) is close or a best fit to the actual acceleration. This will occur when the constant acceleration is some constant K times the actual peak acceleration experienced by the spacecraft at periapsis and is given by

$$A_c = K \, A_p$$

$$A_p = -\frac{C_d \, A}{m} \, q_p$$

$$q_p = \frac{1}{2} \, \rho_p \, v_p^2$$

We may solve the above equations for ΔT_c by equating ΔV_c with the actual integrated velocity change which is approximately ΔV_R and obtain

$$\Delta T_c = \frac{2(1+e)}{v_p^2} \sqrt{\frac{\pi G M h_0}{e}}$$

where, from inspection, the value of K is seen to be a little less than one and has been arbitrarily assigned the value

$$K = \frac{1}{\sqrt{2}}$$

The atmosphere reconstruction procedure thus consists of estimating the total integrated velocity change or ΔV experienced by the spacecraft during passage through the atmosphere and relating this quantity to the parameters of a detailed precise model of the spacecraft and atmosphere.

Consider the following numerical experiment that may be performed on a computer. A trajectory program is initialized with the spacecraft state about 5 min prior to periapsis. The trajectory is integrated through the Venus atmosphere and terminated 5 min after periapsis. A precision integrator is employed and all the significant force models are turned on that perturb the spacecraft. For the first integration, the atmospheric density is set equal to zero and the osculating period at the end of the trajectory is recorded as P_{nom}. We repeat the integration, only this time we turn on the atmosphere and at the end of the trajectory we compute the change in the osculating period which is

$$\Delta P_{int} = P - P_{nom}$$

In order to test the linearity, we may compute the expected osculating period change from

$$\Delta P = \frac{\partial P}{\partial v} \Delta V$$

$$\Delta P_R = \frac{\partial P}{\partial v} \Delta V_R$$

where ΔV is obtained by direct integration of the drag acceleration. The partial derivative is also computed from osculating orbit elements obtained by transformation of the Cartesian state vector at periapsis. We next take the precision integrated ΔV and divide by ΔT_c of 100 s to obtain a constant acceleration A_{100} for comparison. A third precision integrated trajectory is computed using the constant acceleration model and the osculating period (P_{100}) is computed at the reference epoch 5 min after periapsis. A fourth precision integrated trajectory is computed only this time the entire ΔV is applied as an impulse at periapsis resulting in P_{imp}. Finally, this entire experiment is repeated at several atmospheric base densities resulting in ΔV's that range from a fraction of a millimeter per second to over a meter per second and the results are displayed in Table 9.6.

Shown in Table 9.6 is the velocity change (ΔV) obtained by numerical integration and the velocity change (ΔV_R) obtained from the formula. We would expect close agreement since both of these results are based on the same equations of motion and the difference represents only errors in the approximations. Also shown is the period change that would be predicted by the partial derivative assuming linearity of the equations of motion. When compared with the precision integrated period change the error is about 1%. A more relevant comparison is between the period change obtained by integrating the drag assuming an exponential atmosphere and the period change obtained by integrating a constant acceleration. Here, the agreement is generally less than 0.1%. It is reasonable to assume that the actual modeling error attributable to orbit determination software based on a constant acceleration model will also be less than 0.1%. In order to obtain this accuracy the actual integrated drag acceleration must be compared. If the osculating periods are compared, they must be corrected for nonlinearity or the error will be about 1%.

Table 9.6 Drag ΔV and resulting period change

ΔV (m/s)	ΔV_R (m/s)	ΔP_{int} (s)	$\frac{\partial P}{\partial v} \Delta V$ (s)	$\frac{\partial P}{\partial v} \Delta V_R$ (s)	ΔP_{imp} (s)	ΔP_{100} (s)
0.2016×10^{-3}	0.2016×10^{-3}	0.06295	0.06220	0.06221	0.06257	0.06254
0.2016×10^{-2}	0.2016×10^{-2}	0.6226	0.6220	0.6221	0.6224	0.6219
0.2016×10^{-1}	0.2016×10^{-1}	6.2186	6.220	6.2205	6.220	6.2183
0.201677	0.201690	62.1525	62.201	62.205	62.166	62.150
1.008444	1.008467	310.02	311.01	311.03	310.09	310.01

9.3.5 Covariance Analysis Results

In order to determine the orbit determination error, a covariance analysis of all the error sources that contribute must be performed. Since the actual orbit determination error is primarily a function of the mapping of the gravity field error, it is difficult to assign a numerical value. Analysis of residuals indicates that the gravity harmonics contribute about 15–20 ms one sigma to the period estimation error and about 200 m to the periapsis altitude estimation error. Another component of the orbit determination error is data noise which is related to the orbit determination error by the measurement error, number of data points, and observability of the system.

For PVO atmosphere entry we are primarily interested in predicting the spacecraft state at periapsis and reconstructing the drag acceleration. Both of these navigation requirements are related to prediction and estimation of the period (P) and periapsis altitude (H_p). In order to gain some insight into the affect of data noise, a covariance analysis was performed where the amount and geometric placement of the data was varied. The results are shown in Table 9.7.

The first case consisted of processing all the data available form orbits 503 through 505. The period error attributable to data noise is about 6 ms in period and less than a meter in periapsis altitude. If the gravity field were perfectly known, these would have been the orbit determination errors when the experiment was performed back in 1980. However, when the affect of gravity harmonic errors is included, the actual orbit determination error is about 20 ms in period and 200 m in periapsis altitude.

The second case shows the orbit determination error attributable to data noise for the sparse tracking expected in 1992 as shown in Fig. 9.9. The data noise period error of 21 ms is now comparable to the gravity harmonic period error. The periapsis altitude error remains relatively unaffected by the reduced data. Since the orbit determination error is now dominated by data noise we are much more susceptible to degradation from loss of data quality or loss of tracking data.

The third and final case is intended to show the accuracy of predicting the next periapsis based on data taken up to a few hours after the previous periapsis passage. These results indicate that the time of the next periapsis may be predicted to well within the required 1 s and the periapsis altitude can be predicted to about 200 m when gravity harmonics are taken into account. These predictions assume a ballistic arc free of maneuvers and will deteriorate some when maneuvers are included.

Table 9.7 Pioneer orbit determination errors

Data (Arc)	Sigma P (s)	Sigma H_p (km)
Continuous tracking apoapsis 503 to apoapsis 506	5.74×10^{-3}	$.334 \times 10^{-3}$
Sparse tracking apoapsis 503 to apoapsis 506	20.9×10^{-3}	1.20×10^{-3}
Sparse tracking apoapsis 503 to periapsis 505 + 1 h	34.2×10^{-3}	1.21×10^{-3}

9.4 Near Earth Asteroid Rendezvous

Prior to the NEAR mission, little was known about Eros except for its orbit, spin rate, and pole orientation, which could be determined from ground based telescope observations. Radar bounce data provided a rough estimate of the shape of Eros. On December 23, 1998, after an engine misfire, the NEAR spacecraft flew by Eros on a high velocity trajectory that provided a brief glimpse of Eros and allowed for an estimate of the asteroids pole, prime meridian, and mass. This new information, when combined with the ground based observations, provided good *a priori* estimates for processing data in the orbit phase. After a 1 year delay, NEAR orbit operations began when the spacecraft was successfully inserted into a 320×360 km orbit about Eros on February 14, 2000. Since that time, the NEAR spacecraft was in many different types of orbits where radiometric tracking data, optical images, and NEAR Laser Rangefinder (NLR) data allowed a determination of the shape, gravity, and rotational state of Eros. The NLR data, collected predominantly from the 50-km orbit, together with landmark tracking from the optical data has been processed to determine a 24th degree and order shape model. Radiometric tracking data and optical landmark data were used in a separate orbit determination process. As part of this latter process, the spherical harmonic gravity field of Eros was primarily determined from the 10 days in the 35-km orbit . Although the gravity field of Eros has been determined to degree and order 10, differences between the measured gravity field and one determined from a constant density shape model are detected only to degree and order 6. The offset between the center-of-figure and the center-of-mass is only about 30 m indicating a very uniform density (1% variation) on a large scale (35 km). Variations to degree and order 6 (about 6 km) may be partly explained by the existence of a 100 m regolith or by small internal density variations. The best estimate for the J2000 right ascension and declination of the pole of Eros is $\alpha = 11.3692 \pm 0.003°$ and $\delta = 17.2273 \pm 0.006°$, respectively. The rotation rate of Eros is $1639.38922 \pm 0.00015°$/day which gives a rotation period of 5.27025547 h. No wobble for Eros has been detected that is greater than 0.02°. Solar gravity gradient torques would introduce a wobble of at most 0.001°.

The original plan for Eros orbit insertion called for a series of rendezvous burns beginning on December 20, 1998, that would insert the NEAR spacecraft into Eros orbit in January 1999. As a result of an unplanned termination of the first rendezvous burn, NEAR continued on its high-velocity approach trajectory and passed within 3900 km of Eros on December 23, 1998. At this time, it was not possible to place the NEAR spacecraft in orbit about Eros. Instead, a modified rendezvous burn was executed on January 3, 1999, which resulted in the spacecraft being placed on a trajectory that slowly returned to Eros with a subsequent delay of the Eros orbit insertion maneuver until February 2000. The flyby of Eros provided a brief glimpse and allowed for a crude estimate of the pole and prime meridian with an error of 2° along with a 10% mass solution. Orbital operations commenced on February 14, 2000, with an orbit insertion burn that placed the spacecraft into a nearly circular

Table 9.8 Eros orbit segments

Segment	Start date time (UTC)	Length (days)	Orbit (km × km)	Period (days)	Inc. (deg.) ATE[a]	Inc. (deg.) SPOS[b]
1	2/14/00 15:33	10.1	366 × 318	21.8	35	176
2	2/24/00 17:00	8.1	365 × 204	16.5	33	172
3	3/3/00 18:00	29.3	205 × 203	10.0	37	171
4	4/2/00 02:03	9.8	210 × 100	6.6	55	178
5	4/11/00 21:20	10.8	101 × 99	3.4	59	177
6	4/22/00 17:50	8.0	100 × 50	2.2	64	179
7	4/30/00 16:15	68.1	51 × 49	1.2	90	160
8	7/7/00 18:00	6.3	50 × 35	1.0	90	165
9	7/14/00 03:00	10.6	37 × 35	0.7	90	163
10	7/24/00 17:00	7.1	50 × 37	1.0	90	161
11	7/31/00 20:00	8.2	51 × 49	1.2	90	159
12	8/8/00 23:25	18.0	52 × 50	1.2	105	178
13	8/26/00 23:25	10.0	102 × 49	2.3	112	179
14	9/5/00 23:00	37.3	102 × 100	3.5	115	150
15	10/13/00 05:45	7.6	100 × 50	2.2	130	179
16	10/20/00 21:40	5.0	52 × 50	1.2	133	178
17	10/25/00 22:10	0.8	50 × 20	0.7	133	168

[a]*ATE* Asteroid True Equator
[b]*SPOS* Sun Plane of Sky

350 km orbit. A series of propulsive burns lowered the spacecraft orbit to a 50 km and then a 35 km circular orbit where the data acquired allowed precise estimates of Eros physical parameters. Table 9.8 lists the orbit phases for the NEAR mission included in this study from the beginning on February 14, 2000 to the close flyby within 5 km of the surface of Eros on October 25, 2000.

Estimates of the initial attitude and spin rate of Eros, as well as of reference landmark locations used for optical navigation, were obtained from images of the asteroid. In the planned navigation strategy, these initial estimates were used as *a priori* values for a more precise refinement of these parameters by an orbit determination technique which processes optical measurements combined with Doppler and range tracking. Although laser altimetry could be included in the orbit determination process, these data were processed separately using the orbits determined from the optical and radiometric data.

In addition to allowing accurately determined orbits about Eros, the gravity harmonics place constraints on the internal structure of Eros. The shape model was obtained by processing optical landmark and laser altimetry data. This shape model was then integrated over the entire volume, assuming constant density, to produce a predicted gravity field. A comparison of the true gravity field with this predicted gravity field from the shape model then provides insight into Eros' internal structure. The location of the center of mass derived from the first degree harmonic coefficients directly indicates the overall mass distribution. The second degree harmonic coefficients provide insight into the orientation of Eros principal axes. Higher degree harmonics may be compared with surface features to gain additional insight into mass distribution.

9.4.1 Orbit Determination Strategy

Several strategies have been used to determine the NEAR orbits and the physical parameters of Eros. The data types used for determining NEARs orbit are radiometric X-band (8.4 GHz downlink) Doppler and range, and optical imaging of landmarks. A SRIF filter is used to process the data and this sequential filter is designed to handle up to 800 parameters including 18 state parameters. The estimated parameter set includes initial spacecraft state, propulsive maneuvers, solar pressure parameters, stochastic accelerations, Eros ephemeris, Eros attitude and rotation state, and physical parameters that describe the size, shape, and gravity of Eros. Eros physical parameters include gravitational harmonics to degree and order 12, inertia tensor elements, and the location of over 100 landmarks. The solution for nongravitational accelerations presents a particular challenge to the orbit determination filter. These accelerations include attitude control gas leaks and solar pressure. The solar pressure is modeled as a collection of reflecting surfaces with 12 separate parameters. Solar pressure mismodeling and any residual accelerations associated with outgassing from the spacecraft are lumped together and treated both as a constant acceleration and as stochastic accelerations. The stochastic accelerations are modeled as three orthogonal independent exponentially correlated process noise components with an amplitude of 1.0×10^{-12} km/s^2 and a correlation time of 1 day. The total number of estimated parameters for a typical orbit determination solution is about 600.

The differences in the moments of inertia may be determined from the gravity harmonic coefficients, but a particular moment of inertia about any axis cannot be determined from these difference alone. In the orbit determination software, the joint solution for both the gravity and rotational motion of Eros permits a determination of the principal moments of inertia provided the angular acceleration (or wobble about the principal axes) can be detected by the orbit determination filter. The solution strategy involved processing several days of data at a time to converge slowly on the orbit solution. First, about 2 days of data are processed and the solution is fed back to the filter and the data are processed again. This process is repeated until convergence is achieved. At this point several more days of data are introduced to the filter and processed iteratively until another solution is obtained. Additional data are introduced in batches of several days until all the data are processed. Otherwise, processing longer batches of data, especially at the beginning of the mission, resulted in divergence. Once the filtering is complete, the spacecraft trajectory, Eros ephemeris, and Eros attitude files are produced containing

Chebyshev polynomials as a function of time. Gravity harmonic, landmark location, maneuver parameter, and shape harmonic coefficient files are also produced.

Optical tracking of landmarks in the imaging data taken by NEAR's Multispectral Imager (MSI) is a powerful data type for determining NEARs trajectory and the rotation of Eros. Tracking individual landmarks, which are small craters, enables orbit determination accuracies on the order of the camera resolution or several meters. This exceeds the accuracy that can be obtained from radio metric data alone, from fitting limb data or from any measurement scheme that is dependent on developing a precise shape model. We need only develop a database of landmarks and identify the landmarks on more than one image in order to obtain useful information about the spacecraft orbit or Eros' rotation. The procedure of identifying and cataloging landmarks is aided by referring the landmarks to a model of the topographic surface or shape model. The actual identification of individual landmarks depends upon observing them in an image having many landmarks of various sizes to provide a context.

In addition to the one data arc (July 3 to August 7) used to determine the gravity field and rotation of Eros, three other data arcs were used to process the NLR data (April 30 to June 1, June 1 to July 3, and August 7 to September 12). For all the data arcs, the attitude of Eros is fixed to the solution obtained from the gravity solution data arc. This is to maintain consistency when comparing the estimated gravity solution with the shape model gravity solution. Once a good solution was obtained for both the spacecraft trajectory and Eros attitude as a function of time, some additional processing was required to transform the results to a more usable format and to solve for the shape. The solution for the shape of Eros is obtained by processing NLR data in a separate program that reads the spacecraft ephemeris and Eros attitude files.

Although Eros is a very irregular body, the gravitational potential is modeled by a spherical harmonic expansion with normalized coefficients (C_{nm}, S_{nm}) given by

$$U = \frac{GM}{r} \sum_{n=0}^{\infty} \sum_{m=0}^{n} \left(\frac{r_0}{r}\right)^n P_{nm}(\sin(\phi) \left[C_{nm} \cos(m\lambda) + S_{nm} \sin(m\lambda)\right]$$

where n is the degree and m is the order, P_{nm} are the Legendre polynomials and associated functions, r_0 is the reference radius of Eros (16.0 km), ϕ is the latitude, and λ is the longitude. The harmonic coefficients of degree one are set to zero since the origin of the coordinate system is chosen to be the center of mass of the body. This expansion converges outside the smallest sphere enclosing Eros. All the NEAR data employed are outside this sphere and so spherical harmonics is the simplest way to compare the gravity and shape models. All gravity and shape results are mapped onto a sphere of radius 16 km. For mapping the gravity field to the surface of Eros, one must use alternative methods such as direct integration over the volume of Eros defined by the shape model.

9.4.2 *Eros* A Priori *Physical Model*

Determination of the spacecraft orbit about Eros is intimately associated with the development of an accurate physical model of Eros. Eros is the principal source of perturbations on the spacecraft's trajectory and the principal source of data for determining the orbit. The model of Eros used for orbit determination is similar to the model used for science investigation. The major difference is in emphasis of detail.

During a particularly close Earth approach (0.15 AU) in January 1975, there was a coordinated ground-based observation campaign to characterize the physical nature of Eros. Photometric, spectroscopic, and radar measurements provided a diverse data set that allowed the asteroid's size, shape and spectral class to be determined. Eros is an S-class asteroid with a geometric albedo of about 0.27. The absolute magnitude of Eros (at zero phase angle and one AU from both the sun and Earth) is 11.16. From the light curve, which reaches 1.47 magnitudes in amplitude, the rotation period and pole direction were determined.

During the December 1998 flyby, a crude estimate of Eros mass and pole location was obtained. The pole location confirmed ground-based measurements to an accuracy of about 2°. Observation of the lit portions of Eros by the Multispectral Imager (MSI) permitted a rough shape determination. The gravity harmonic coefficients were computed from this shape determination, by numerical integration assuming constant density. Light curve data obtained during the flyby yielded a precise rotation rate for Eros and enabled location of the prime meridian with respect to a large crater discernible in the images. This information was used as *a priori* data for the orbit phase solution.

9.4.3 *Orbit Determination Solution*

In addition to the models describing the estimated parameters, calibrations obtained from other models are applied to the Doppler, range and optical data. The calibration data included seasonal and daily troposphere and ionosphere models based upon on-site GPS and weather measurements and a solar plasma model. DSN station locations are modeled to about 4 cm accuracy with Earth precession, nutation, polar motion, ocean tidal loading, solid Earth tide, and tectonic plate motion. A landmark file consisting of *a priori* landmark locations and unique identification numbers was assembled along with a picture sequence file that contained camera pointing and image coordinates for each landmark that was identified. Additional models that were needed for parameter estimation include the spacecraft clock model, a solar pressure model, propulsive maneuvers, and initial state vectors for the equations describing the motion of the spacecraft, planets and Eros.

The gravity and pole solution data arc, which included Doppler, range and optical imaging of landmarks, extended from July 3, 2000 through August 7, 2000. Nearly continuous Doppler data were processed and the post fit residuals for this

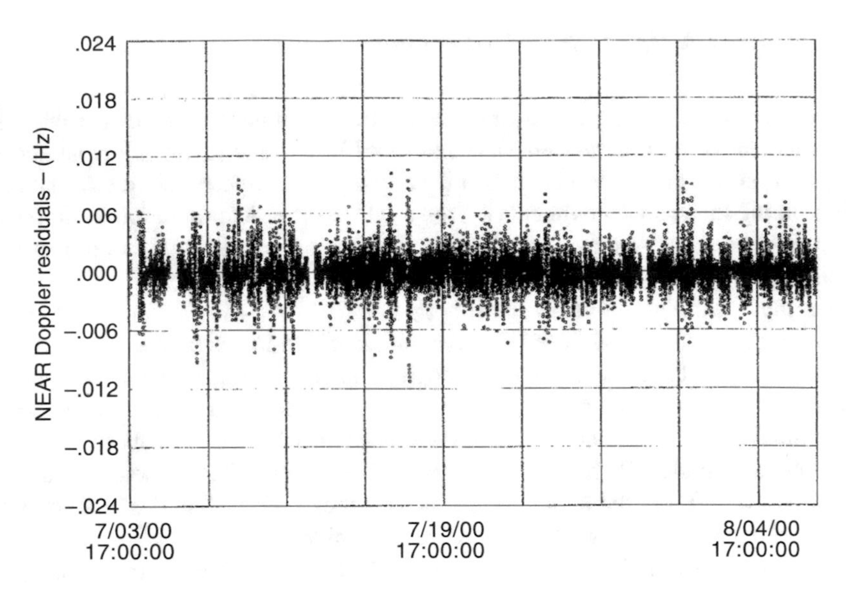

Fig. 9.11 NEAR Doppler residuals

solution are shown in Fig. 9.11. The ordinate is the measurement residual in hertz. A spacecraft radial velocity component of 1 mm/s measured along the line of sight from a particular DSN tracking station corresponds to approximately 0.054 Hz of Doppler phase shift over the count time interval which is typically 60 s. The Doppler signature shown in Fig. 9.11 reveals noise with a periodic amplitude of 0.002 Hz (0.03 mm/s) rms.

Optical data residuals are shown in Fig. 9.12. The ordinate of this figure is the measurement error in pixel(x) or line (y) direction in an image. One line subtends 165 μrad and one pixel 95 μrad. In a 50 km orbit, the line and pixel measurement error translate to 5.6 m and 3.2 m, respectively, when observing landmarks on the ends of Eros. The rms of the measurement error is about two lines and pixels and permits sub-meter accuracy when more than 3000 optical observations are processed by the orbit determination filter. High-precision orbits are obtained by processing optical data since individual landmarks may be located with an accuracy of a few meters with respect to the center of mass. With the optical data giving highly accurate orbits with an accuracy of about 1.5 m, the range data are able to tie down the Eros ephemeris.

Including the NLR data in the orbit determination solution does not improve the spacecraft orbit. The shape model has errors on the order of a hundred meters which is far greater than the orbit error. However, the NLR data were useful for determining a shape model that is accurate to about 100 m. This was accomplished by processing a high precision spacecraft ephemeris file and Eros attitude file, obtained from the orbit determination solution, in a separate program that solves only for the shape model harmonic coefficients and NLR bias parameters. The post fit residuals are shown in Fig. 9.13.

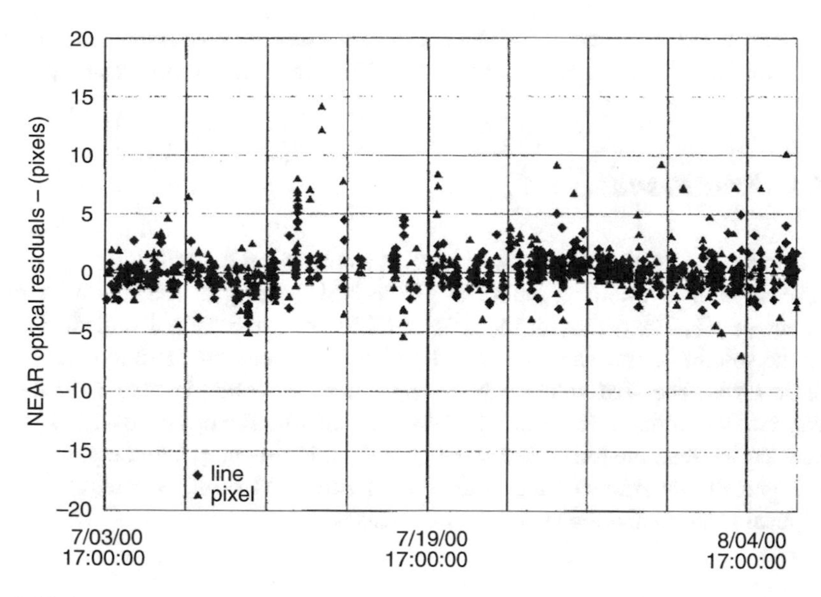

Fig. 9.12 NEAR optical residuals

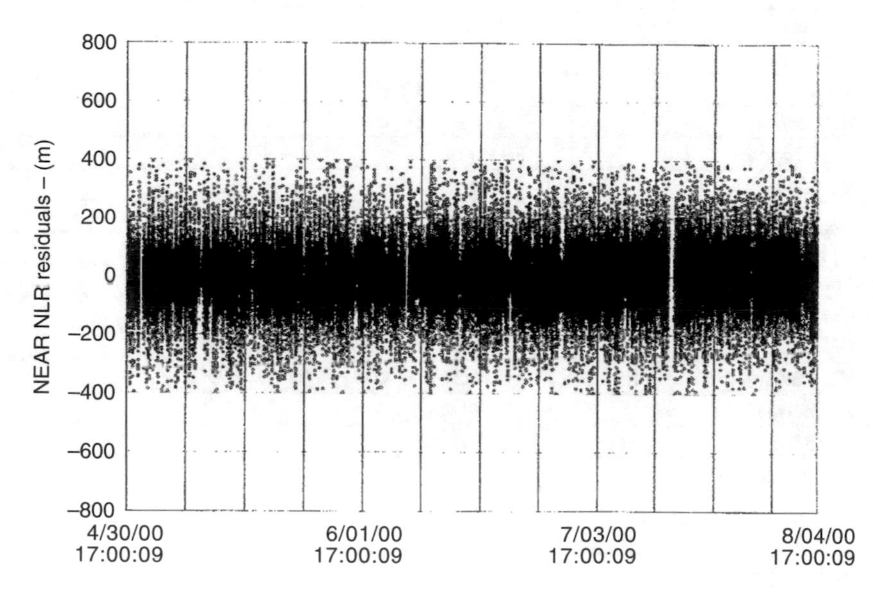

Fig. 9.13 NLR residuals

The rms error is 109 m and measurement errors greater than 400 m were rejected by the filter. The large rms error of the NLR measurement belies the accuracy of the NLR since this residual error is dominated by modeling errors in determining the shape. The instrument error is only a few meters. Since the modeling error is unbiased, a considerable reduction in the determination of the mean radius of Eros may be expected when the 263,490 NLR observations are processed. Eros volume may be estimated to an accuracy of less than 1% using this solution strategy.

9.4.4 Eros Results

The solution for Eros' physical parameters is summarized in Table 9.9.

The mass and volume combine to give a bulk density of 2.67 g/cm^3 with an accuracy of 1%. The errors for the pole and GM are at the formal statistical errors from the solution covariance but scaled higher by a factor of three to give a more realistic error. The GM solution based upon the radio metric data only is very sensitive to the initial pole value. The GM solution with the optical data is 10 times more accurate than the radio metric only solution. However, when the pole is fixed to the optically determined values, the radio metric GM solution agrees well with the optical solution and the uncertainty decreases.

Table 9.9 Eros physical parameters

Parameters	Values
Size and density	
Volume	2503 ± 25 km^3
Bulk density	2.67 ± 0.03 g/cm^3
X_{cg} of figure	−9.7 m
Y_{cg} of figure	2.4 m
Z_{cg} of figure	32.6 m
Mass properties	
Mass	$(6.6904 \pm 0.003) \times 10^{15}$ kg
GM (optical radiometric)	$(4.4631 \pm 0.0003) \times 10^{-4}$ km^3/s^2
GM (radiometric)	$(4.4584 \pm 0.0030) \times 10^{-4}$ km^3/s^2
GM (radiometric and optical pole)	$(4.4621 \pm 0.0015) \times 10^{-4}$ km^3/s^2
I_{xx} (normalized)	17.09 km^2
I_{yy} (normalized)	71.79 km^2
I_{zz} (normalized)	74.49 km^2
X principal axis	9.29° East (definition)
Pole (optical)	
Right ascension	$11.369 \pm 0.003°$
Declination	$17.227 \pm 0.006°$
Rotation rate	$1639.38885 \pm 0.0005°$/day
Prime meridian	326.06° (at epoch and equinox J2000)
Pole (radiometric)	
Right ascension	$11.363 \pm 0.01°$
Declination	$17.230 \pm 0.02°$
Rotation rate	$1639.38922 \pm 0.0002°$/day

9.4.5 Shape Model

The Eros shape model obtained from the NLR data is in the form of harmonic coefficients through degree and order 24. One may use the harmonic expansion to compute the radius of Eros as a function of latitude and longitude. The resulting topographic map, shown in Fig. 9.14, reveals two mountainous looking features about the size of Mount Everest. This is an illusion since these features are simply the elongated ends of Eros. The contour lines shown are accurate to about 100 m, and this can be verified by comparing the shape of Eros projected into two dimensions to actual images of Eros taken by the MSI. Where the curvature is high, the shape model error is as high as 200 m.

The accuracy of the shape model may also be confirmed by computing the radius vectors for reference landmarks whose locations have been determined to about 5 m. The locations of about 43 landmarks were confirmed to be on the shape model surface with an rms error of about 50 m. A few of the landmarks were above or below the shape model surface by as much as 200 m in the regions of Eros where the nadir pointed NLR intersected the surface at a high incidence angle.

Even though the local variation in the shape model error suggests an accuracy of about 100 m, the error in determining the average radius integrated over the entire surface is much smaller. The trajectory error and instrument measurement error combined are about 10 m. Since the shape model error associated with the harmonic coefficients is unbiased, the error in determining the average radius, which is directly related to the volume determination, may be reduced considerably by taking many measurements and statistically averaging. This averaging, which is implicitly performed by the orbit determination filter, is effective when a large number of measurements are processed, since the error in the average radius is reduced by the square root of the number of measurements. For 263,490 NLR

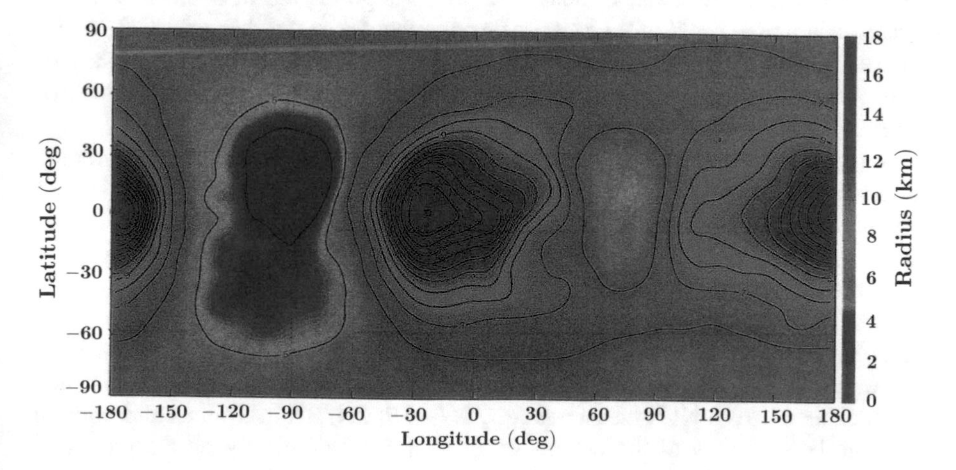

Fig. 9.14 Eros shape model

measurements, the modeling error may be reduced by a factor of about 500, which is well below the level where systematic errors dominate. Thus the volume of Eros and the low order shape harmonic coefficients may be determined to an accuracy of 1% provided the surface is sampled randomly and systematic errors associated with the trajectory and instrument biases are about 10 m. The NLR data acquisition strategy for NEAR resulted in fairly uniform coverage of Eros, owing to the circular 50 km polar orbit and the relatively rapid rotation of Eros. The random character of the sampling is at very small scales. Statistically, NLR samples that measure the surface at the top of boulders are compensated by samples that fall in craters. This mathematical property of laser altimetry gives this method a distinct advantage over optical imaging where shadows tend to obscure the surface at small scales.

9.4.6 Gravity Harmonics

Determination of the gravity harmonic coefficients of Eros is a direct result of the spacecraft orbit determination process necessary to navigate the spacecraft. The harmonic coefficients are estimated by observing the acceleration of the spacecraft in orbit. As the spacecraft is maneuvered closer to Eros, the degree of the harmonic expansion must be increased in order to provide the required accuracy for orbit prediction. This results in determining Eros gravity field to fairly high precision. At degree 10, the uncertainty or noise in the gravity field is roughly equal to the signal as given by the rms of the coefficients.

An *a priori* gravity model can be developed by integrating the potential function over the shape model determined by NLR or MSI observations assuming constant density. The results are shown in Table 9.10 for comparison. The close agreement of the gravity coefficients obtained from spacecraft dynamics and those obtained from the NLR-derived shape model provides a high degree of confidence in the results when used for NEAR spacecraft navigation. Since the shape-derived gravity coefficients assumed a constant density, the closeness of the agreement for the two sets of coefficient values (Table 9.17) indicates that the material within the interior of Eros is nearly of uniform density.

Of particular interest are the first degree and order terms of the harmonic expansions. For the spacecraft orbit solution, these terms were explicitly set to zero forcing the center of the coordinate system to coincide with the center of mass of Eros. Thus, the values of these coefficients from the shape model provide a direct measure of the offset of the center of figure from Eros center of mass, since the vector from the origin to the center of mass may be determined by multiplying the first degree and order coefficients by the reference radius (16 km). The coefficients shown in Table 9.17 reveal that the center of figure offset vector for Eros, obtained independently from NLR measurements, is (−9.7, 2.4, 32.6). This result indicates that the bulk density of the octants of Eros, defined arbitrarily by the planes of the reference coordinate axes, agrees within one percent. This is another strong indication of the uniformity of Eros internal structure.

Table 9.10 Eros gravity harmonic coefficients

Coefficient (R_0 =16.0 km)	Solution spacecraft dynamics	Solution shape model integration
C_{10}	0	0.001175
C_{11}	0	−0.000348
S_{11}	0	0.000088
C_{20}	−0.052478 (0.000051)	−0.052851
C_{21}	0	0.000102
S_{21}	0	0.000012
C_{22}	0.082483 (0.000061)	0.083148
S_{22}	−0.027909 (0.000035)	−0.028197
C_{30}	−0.001400 (0.000030)	−0.001747
C_{31}	0.004059 (0.000006)	0.004086
S_{31}	0.003375 (0.000006)	0.003401
C_{32}	0.001791 (0.000016)	0.002127
S_{32}	−0.000691 (0.000016)	−0.000840
C_{33}	−0.010373 (0.000027)	−0.010492
S_{33}	−0.012104 (0.000027)	−0.012216
C_{40}	0.012900 (0.000070)	0.013077
C_{41}	−0.000106 (0.000014)	−0.000145
S_{41}	0.000136 (0.000015)	0.000165
C_{42}	−0.017488 (0.000035)	−0.017647
S_{42}	0.004577 (0.000030)	0.004624
C_{43}	−0.000320 (0.000044)	−0.000313
S_{43}	−0.000141 (0.000044)	−0.000194
C_{44}	0.017552 (0.000062)	0.017694
S_{44}	−0.009009 (0.000061)	−0.009118

The gravity field of Eros as a function of latitude and longitude is shown in Fig. 9.15 for harmonics up to degree 8. The gravity field is displayed in milligals (1 gal = 1 cm/s) on a sphere with a radius of 16 km. The central body (GM) term of the harmonic expansion is not included in computing the acceleration and this accounts for the negative values. Comparison of the gravity field map with the topographic map shown in Fig. 9.14 does not reveal a high degree of correlation. The ends of Eros stand out, but surface features on a smaller scale are not seen. This is because the surfaces of constant gravity potential do not conform well to the shape of Eros and, when displayed on a sphere, the ends of Eros are given more weight than the central part. The advantage, however, for displaying the gravity field on a sphere is that the formal gravity uncertainty is very nearly uniform and is about 0.3 mgals for coefficients to degree 6 and 2.0 mgals to degree 8. Instead of using Fig. 9.14, a more meaningful comparison is to compare the Eros gravity map with the gravity map obtained from the Eros shape model assuming constant density. Since the gravity map from Eros shape would look very much like the actual Eros gravity map, the difference between the two maps is plotted as a function of latitude

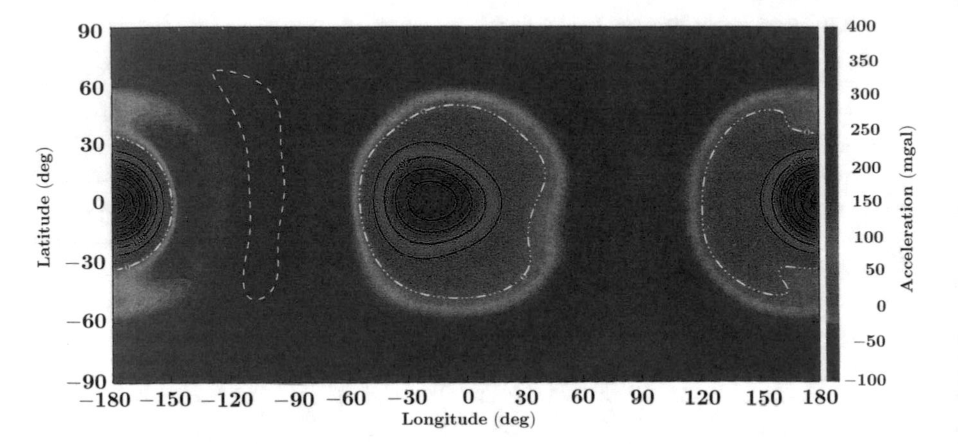

Fig. 9.15 Radial acceleration

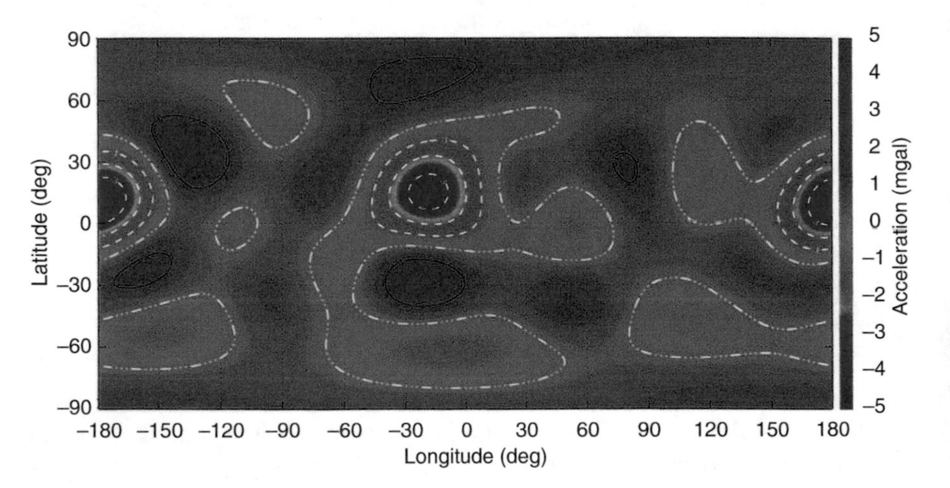

Fig. 9.16 Bouguer map

and longitude. This difference map (estimated gravity minus gravity from shape integration) is known as a Bouguer map and is shown in Fig. 9.16 for coefficients through degree 6. Differences in the gravity field reveal peaks and valleys uniformly distributed over Eros with maxima and minima of 3–4 mgal. The Bouguer variations are about ten times the formal uncertainty derived from the gravity covariance and these differences are about 1–2% of the maximum gravity amplitude.

Given the Bouguer map, there is no unique solution for the mass distribution of Eros. Several possible explanations for the observed mass deficiency at the ends of Eros include a less dense regolith covering on the order of 100 m distributed perhaps uniformly over the surface of Eros or a more dense concentration of material near the center of Eros. At degree 6, a 100 m uniform covering with a density contrast of 0.6 g/cm^3 produces a signature of -1.0 and -0.4 mgal at the asteroid ends. The observed signature therefore requires a higher density contrast, thicker regolith, or a variable thickness regolith that may be correlated with greater thicknesses for the highest potential areas. The Bouguer map also displays a shift of the negative anomaly to the northern hemisphere indicating less dense material. This may be related to higher potential areas also being shifted to the north where less dense regolith may accumulate. An increase in density of 5% for the central part of Eros, in the form of a sphere with 20% of the volume of Eros, results in a −3.0 mgal signature in the Bouguer map and very nearly matches the observed variation in Fig. 9.14.

9.4.7 Polar Motion

An important result that may be obtained from the NEAR data is an estimate of the moments of inertia about the principal axes. As described earlier, the moments of inertia provide insight into the radial distribution of mass. Estimates of the moments of inertia cannot be obtained if Eros is in principal axis rotation and there is no free precession. Therefore, one of the priorities of the NEAR mission is to measure the free precession of Eros. Precession results from disturbances of Eros' rotational motion from quakes, impacts, or gravitational torques. The free precession resulting from distinct events will damp out depending on the rate of internal energy dissipation. The forced precession from external gravity sources persists, but is low in amplitude. The Sun's gravity gradient produces a small forced precession and nutation.

The response of Eros to the solar gravity gradient torque depends on the orbit of Eros, the attitude and body fixed spin vector of Eros at some reference epoch, and the inertia tensor. All of these parameters may be solved for with high precision except the diagonal elements of the inertia tensor and the components of spin in body fixed coordinates normal to the spin axis. The second degree gravity harmonic coefficients provide the differences in the values of the diagonal elements of the inertia tensor, but the trace or any one diagonal element is needed to complete the inertia tensor. The complete inertia tensor may be obtained by numerical integration of the shape model. In order to minimize the error, only the smallest diagonal element is needed to complete the gravity harmonic based inertia tensor. Thus, the Ixx term of the shape model inertia tensor is used to construct the gravity based inertia tensor.

The determination of the spin vector components normal to the spin vector is needed to completely determine the free precession of Eros. These spin vector components place the angular momentum vector in Eros body-fixed coordinates. The normal spin vector components are too small to be resolved; however, the magnitude of the spin vector can be determined with very high precision. This high precision measurement is obtained by observing for several weeks small craters near the ends of Eros whose motions can be observed at the one meter level. The motion of the principal z axis projected onto the sky is shown in the top plot in Fig. 9.17 as a function of right ascension and declination. The amplitude of the free precession

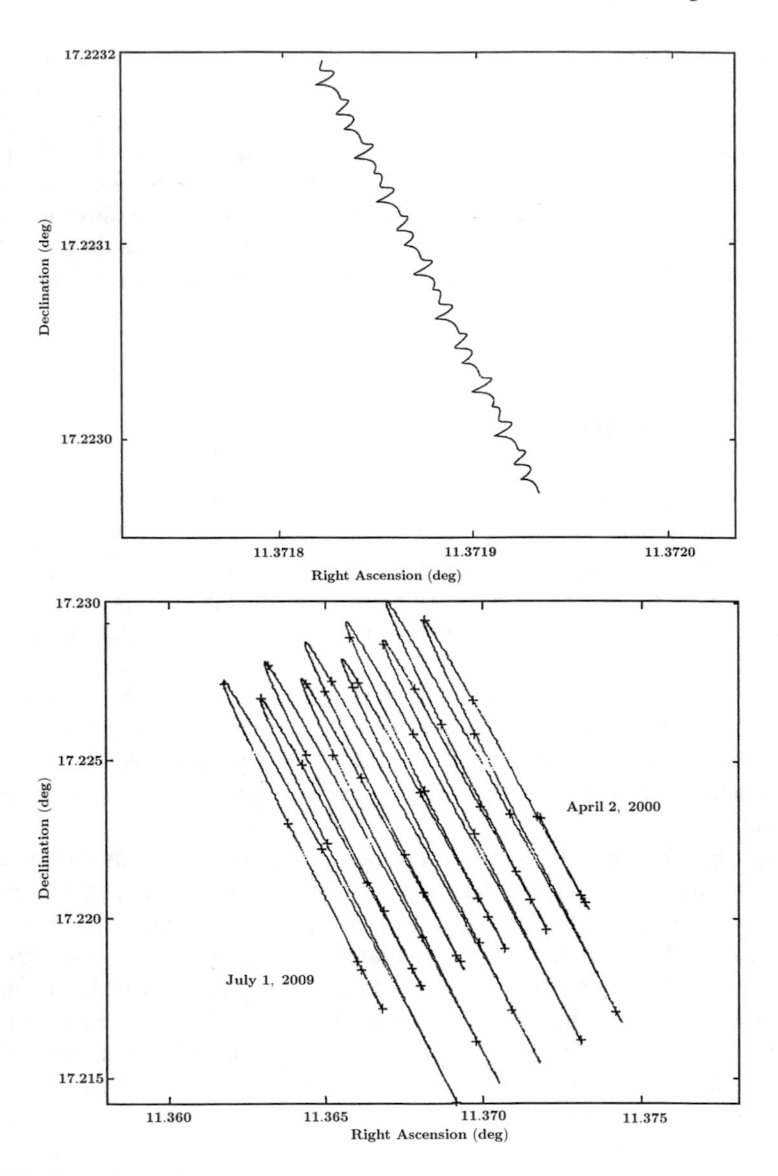

Fig. 9.17 Eros polar motion

is about 36 milliarcseconds and the forced precession over 3 days moves the pole and angular momentum vector about one arc second. This motion is too small to detect by the orbit determination filter. The precession is about 0.01° over 9 years, which is well beyond the lifetime of the NEAR mission. However, the short-period nutation has an amplitude of 0.02° over 6 months, as shown on the bottom plot of Fig. 9.17, and this could be detected. The solution obtained by processing 6 weeks of navigation data does not indicate free precession above that induced by the Suns gravity. However, this does not rule out free precession from other sources with amplitudes up to as much as 0.02°.

9.5 MESSENGER

For a month after the launch of the MESSENGER mission to Mercury, the spacecraft trajectory was perturbed by nongravitational accelerations that resulted in a migration of several thousand kilometers in the target B-plane. It is speculated that the accelerations were due to outgassing of water trapped in the composite materials of the spacecraft. Nongravitational accelerations are difficult to model, leading to inconsistent solutions for the spacecraft state from Doppler and range data. These nongravitational accelerations may be modeled as a sum of exponentially decaying stochastic vectors with different correlation times.

Immediately after a spacecraft is launched on an interplanetary trajectory, tracking data is acquired, and an orbit determination solution is computed for the outgoing trajectory and mapped to the target planet. The initial solution includes estimates of the spacecraft state, and of small turbulent accelerations that act on the spacecraft during the first few weeks, referred to as nongravitational accelerations. These accelerations are usually attributed to errors in the solar pressure or propulsion models. Although nongravitational accelerations are very small, on the order of 10^{-11} km/s^2, the orbit determination process is very sensitive to them. When they are not estimated accurately, it is difficult to obtain convergence, and the Doppler and range solutions are inconsistent. After a few weeks, the nongravitational accelerations diminish, and good solutions are obtained. However, after the Viking spacecraft was launched, significant nongravitational accelerations attributed to air trapped in the lander parachute were observed for several weeks.

For a month after the launch of the MESSENGER mission to Mercury, the spacecraft was perturbed by nongravitational accelerations that resulted in a migration of several thousand kilometers in the target B-plane. These accelerations were greater in magnitude and lasted longer than usual. It is conjectured that they may be attributed to outgassing of water vapor and other gases trapped in the newer composite materials of the spacecraft.

It is difficult to model accelerations due to volatile elements escaping from the spacecraft, which are assumed to be comprised of water vapor, and radiate in all directions. Only the total acceleration of the spacecraft can be estimated from Doppler and range data. Most volatile elements evaporate after a few days. Isotropic radiation (in all radial directions equally) results in negligible net acceleration of the spacecraft, and would not be observed. However, differences in the surface temperature of the spacecraft would result in anisotropic radiation. Initially, more gas would radiate toward the Sun, resulting in net accelerations away from the Sun. Later, because the same amount of gas is present on both sides, the net acceleration would be toward the Sun. The correlation time of surfaces exposed to the Sun would be greater than that of surfaces in shade. A stochastic model of gas radiated with several correlation times seems appropriate.

9.5.1 Initial Post-Launch Orbit Determination

After launch, the data types were X-band Doppler and range, with one sigma errors of 0.1 mm/s and 0.7 m, respectively. Figure 9.18 displays Doppler and range residuals from initial orbit determination results for the first month after launch of the MESSENGER mission. Only one stochastic vector is used to represent

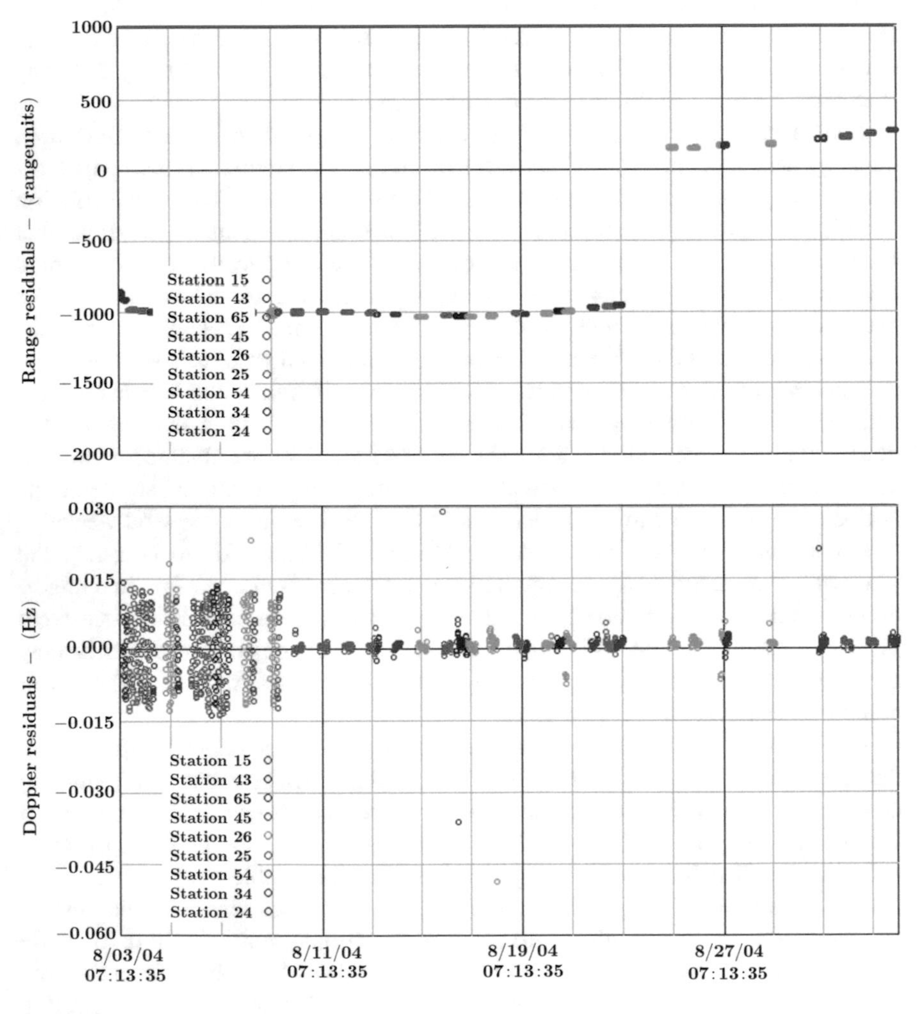

Fig. 9.18 Doppler data only solution

nongravitational accelerations. The data arc extends from shortly after launch on 8/3/04 07:13:35 GMT and extends to 9/4/2004. The estimated parameters are the initial state, Earth ephemeris, a propulsive maneuver on 8/24/2004, solar pressure model parameters and one stochastic acceleration vector $Ae^{-t/\tau}$, where A is a stochastic three-dimensional vector with white noise components of one sigma 1×10^{-12} km/s^2 (one nanometer/s^2), and τ is a 2-day correlation time (172,800 s). One such vector has usually been adequate to model nongravitational accelerations after launch. The data consisted of radiometric Doppler weighted at 6 mHz and loosely weighted range. It shows an apparent range bias of 1000 range units (about 143 m) until the first maneuver and about 200 range units after the maneuver. The Doppler residuals show an rms error of about 10 mHz (0.175 mm/s) for the first 9 days decreasing to about 6 mHz there after. The early sinusoidal signature in the Doppler data can be attributed to a slow spacecraft spin which was not modeled.

Figure 9.19 shows the same data arc and estimation strategy, but the Doppler data is loosely weighted and the range data is weighted at 500 range units. Now the range solution fits well, but the Doppler signature shows scalloping that might be attributed to a time error. It is unlikely that the discrepancy between Figs. 9.18 and 9.19 can be explained by problems with the data. The observed error must be attributable to a common source, which could not be a tracking station, since tracking stations around the Earth all give consistent results. The likely cause of this discrepancy is unmodeled accelerations acting on the spacecraft. A more accurate model for the accelerations is required.

One way to model a stochastic acceleration would be to represent each component as an orthogonal series, such as a Fourier sine series or a set of Tchebyshev polynomials. The coefficients would be estimated as stochastic parameters. The interval of the series expansion would move forward in time as batches of data are processed, and as noise introduced to the filter modulates the coefficients.

Here an alternate approach is applied. The orbit determination filter is already designed to process one exponentially correlated stochastic vector. Instead, a sum of several exponentially correlated stochastic acceleration vectors with generic integer correlation times is used. For five sets of stochastic acceleration vectors, a total of 15 components are estimated. To illustrate the method, MESSENGER Doppler and range tracking data are processed from launch through the first 4 months of the mission. The resulting time history of nongravitational accelerations is analyzed to verify the reasonableness of the results. The suspected outgassing could come from several sources. A single stochastic acceleration vector may not suffice because the time constant of the exponential decay is unknown and more than one source may be present. This suggests a stochastic model consisting of a sum of exponentially decaying stochastic acceleration vectors with varying time constants. While one might expect such a series to model effectively only processes that are exponentially decaying, it will be shown later that such a model can represent any function over a reasonable time interval much the same as a half range Fourier expansion can be used. Figure 9.20 shows the result for the same data arc where the number of stochastic acceleration vectors was increased from one to five with correlation times

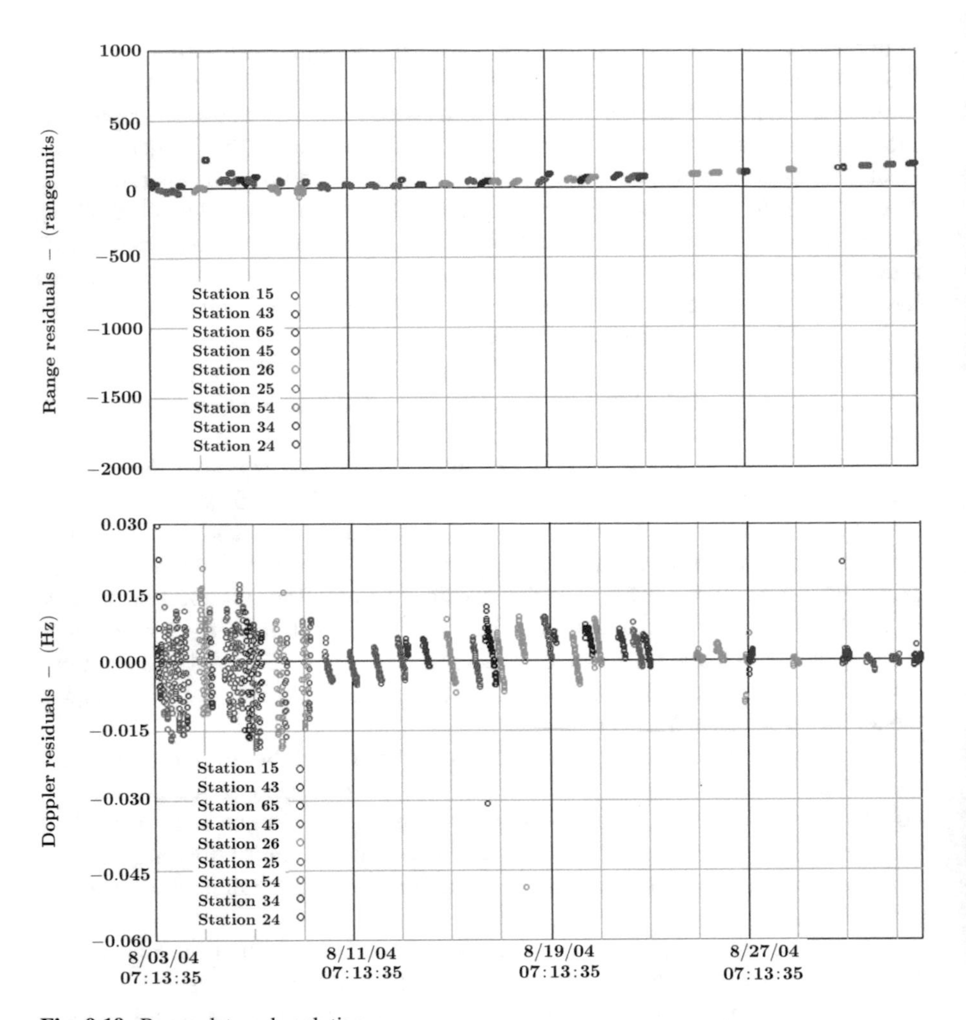

Fig. 9.19 Range data only solution

of 1, 2, 4, 8, and 16 days. A much better fit to the data is obtained. The position and velocity errors along the line-of-sight from the tracking stations are less than 20 m and 0.1 mm/s, respectively.

In Fig. 9.21, the time history of stochastic nongravitational accelerations attributable to outgassing is broken down by components. Ax is the component of the acceleration along the spacecraft-Sun line. The Ay and Az acceleration components are normal to the Sun line with Ay in the orbit plane and Az normal to the orbit plane. Solar pressure acceleration is not included. It would contribute another 60 nm/s^2 in the Ax direction. The accelerations were obtained by summing the five stochastic acceleration vectors described above. The dominant component is along the Sun line away from the Sun as one might expect. The volatile elements on the Sun side of the spacecraft would be heated to a higher temperature and

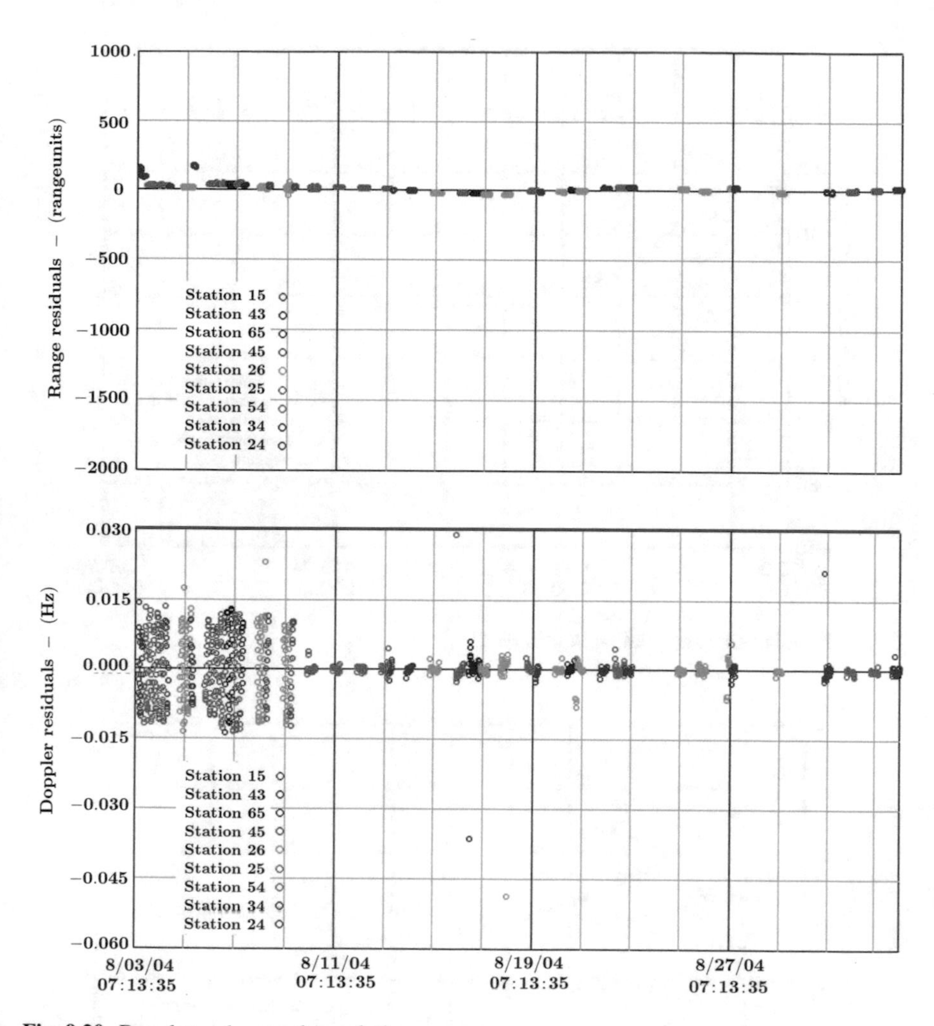

Fig. 9.20 Doppler and range data solution

consequently apply more force in this direction. The reversal in direction of the normal components would be explained by the presence of different sources outgassing in a given direction. One source would expel gas with a short time constant and be over taken by another source with a longer time constant.

The individual components of the acceleration along the Sun line due to the exponential model with various correlation times are shown in Fig. 9.22. Initially, the acceleration is dominated by components with 1-day and 8-day correlation times. After several weeks, the 8-day and 16-day components dominate. The migration from high frequency to low frequency components is consistent with the initial outgassing of the more highly volatile elements with short time constants.

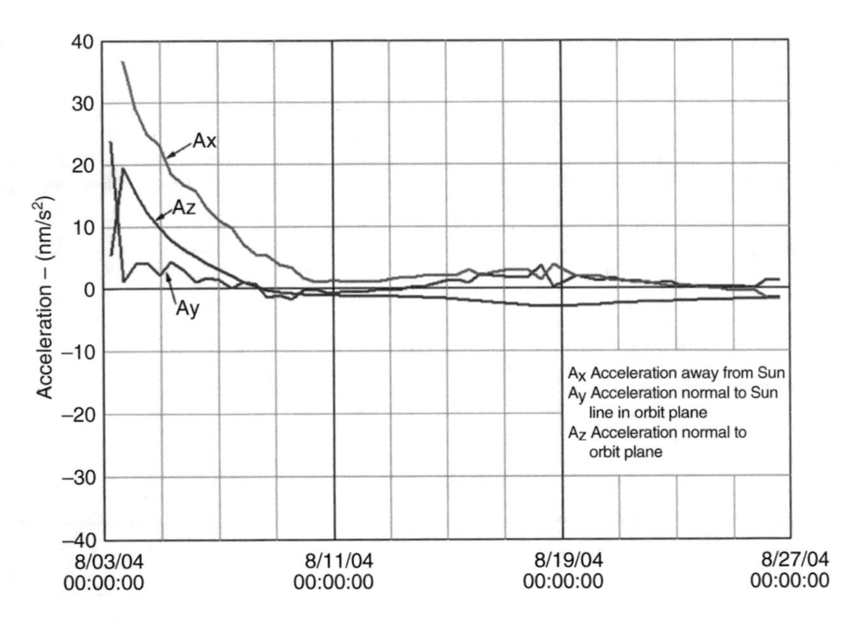

Fig. 9.21 Outgassing acceleration time history

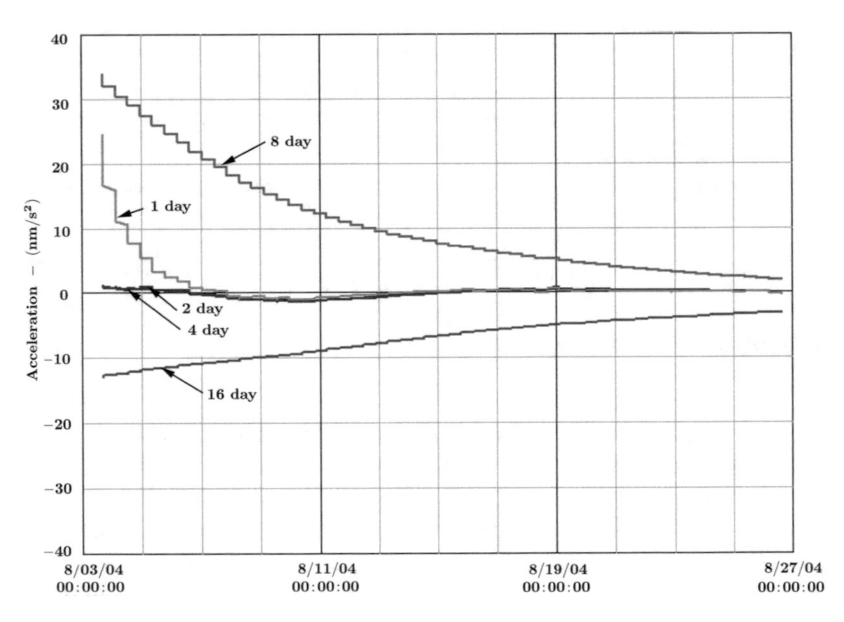

Fig. 9.22 Stochastic acceleration components along the sun line

After the supply of gas is largely exhausted, the longer time constant, less volatile elements then dominate, but at a diminished amplitude.

The filter can go unstable if too much stochastic white noise is introduced. It is desirable to have a stochastic model with coefficients that vary slowly as the acceleration is changing. Since the correlation time of each component is fixed, the filter adjusts the amplitude of each component at the beginning of each batch of data that is processed. This adjustment must be small since the *a priori* white noise is only 1.0 nm/s^2 while the observed acceleration is as high as 40.0 nm/s^2. The exponential character of each of the components confirms that small adjustments are taking place.

Figure 9.23 shows Doppler and range residuals for a long arc solution from launch to December 9, 2004. The filter estimation strategy is the same, but two additional maneuvers on September 24, 2004 18:00:46 GMT and November 18, 2004 19:31:04 GMT are included. The rms range error is less than 70 range units or 10 m over the entire data arc. The solution mapped to the B-plane at Earth return in August of 2005 is within a few hundred km of short arc solutions with data through May of 2005.

Figure 9.24 shows the outgassing acceleration history for the long arc solution. The solution is good through mid November where accelerations of about 5.0×10^{-12} nm/s^2 appear. These are probably associated with some unmodeled accelerations related to attitude and solar pressure mismodeling. Attempts to extend the long arc solution beyond December 9, 2004 into January of 2005 resulted in the solution migrating a few thousand km in the b-plane at Earth return on August 3, 2005. While this error is small for this time in the mission, orbit determination and long term prediction require high precision modeling of small accelerations on the order of 1.0×10^{-12} nm/s^2.

9.5.2 *Estimated Accelerations from Assumed Water Vapor*

Figure 9.24 indicated that the magnitude of the unmodeled accelerations was approximately 40.0 nm/s^2. If water vapor is the dominant source, it would be of interest to compute the amount of water necessary to cause the observed accelerations. The kinetic theory of gases will be used to determine the net thrust. Gas vented to space applies an acceleration to a spacecraft in a manner similar to that of a rocket engine. A small amount of gas released to space from a spacecraft will expand to the vacuum of space and exert a pressure against the spacecraft. Gas confined to a closed container would produce no net force. Gas evaporating from materials in the side of the spacecraft would expand against the side of the spacecraft, but be unobstructed in the direction away from the spacecraft, resulting in a net pressure transient and acceleration of the spacecraft. From Newton's laws, the pressure of a gas on a spacecraft is equal to the rate of change of momentum associated with the gas molecules striking the spacecraft.

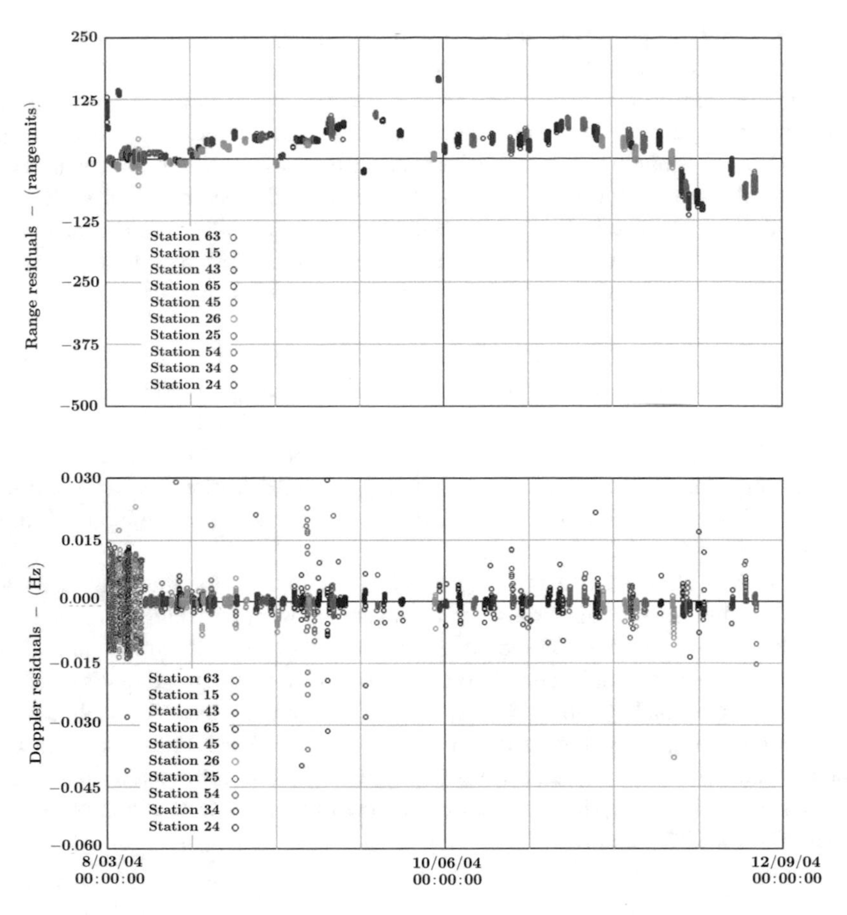

Fig. 9.23 Filter performance over 4 months

The velocity of a gas molecule may be obtained from the kinetic theory of gasses (Eq. 1.46) and is given by

$$\bar{v}^2 = \frac{3RT}{M}$$

If it is assumed that the gas is vented at a constant rate, then from Newton's law,

$$F = \bar{v}\frac{dm}{dt} = M_{sc}A_{sc}$$

where m is the mass of gas vented, M_{sc} is the spacecraft mass, and A_{sc} is the spacecraft acceleration. Solving for the gas mass flow rate,

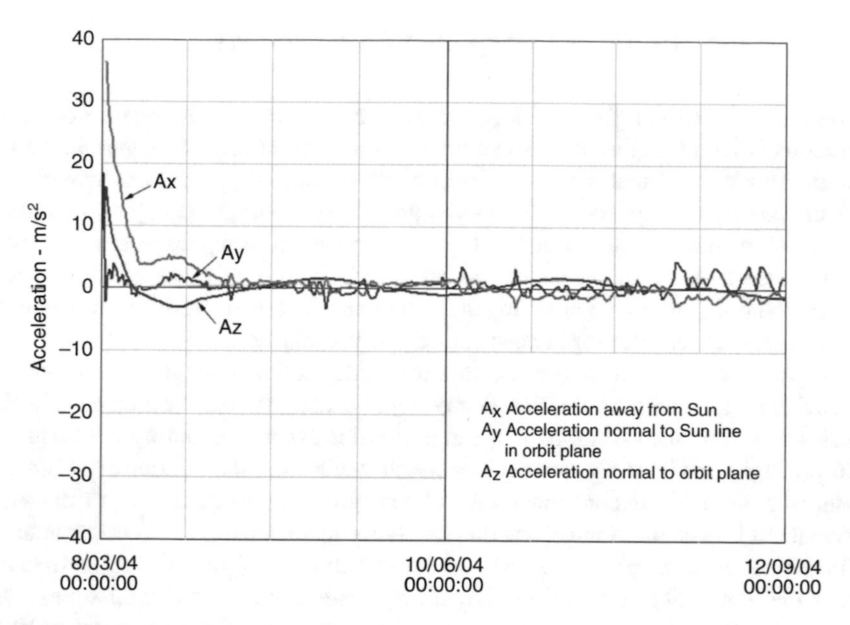

Fig. 9.24 Outgassing solution over 4 months

$$\frac{dm}{dt} = \frac{M_{sc}}{\sqrt{\frac{3RT}{M}}} A_{sc}$$

From Fig. 9.24, the spacecraft acceleration component attributable to outgassing is observed to decay exponentially after launch with a time constant (τ) of about 5 days.

$$A_{sc} = A_0 e^{\frac{-t}{\tau}}$$

where $A_0 = 4.0 \times 10^{-8}$ m/s and $\tau = 432,000$ s. After substituting the spacecraft acceleration and integrating from 0 to ∞, the total mass of gas expended is

$$\Delta m = \frac{M_{sc} A_0 \tau}{\sqrt{\frac{3RT}{M}}}$$

If the gas expended is water vapor ($M = 0.018$ kg/mol) at room temperature (293° K), the total amount of water vapor expended is approximately 0.014 kg. For this calculation, the gas constant (R) is 8314 kg $\cdot$ m^2/(s$^2 \cdot$ mol $\cdot$ K), and the mass of the spacecraft (M_{sc}) is 512 kg. Using 61 cubic inches of water per kilogram, less than one strategically placed cubic inch of water can cause the observed acceleration. Somewhat more water would be needed if distributed over the spacecraft.

9.5.3 Curve Fitting with Exponential Functions

The functions chosen for modeling are stochastic exponentials because one such function is built into the stochastic model and filter used for orbit determination. It might be expected that a series of exponentially decaying functions would only be effective when applied to the modeling of exponentially decaying processes. However, a truncated series of this type can be shown to represent any piecewise continuous function on an interval as effectively as that of other truncated series representations by common orthogonal functions that arise from Sturm-Liouville problems, such as half-range expansions or polynomial series.

Figure 9.25 shows a comparison of a triangle function, over an interval normalized from 0 to 1, with several fifth degree representations. The coefficients of a sine series was computed from the half range Fourier expansion and by least squares. Surprisingly, the least squares fit does much better than the Fourier expansion. A sum of 5 decaying exponentials with varying time constants was fit to the same triangle. It is even more surprising that the decaying exponentials did better than the Fourier series. The apparent explanation is that the least square fit is able to better alias the higher degree terms that are simply truncated by the Fourier series. One reason for the difference is the criterion for best estimate of f(t) on the interval [0,1]. The least squares best estimate $\hat{f}$ of the form

$$\hat{f}(x) = \sum_{n=1}^{N} C_n sin(nx)$$

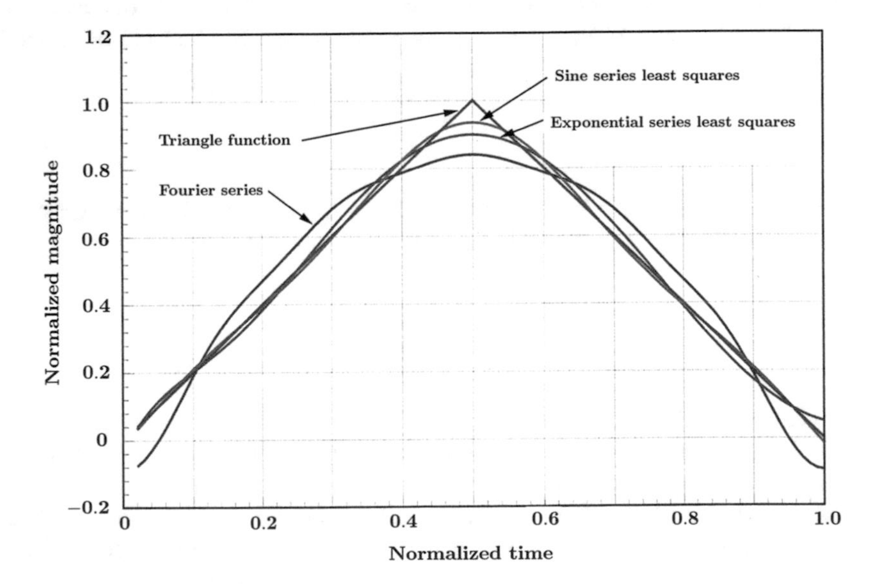

Fig. 9.25 Comparison of triangle function with various representations

is to choose the constants C_i to minimize

$$\sum_{i=1}^{m}\left(\hat{f}(x_i) - f(x_i)\right)^2$$

where $x_1, \ldots x_m$ are measurement points in the interval [0, 1]. The constants in the Fourier series estimate for f(x) are computed by integration on [0,1]

$$C_n = \int_0^1 f(x)\, sin(nx)dx$$

and minimize

$$\int_0^1 |\hat{f}(x) - f(x)|^2\, dt$$

In order to gain some insight into the nature of curve fitting versus expansions based on orthogonal functions, consider the ramp function shown in Fig. 9.26. The first five terms of the convergent Fourier half range expansion is compared with a least square fit to the same sine function series, a least square fit to a sum of decaying exponential functions and power series obtained from both the sine function and exponential function sum. The curve fitting and power series methods result in a high precision fit over the interval [0,1], while the Fourier series, by comparison, does not perform very well.

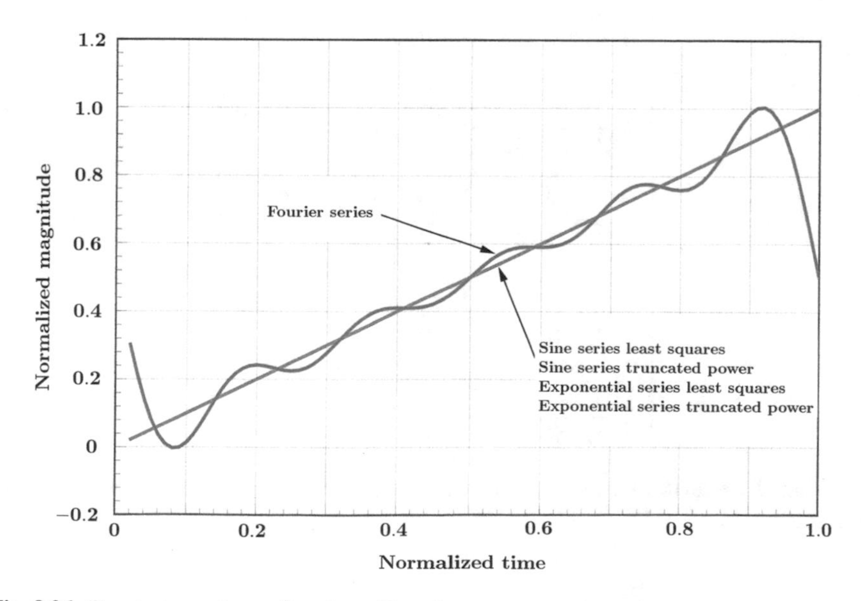

Fig. 9.26 Comparison of ramp function with various representations

The reason is obvious from close examination of the power series method for the decaying exponentials. We approximate each exponential function by the first five terms of its power series expansion.

$$e^{-xn} \approx 1 - nx + \frac{n^2}{2!}x^2 - \frac{n^3}{3!}x^3 + \frac{n^4}{4!}x^4 - \frac{n^5}{5!}x^5$$

Suppose an arbitrary function of x is represented as a sum of five exponentially decaying functions.

$$f(x) \approx \sum_{n=1}^{5} C_n\, e^{-xn}$$

so that

$$f(x) = \sum_{n=1}^{5} C_n (1 - nx + \frac{n^2}{2!}x^2 - \frac{n^3}{3!}x^3 + \frac{n^4}{4!}x^4 - \frac{n^5}{5!}x^5)$$

To fit the ramp function, the $\sum n\ C_n$ must be 1 and all the C_n associated with powers of x multiplied by the appropriate factors in the exponential series must sum to 0. The ramp function has only the linear term. All the other powers of x must be annihilated. In matrix notation the above equation becomes

$$\begin{bmatrix} 1 & 1 & 1 & 1 & 1 & 1 \\ 0 & -1 & -2 & -3 & -4 & -5 \\ 0 & \frac{1}{2!} & \frac{2^2}{2!} & \frac{3^2}{2!} & \frac{4^2}{2!} & \frac{5^2}{2!} \\ 0 & \frac{1}{3!} & \frac{2^3}{3!} & \frac{3^3}{3!} & \frac{4^3}{3!} & \frac{5^3}{3!} \\ 0 & \frac{1}{4!} & \frac{2^4}{4!} & \frac{3^4}{4!} & \frac{4^4}{4!} & \frac{5^4}{4!} \\ 0 & \frac{1}{5!} & \frac{2^5}{5!} & \frac{3^5}{5!} & \frac{4^5}{5!} & \frac{5^5}{5!} \end{bmatrix} \begin{bmatrix} C_0 \\ C_1 \\ C_2 \\ C_3 \\ C_4 \\ C_5 \end{bmatrix} = \begin{bmatrix} 0 \\ 1 \\ 0 \\ 0 \\ 0 \\ 0 \end{bmatrix}$$

Inverting the matrix and multiplying times the right side results in the following coefficients.

$$\{C_n\} = \{2\frac{17}{60},\ -5,\ 5,\ -3\frac{1}{3},\ 1\frac{1}{4},\ -\frac{1}{5}\}$$

The error in the power series expansion is caused by the higher than 5° terms of the exponential series that have been truncated. If a least square fit of the data over the interval 0–1 is performed, the error will be less because the truncation error is aliased. The coefficients for the least square fit that are directly comparable to C_n are

$$\{C_n\} = \{2.78, \ -8.15, \ 12.99, \ -13.50, 7.73, \ -1.85\}$$

Observe that the least square coefficients $\{C_n\}$ are greater than the power series coefficients $\{C_n\}$. Given any collection of functions with enough available powers of x, the least squares solution will effectively generate a power series representation of the function to be modeled.

A least squares series solution using a small number of functions is clearly better able to alias the error from truncating higher order terms, but as more terms are added to a least squares solution, all lower order constants C_i must be recomputed, and experimentation indicates the constants C_i can grow without bound. This observation is more apparent for the triangle function. The properties of orthogonal series of eigenfunctions such a Fourier series are well known, and may be found in any reference that discusses Sturm-Liouville theory and boundary value problems.

9.6 New Horizons

Navigation of the New Horizons spacecraft during approach to Pluto and its satellite Charon presented several new challenges related to the distance from the Earth and Sun and the dynamics of two body motion when the mass ratio results in the barycenter being outside the radius of the primary body. Since the Earth is about 30 a.u. from the spacecraft during the approach to Pluto and Charon, the round trip light time is greater than 8 h making two-way Doppler tracking difficult. The great distance from the Sun also reduces the visibility of Pluto since Pluto receives about 1/900 of the solar radiation as the Earth. The two body motion involves Pluto and Charon moving in elliptic orbits about each other, and the system mass is a simple function of the period and semi-major axis of the orbit. The period can be measured to high precision from Earth based telescope observations and the orbit diameter can be measured to a precision of perhaps 100 km enabling the system mass to be determined within 1 %.

The mass ratio or the allocation of mass between Pluto and Charon is more difficult to discern from Earth based observations. Pluto and Charon orbit about their barycenter in elliptical orbits whose semi-major axes are inversely proportional to their mass. Therefore, the mass ratio can only be determined by observing the motion on a star background over some time and removing the heliocentric orbital motion. Since these measurements are difficult to make from Earth based telescopes, it is expected that the mass ratio and corresponding orbit sizes about the barycenter will not be determined accurately until spacecraft based optical measurements are obtained during approach.

The New Horizons approach navigation strategy must be designed to enable precision determination of the Charon orbit about Pluto and the spacecraft orbit relative to Pluto and Charon as well as to control the spacecraft approach trajectory and deliver the spacecraft to a position for science observations. The initial navigation activity after detection of Pluto is to separate the orbit of Charon from the orbit of Pluto about their common barycenter. In order for these observations to be useful, Charon must be separated from Pluto by more than 100 pixels. As a by product, a more precise estimate of the system mass and spacecraft trajectory is obtained. As the spacecraft approaches the planetary system, the aim point relative to Pluto and the timing of Charon in its orbit is determined to an accuracy that permits an orbit correction maneuver to be executed so that the spacecraft is placed on the correct trajectory for science observations. If a substantial time adjustment is necessary to intercept Charon in its orbit about Pluto, it is important that this maneuver be performed as early as possible. Time change maneuvers are expensive to perform when the spacecraft is close to the Pluto/Charon system. During the approach to Pluto and Charon, the time of closest approach is not well determined. The error in the time of closest approach is proportional to the error in the range from Earth which is dominated by the Pluto and Charon ephemeris error. The distance of the spacecraft from Pluto and Charon cannot be determined with high precision until the spacecraft is close enough to observe the position parallax. This occurs during the final 1–2 days before encounter for Pluto only observations or during the final 2–4 days before encounter for observations of both Pluto and Charon. The timing of science observations during flyby is critically dependent on knowledge of the time of closest approach. A late update of the encounter sequence timing is planned based on optical navigation images acquired during approach.

9.6.1 Pluto and Charon Approach

The Pluto/Charon approach phase begins at about 120 days prior to Pluto encounter. Navigation activities that are performed during the approach phase include initial detection of Pluto, search for co-orbitals, ephemeris refinement, and a sequence of approach maneuvers that are designed to place the spacecraft on a trajectory that is optimum for science observations. The detection of Pluto as early as possible is advantageous from the standpoint of ephemeris verification and improvement to assure early tracking in support of the initial approach TCM and subsequent maneuvers. The optical measurement is obtained from an image of either Pluto or Charon using the LORRI or MVIC camera. The accuracy of this data type is a function of the picture element (pixel) spacing and the focal length of the camera optics. For the LORRI camera, the resolution is about 5 μrad per pixel and for the MVIC the resolution is about 20 μrad per pixel. Detection depends on Pluto's brightness as seen from the spacecraft and the imaging cameras sensitivity. The sensitivity of the camera depends on its light gathering capability (i.e., its aperture), the lens/filter/sensor light transfer and conversion efficiency, and the

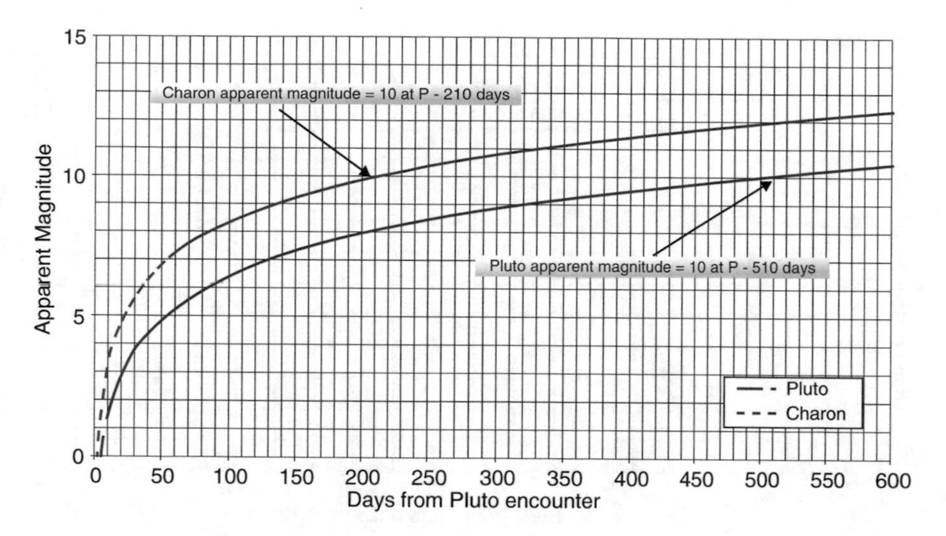

Fig. 9.27 Pluto/Charon apparent magnitude

various electronic processes that produce image noise. The MVIC and LORRI cameras are able to detect an object when the brightness is greater than magnitude 10. Figure 9.27 gives the apparent magnitude of Pluto and Charon as the spacecraft approaches the system. The figure shows the point on approach at which the theoretical apparent magnitude is brighter than magnitude 10 for Pluto and Charon. Pluto reaches apparent magnitude 10 at about 510 days prior to closest approach and Charon at about 210 days prior to closest approach.

As the spacecraft range closes, Pluto's image will become brighter and expand thus improving the optical navigation image location accuracy. Optical navigation begins to exceed the performance of Earth-based observations approximately when the spacecraft camera resolution exceeds that of Earth-based telescopes. Figure 9.28 shows the resolution of the two spacecraft cameras as a function of time from Pluto. A value of 0.043 arcsec/pixel is used as the reference Hubble resolution of the Wide Field Planetary Camera. The plot shows that resolution of the primary MVIC camera becomes better than that of Earth-based observations at 44 days before the encounter, whereas the LORRI camera resolution exceeds the Earth-based resolution at 170 days before the encounter.

9.6.2 Pluto Approach Time-of-Flight Determination

As the spacecraft approaches Pluto and Charon from a great distance, the orbit determination error relative to Pluto is a statistical combination of the independently determined spacecraft and Pluto ephemeris errors. The spacecraft is initially too far from Pluto to make use of direct observation of Pluto. As the spacecraft enters

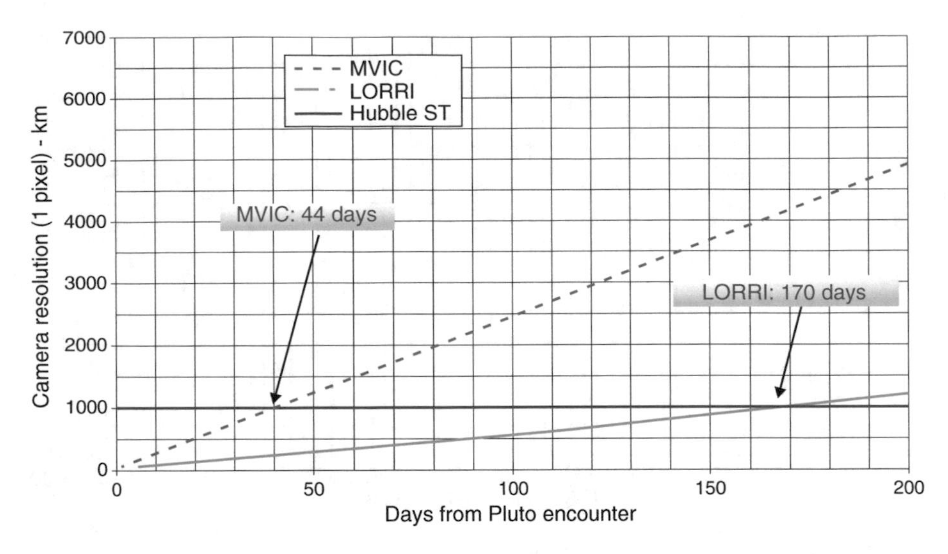

Fig. 9.28 Camera resolution versus hubble space telescope

Pluto's sphere of influence, the Doppler and range measurements are able to detect the gravitational acceleration of the Pluto/Charon system and the onboard optical navigation camera is able to detect the Pluto and Charon angular position on the star background. These measurements, either processed separately or in combination, can determine the approach velocity and position in the B-plane with high precision. The velocity determination error is within 1 mm/s in all three Cartesian components and the position error is about 5 μrad times the range from the spacecraft to Pluto for the two Cartesian components in the B-plane. The third component of position, along the down track or time-of-flight direction, is not very well determined. The time-of-flight error is determined by observation of the Pluto gravitational acceleration by the Doppler and range data or the position parallax associated with the angular motion of Pluto and Charon on the star background.

For Doppler data, an approximate analytic formula for the time-of-flight error may be derived that provides insight into the problem of time-of-flight or range-to-go determination. The time-of-flight error is simply the range-to-go distance error times the approach velocity (V_∞). The geometry is illustrated in Fig. 9.29. As the spacecraft approaches Pluto, it is accelerated by Pluto's gravity. The approach velocity magnitude and direction is known to very high precision as a result of tracking the spacecraft and observing Pluto's motion for years. The velocity along the line-of-sight from Earth ($\dot{\rho}$) can also be measured with high precision by the DSN. The change in velocity magnitude is given by

$$\Delta v = \dot{r} - V_\infty$$

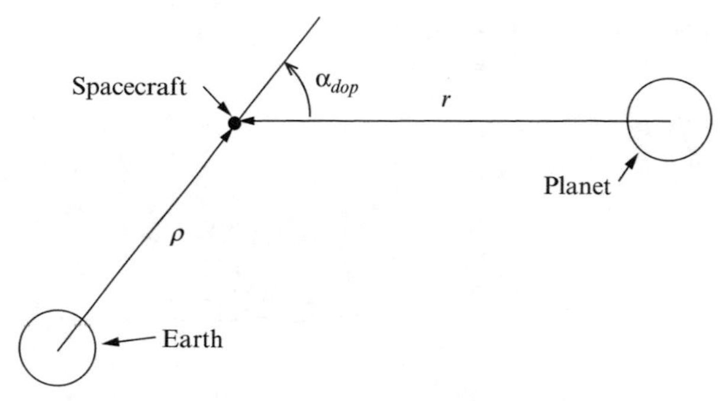

Fig. 9.29 Planetary approach doppler measurement accuracy

The velocity change can be obtained by integrating the gravitational acceleration during approach to Pluto.

$$\Delta v = \int_{-\infty}^{t} \frac{GM_p}{r^2}\, dt$$

Since the integrated acceleration is small relative to V_∞, the range may be approximated by

$$r \approx V_\infty\, t$$

and

$$\Delta v \approx \int_{-\infty}^{t} \frac{GM_p}{V_\infty^2\, t^2}\, dt$$

resulting in

$$\Delta v \approx \frac{-GM_p}{V_\infty^2\, t}$$

The sensitivity of the approach velocity with respect to time-of-fight variation is obtained by taking the partial derivative.

$$\delta \Delta v \approx \frac{GM_p}{V_\infty^2\, t^2}\, \delta t$$

The sensitivity of the Earth line-of-sight range rate to approach velocity is simply its projection onto the approach velocity vector and

$$\delta \dot{\rho} \approx \frac{GM_p \cos \alpha_{dop}}{V_\infty^2\, t^2}\, \delta t$$

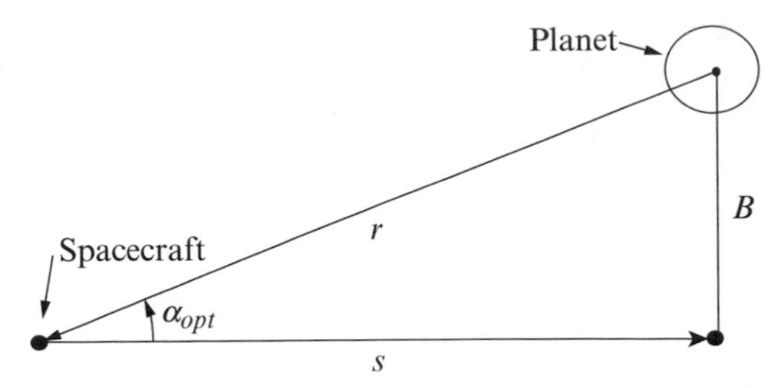

Fig. 9.30 Planetary approach optical measurement accuracy

A formula for the time-of-flight error ($\sigma(t)$) as a function of time from Pluto and Doppler measurement error is then given by

$$\sigma(t) \approx \frac{V_\infty^2 \, t^2}{GM_p \, \cos\alpha_{dop}} \, \sigma(\dot{\rho})$$

For optical data, another approximate formula may be derived for the time-of-flight error. The geometry is illustrated on Fig. 9.30. As the spacecraft approaches Pluto from a great distance, images of Pluto on a star background provide a strong determination of the direction of the approach asymptote. As the spacecraft approaches Pluto, the angular position of Pluto on the star background will begin to move away from the approach asymptote direction because of position parallax. The observation of Pluto's motion on the star background may be used to determine the range-to-go and time-of-flight.

From the geometry, the tangent of the angle between Pluto and the approach velocity vector is given by

$$\tan\alpha_{opt} = \frac{B}{s}$$

The distance (s) from the spacecraft to the B-plane may be approximated by

$$s \approx V_\infty \, t$$

Taking the partial derivative of α_{opt} with respect to t gives

$$\sec^2\alpha_{opt} \, \delta\alpha_{opt} \approx \frac{-B}{V_\infty \, t^2} \, \delta t$$

The formula for the time-of-flight error as a function of time from Pluto closest approach and optical measurement error is then given by

$$\sigma(t) \approx \frac{V_\infty \, t^2}{B \cos^2\alpha_{opt}} \, \sigma(\alpha_{opt})$$

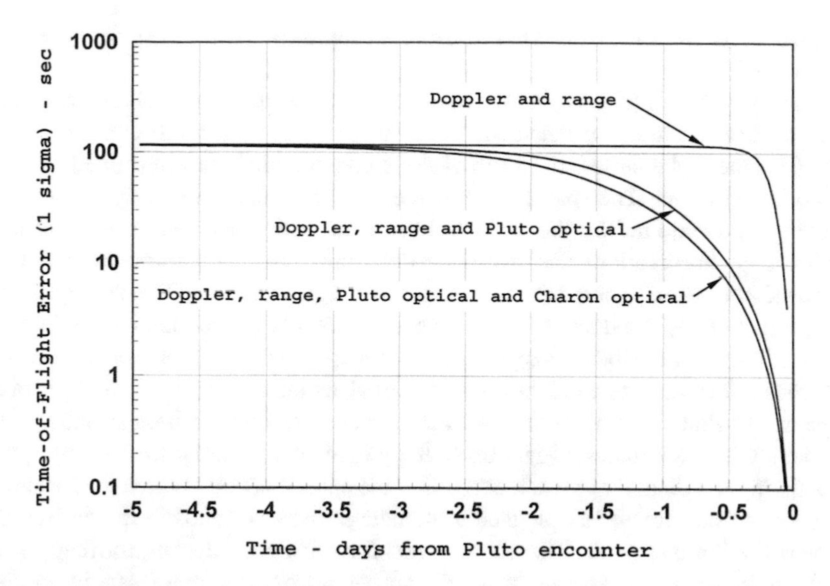

Fig. 9.31 Analytic Pluto approach time-of-flight error

The error in the time-of-flight estimation as a function of time from Pluto closest approach is shown in Fig. 9.31 for Doppler data, Doppler and Pluto optical data and Doppler, Pluto optical and Charon optical data. For this analysis V_{∞} is 13.7 km/s, B is 13,000 km for the aim point and 19,000 km for the radius of Charon's orbit and α_{dop} is 14°. The Doppler and optical measurement errors are 1.1 mm/s and 5 μrad, respectively, assuming the LORRI camera. Pluto observation science requires knowledge of the time-of-flight to be less than 100 s which corresponds to about 1300 km down track error. As shown in Fig. 9.29, the Doppler only orbit determination error does not decrease below 100 s until about 6 h before closest approach, too late to be of use for a science instrument pointing update. With optical data, the time-of-flight error is about 39 s 1 day before Pluto closest approach. The addition of Charon optical data decreases the error about 50% from that obtained with only Pluto optical data. The Charon orbit baseline is about 50% greater than the baseline provided by the approach asymptote aim point and Pluto.

9.6.3 *Pluto and Charon Approach Covariance Analysis*

A detailed covariance analysis was performed of navigation and orbit errors during approach to Pluto and Charon. This analysis included all the error sources that affect navigation accuracy and the data acquisition strategy that will be used. The orbit determination error is determined by filtering simulated data using the same square root information filter that will be used for flight operations. The filtered best estimate of the orbit is mapped to Pluto closest approach to provide a common basis for comparison.

During approach to the Pluto/Charon system, the spacecraft orbit determination error relative to Pluto is a statistical combination of the spacecraft ephemeris error and the planet ephemeris error. Both of these ephemerides are determined by Earth based observations. The spacecraft ephemeris error is determined by radio metric tracking data acquired by the DSN and the planet ephemeris error is determined by telescope observations. Both of these determinations are accurate to about one thousand kilometers. For the Pluto ephemeris error, it is assumed that an observation campaign will be conducted about 1 year before Pluto encounter to reduce the effect of long-term velocity mapping errors. An approach ephemeris error of several thousand kilometers is sufficient to ensure initial acquisition of Pluto and control the approach to Pluto/Charon until optical data is acquired by the imager onboard the spacecraft. The approach navigation strategy is to acquire radio metric and optical data during the distant approach and refine the spacecraft orbit relative to Pluto. A sequence of maneuvers are planned to maneuver the spacecraft to the desired aim point in the Pluto B-plane. These maneuvers are scheduled after performing a trade between the improved knowledge of the spacecraft orbit as more data is acquired and the cost in propellant of delaying the adjustment to the aim point.

The estimated parameters during approach to Pluto and Charon include spacecraft state, propulsive maneuver components, solar pressure model parameters, stochastic accelerations and Pluto and Charon ephemerides, gravity, pole, prime meridian, and rotation rate. For navigation, the accuracy of spacecraft ephemeris estimation is of prime interest. The sensitivity of the spacecraft ephemeris estimation error for various data acquisition strategies in shown in Figs. 9.32 through 9.34. On these figures, the spacecraft position error is shown as a function of time from Pluto closest approach mapped to the Pluto B-plane. Figure 9.32 shows the spacecraft orbit determination error for Doppler and range data starting at 20 days before Pluto encounter and continuing to encounter. Prior to encounter minus 20 days, the Doppler and range data cannot measure the Pluto/Charon gravitational acceleration and the orbit is determined from Earth based observations. As can be seen in Fig. 9.32, the Doppler and range orbit determination error does not improve until a few hours before encounter which is consistent with the analytic result shown in Fig. 9.31 for the time-of-flight error. Studies of planetary approach orbit determination show that the B-plane position errors are theoretically one half the down track position or time-of-flight error. Since the down track (s) position error is equal to V_∞ times the time-of-flight error the numerical values shown in Fig. 9.32 a few hours before encounter for the B-plane position are roughly a factor of 26 times the time-of-flight numerical values allowing for the mixed units.

Figure 9.33 shows the spacecraft orbit determination error when optical observations of Pluto are included with the radio metric data. The components of the spacecraft position error in the B-plane, the plane normal to the approach velocity vector, are reduced proportional to the range from Pluto times the angular measurement error. The angular measurement error is 5 μrad associated with the LORRI camera. For the MVIC camera, these results should be inflated by about

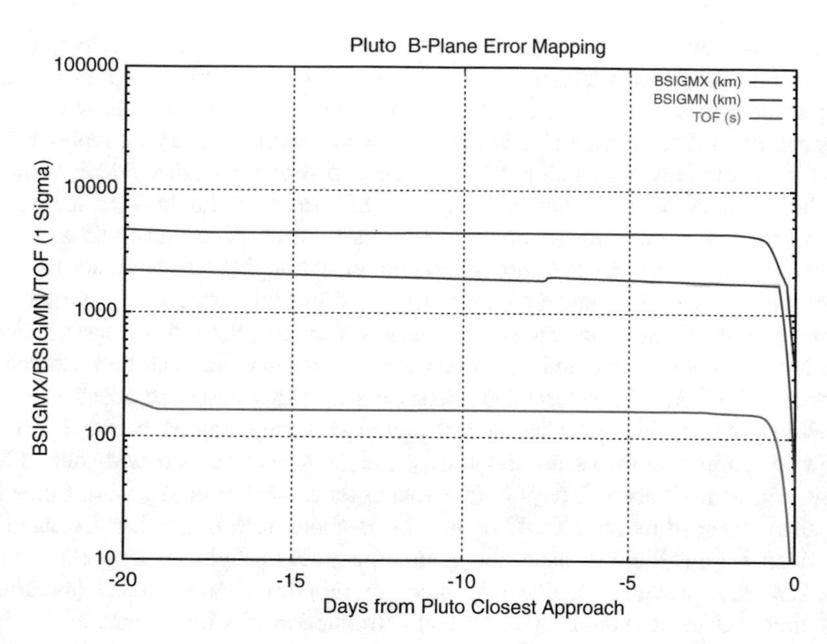

Fig. 9.32 Pluto approach—Doppler and range data only

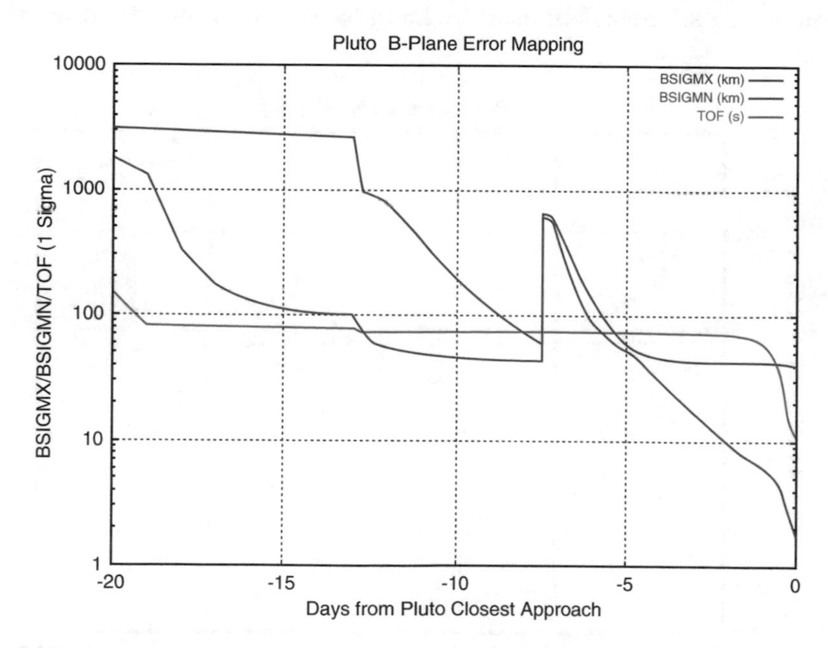

Fig. 9.33 Pluto approach—Doppler, range and optical imaging of Pluto

a factor of four (20 μrad). Several events occur during approach that temporarily distort this simplified analysis. Most notably, a propulsive maneuver executed at encounter minus 7 days inflates the mapped orbit determination error until the Doppler and range data are able to resolve the spacecraft velocity a couple of days later. At about encounter minus 13 days, the B-plane errors are suddenly reduced by the introduction of new optical images. A uniform reduction in B-plane position error occurs when the images are acquired on a uniform time schedule as occurs from encounter minus 13 days through encounter. Throughout the approach to Pluto, until a few days before encounter, the time-of-flight error remains essentially the same as for the Doppler and range only case shown in Fig. 9.32. As discussed above, the time-of-flight error is reduced when the position parallax can be observed as shown in Fig. 9.31. This occurs about 1 day before Pluto closest approach.

When observations of Charon are included during approach, the approach orbit determination errors are essentially the same as obtained with only Pluto observations until about 2 days before encounter as shown in Fig. 9.34. Since the baseline provided by the Pluto/Charon orbit is about 50% larger than the baseline provided by the Pluto/B-plane aim point, the time-of-flight error is about 50% smaller. This provides significantly more margin for a late science instrument pointing update. It should be noted that introduction of Charon data adds some complexity to the approach orbit estimation strategy. The orbit determination filter must be able to solve for the Charon orbit and gravity with considerably more accuracy than has been determined by Earth based observations. However, even

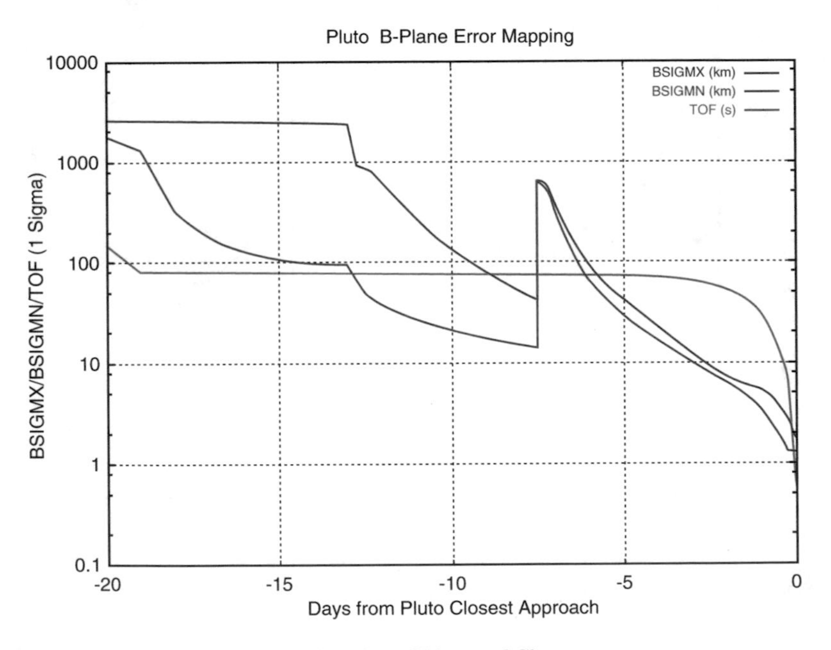

Fig. 9.34 Doppler, range and optical imaging of Pluto and Charon

with Pluto only observations, the dynamics of Pluto and Charon's orbit must be included in the solution. The barycenter of Pluto/Charon is outside the surface of Pluto and uniform motion of Pluto on a star background during approach cannot be assumed.

9.6.4 Spacecraft Orbit Reconstruction

As the spacecraft encounters the Pluto/Charon system, the gravitational perturbation of the spacecraft and observation of craters on both Pluto and Charon permit an accurate determination of the spacecraft trajectory and certain physical parameters that characterize both Pluto and Charon. The improvement in the spacecraft ephemeris during the encounter phase is useful for determining where the spacecraft is headed after encounter. An extended mission to the Kuiper belt is planned. The determination of Pluto and Charon physical parameters is useful for future missions to Pluto/Charon and is of interest for science investigations.

During the encounter phase, radio metric data and many images of Pluto and Charon are obtained. Table 9.11 shows the results of processing this data from before encounter until several days after encounter. In addition to the parameters estimated for approach navigation, some additional parameters are estimated that describe Pluto and Charon. These include the pole and prime meridian angles, rotation rate, gravitational parameters, and the location of craters on the surface of Pluto and Charon. Optical imaging of Pluto and Charon when combined with Doppler tracking data should enable determination of the location of craters relative to the respective centers of mass to an accuracy of about 100 m. The observations that are most useful are obtained within several hours of encounter. During this time interval, the bodies rotate several degrees enabling a determination of the poles, prime meridians and rotation rates. The accuracy is optimistically estimated to be about 0.05° as given in Table 9.11. The *a priori* values for the poles and rotation

Table 9.11 Pluto and Charon parameter estimation errors

		Error (1 sigma)	
Parameters	Pluto/Charon nominal values	Pluto	Charon
Pole and prime meridian			
α (deg)	313.02	0.05	0.075
δ (deg)	9.09	0.05	0.075
W (deg)	236.77	0.05	0.075
$\dot{W}$ (deg/day)	−56.3623195	3.0×10^{-4}	4.5×10^{-4}
Mass properties			
GM (km^3/s^2)	874.05/73.16	0.41×10^{-3}	0.33×10^{-2}
Gravity harmonics			
C_{20}	0/0	0.81×10^{-2}	Not available

rates assume that both Pluto and Charon rotate at the orbital period of the Charon orbit about Pluto and are thus in gravity lock. This assumption is not necessary for determination of these parameters since the data obtained during encounter will be much more powerful than Earth based observational data including the Hubble space telescope. The solution will not be *a priori* limited. If Pluto and Charon are in gravity lock and the orbit is not exactly circular, the pole and prime meridian angles will librate about their nominal values. Further study is needed to determine if the libration angles will be large enough to be detected.

Pluto and Charon gravitational parameters are determined from the Doppler tracking data acquired during encounter. The values for the errors given in Table 9.11 were obtained by detailed covariance analysis of simulated data. When combined with the volume of the bodies obtained from optical imaging, a refined estimate of the bulk densities may be obtained that will be at least an order of magnitude better than obtained from Earth based observations. Also, observation of the Charon orbit about Pluto will enable an even more precise determination of the system mass, an artifact of the equations of motion, that is of marginal interest to physical science but is essential for accurate ephemeris development. The normalized gravitational harmonic C_{20}, which is related to the oblateness, can be resolved to an accuracy of about 0.008 which is at least an order of magnitude greater than the nominal value expected from analysis of the spin and hydrodynamics. The flyby distances will probably be too great for the Doppler data to detect gravity harmonics.

In addition to the physical parameters of Pluto and Charon, the orbit of Charon about Pluto and the heliocentric orbit of the Pluto/Charon barycenter will be determined. This "normal point" will permit an improved ephemeris for these bodies.

9.7 Phobos

The Phobos gravity field provides useful insight into the physical makeup of Phobos and is needed for determination of the orbit of a spacecraft in the vicinity of Phobos. When combined with a figure model and observations of the forced libration of Phobos in its orbit about Mars, certain physical parameters such as mean density and moments of inertia may be determined.

Direct observation of the gravity field of Phobos is limited. The most accurate data that has been obtained is from tracking spacecraft that have flown by or orbited near Phobos. These observations have yielded an important determination of the mass. Indirect determination of the gravity field may be obtained from the figure model. An extensive map of the Phobos topography has been obtained from stereo imaging of the surface by the Mariner 9 and Viking missions. The resulting figure may be integrated to obtain gravity harmonic coefficients, the inertia tensor, and volume assuming constant density. The rotational motion of Phobos may then be integrated around one Martian orbit and the results compared with observation of the forced libration to obtain some insight into the internal structure of Phobos.

9.7.1 Phobos Inertial Properties

The combination of a photographic determination of the figure of Phobos with observations of spacecraft and Phobos dynamics provides an estimate of some of the inertial properties of Phobos. With this limited knowledge, some insight of the internal structure of Phobos may be inferred. The most useful information is contained in the low degree gravity harmonics and moments of inertia.

The key mass properties that have been determined are given in Table 9.12. The gravitation constant (GM) has been determined by tracking spacecraft that have flown in the immediate vicinity of Phobos, most recently the Soviet Phobos Mission, and the mass is obtained by simply dividing by the universal gravitation constant. Integration over the observed surface of Phobos gives the volume (V) and the mean density is simply the mass divided by the volume.
Integration of the first moment over the volume of Phobos gives the center of figure relative to the center of the planetocentric coordinate system. The center of the planetocentric coordinate system is the center of mass as determined by observation of spacecraft and Phobos dynamics. If the density of Phobos is uniform, the center of figure as defined by the above integration is also the center of mass. Thus, the difference may be attributed to inhomogeneity of Phobos or the accuracy of the data. The observed offset from the Viking measurements was too small to be significant and was incorporated into the reduction of the images. This supports a fairly uniform mass distribution for Phobos as far as can be determined from the first moment.

Another perspective of the mass distribution of Phobos may be obtained from the integration of the second moment over the volume of Phobos. The results of this integration for the inertia tensor are given in Table 9.13. The orientation of the Phobos centered coordinate axes are also defined by observation of spacecraft and Phobos dynamics. The x axis of Phobos points toward the center of Mars on the average and the z axis is normal to the Phobos orbit plane. Therefore, the Phobos centered body fixed axes should also be principal axes since this is the orientation that is attained in the steady state over many revolutions. From the cross products of inertia given in Table 9.13, we may compute the location of the principal axes of Phobos figure and these are given in Table 9.14. The offset of the

Table 9.12 Phobos mass properties

Parameter	Value	Units	Definition
GM	7.22×10^{-4}	km^3/s^2	Gravitation parameter
M	1.082×10^{16}	kg	Mass
V	5.673×10^{12}	m^3	Volume
ρ	1.91	g/cm^3	Mean density (cgs)

Table 9.13 Phobos inertia tensor

$$I = \begin{bmatrix} I_{xx} & I_{xy} & I_{xz} \\ I_{xy} & I_{yy} & I_{yz} \\ I_{xz} & I_{yz} & I_{zz} \end{bmatrix}$$

Phobos centered body fixed coordinates

$$I = \begin{bmatrix} 4.72 & -.0338 & -.0341 \\ -.0338 & 5.50 & -.0280 \\ -.0341 & -.0280 & 6.48 \end{bmatrix} \times 10^{23}\, kg - m^2$$

Phobos centered principal axes

$$I = \begin{bmatrix} 4.72 & 0 & 0 \\ 0 & 5.50 & 0 \\ 0 & 0 & 6.48 \end{bmatrix} \times 10^{23}\, kg - m^2$$

Table 9.14 Phobos principal axes of inertia directions

Axis	Latitude[a] (deg)	Longitude[a] (deg)
x_p	1.14	2.53
y_p	1.53	92.5
z_p	88.1	235.8

[a] Phobos centered inertial

figure principal axes of inertia from the Planetocentric coordinate system may be attributed to asymmetric mass distribution or accuracy of the data. The small offset that is observed also supports a uniform distribution of Phobos density within the accuracy of the observations.

9.7.2 Phobos Gravity Field

The gravity harmonic coefficients may also be obtained by integration over the figure of Phobos assuming constant density. The results of this integration are given in Table 9.15. The harmonic coefficients through degree and order two may be directly related to the center of gravity and part of the inertia tensor and thus provide essentially the same insight into the mass distribution as has been discussed above. For the center of gravity, these relationships are

$$x_{cg} = C_{11} r_0$$
$$y_{cg} = S_{11} r_0$$
$$z_{cg} = C_{10} r_0$$

and for the inertia tensor

$$I_{xx} - I_{yy} = -4 M r_o^2 C_{22}$$

Table 9.15 Phobos gravity field

$GM_p = 7.22 \times 10^{-4}\,\text{km}^3/\text{s}^2$, $r_0 = 11.1\,\text{km}$					
Coefficient	m = 0	m = 1	m = 2	m = 3	m = 4
C_{2m}	−0.1035E+00	0.2556E-02	0.1457E-01		
S_{2m}		0.2098E-02	0.1269E-02		
C_{3m}	0.8243E-02	−0.4800E-02	−0.3091E-02	0.6721E-03	
S_{3m}		0.2093E-02	−0.3110E-03	−0.1323E-02	
C_{4m}	0.1892E-01	0.1949E-02	−0.4758E-03	−0.2116E-03	0.6139E-04
S_{4m}		−0.9897E-03	−0.5637E-03	0.1983E-03	−0.3234E-04

$$
\begin{aligned}
I_{yy} - I_{zz} &= M r_o^2 (C_{20} + 2C_{22}) \\
I_{zz} - I_{xx} &= -M r_o^2 (C_{20} - 2C_{22}) \\
I_{xy} &= -2M r_o^2 S_{22} \\
I_{yz} &= -M r_o^2 S_{21} \\
I_{xz} &= -M r_o^2 C_{21}
\end{aligned}
\tag{9.1}
$$

The above six equations for the inertia tensor elements place constraints on the relationship between the inertia tensor and gravity harmonic coefficients. However, only five of these equations are independent. The third equation may be obtained by adding the first two equations. For this reason, it is not possible to completely describe the inertia tensor from gravity measurements alone.

The higher degree terms of the gravity field expansion are useful for predicting the acceleration of a spacecraft that orbits near Phobos. Thus, they are vital for precision navigation. In their own right, the higher order gravity harmonics provide some insight into the homogeneity of Phobos when they are compared with those obtained from the figure assuming constant density (i.e., the coefficients in Table 9.15). At this time, a direct determination of the higher order harmonic coefficients that would be obtained by tracking spacecraft is not available.

9.7.3 Phobos Rotational Dynamics

Consider the rotational equations of motion that relate the observed angular acceleration $\dot{\mathbf{\Omega}}$ and body-fixed spin rates to the applied moment ($\mathbf{M}$).

$$
\begin{aligned}
\mathbf{M} &= I\,\dot{\mathbf{\Omega}} + \mathbf{\Omega} \times \mathbf{H} \\
\mathbf{H} &= I\,\mathbf{\Omega}
\end{aligned}
$$

Table 9.16 Mars gravity field

$GM_m = 42828.44\,\text{km}^3/\text{s}^2,\quad r_m = 3,394.\,\text{km}$					
Coefficient	m = 0	m = 1	m = 2	m = 3	m = 4
C_{2m}	−0.1960E-02	0	−0.5473E-04		
S_{2m}		0	0.3140E-04		
C_{3m}	0.3145E-04	0.4477E-05	−0.5579E-05	0.4845E-05	
S_{3m}		0.2690E-04	0.2895E-05	0.3607E-05	
C_{4m}	−0.1889E-04	0.3494E-05	−0.2077E-06	0.4175E-06	−0.3614E-08
S_{4m}		0.3990E-05	−0.2199E-05	0.1625E-07	−0.2765E-06

As Phobos orbits Mars, Phobos is subjected to a torque from the gradient of the Mars gravity field. The total applied moment to Phobos is obtained by integrating this force times the moment arm over the density and volume of Phobos.

$$\mathbf{M} = \iiint_V \left(\mathbf{r} \times \frac{\mathbf{dF}}{\text{dm}}\right) \rho(\text{r}, \lambda, \phi)\,\text{dV} \tag{9.2}$$

The gravitational force (F) is exerted on an elemental volume element of mass dm. This force is obtained from the Mars gravity field and is given by

$$\frac{d\mathbf{F}}{\text{dm}} = f_g(\mathbf{r}_p, \alpha, \delta, W, GM_m, r_m, C_{Mnm}, S_{Mnm})$$

where $\mathbf{r}_p$ is the vector from the center of Mars to an elementary volume element of Phobos, GM_m is the gravitational constant of Mars, r_m is the reference radius of Mars, and C_{Mnm} and S_{Mnm} are the Mars gravity coefficients as determined by Balmino and are given in Table 9.16.

For the orbit of Phobos about Mars, a state vector was computed at Mars periapsis from the orbit elements given in Table 9.17. The above rotational equations of motion were integrated over one complete orbit of Mars in conjunction with the translational equations of motion and the moment obtained by simultaneous repeated integrations over Phobos's volume. Of particular interest is the forced libration in longitude of Phobos. This is simply the inertial attitude about the z axis minus the mean rotation about the same axis. The amplitude of 0.994° obtained by numerical integration provides some insight into the radial distribution of density.

9.7.4 Analytic Approximation of Forced Libration

The moment about Phobos's coordinate axes may be approximated by assuming that the Mars gravity gradient is constant over the entire volume of Phobos. With this assumption, the moment about Phobos is determined by the second degree gravity harmonics of Phobos which are related to a certain ratio of the moments of inertia.

Table 9.17 Initial conditions

Parameters	Values
Phobos ephemeris	Phobos-centered
$\mathbf{X_o} = \begin{bmatrix} a_{P_o} \\ e_{P_o} \\ \Omega_{P_o} \\ i_{P_o} \\ \omega_{P_o} \end{bmatrix}$	$\mathbf{X_o} = \begin{bmatrix} 9378.5 \text{ km} \\ .015364 \\ 242.703 \text{ deg} \\ 1.0324 \text{ deg} \\ 227.073 \text{ deg} \end{bmatrix}$
Phobos attitude and rates	Rotations from Phobos centered frame
$\mathbf{\Phi_o} = \begin{bmatrix} \alpha_o \\ \delta_o \\ W_o \\ \omega_{xo} \\ \omega_{yo} \\ \omega_{zo} \end{bmatrix}$	$\mathbf{\Phi_o} = \begin{bmatrix} 0 \text{ deg} \\ 90 \text{ deg} \\ 0 \text{ deg} \\ 1.077 \times 10^{-4} \text{ rad/s} \\ 1.077 \times 10^{-4} \text{ rad/s} \\ 8.725 \times 10^{-3} \text{ rad/s} \end{bmatrix}$

The gravitational force of Mars on an elementary mass element of Phobos may be approximated by

$$d\mathbf{F} = \frac{GM_m}{|\mathbf{r}_p - \mathbf{r}|^3}(\mathbf{r}_p - \mathbf{r})\, dm$$

where $\mathbf{r_p}$ is the vector from the center of Mars to the center of Phobos. Substituting the gravity force into the moment equation we obtain

$$\mathbf{M} = GM_m \iiint_V \frac{\mathbf{r} \times (\mathbf{r}_p - \mathbf{r})}{|\mathbf{r_p} - \mathbf{r}|^3} \rho(r, \lambda, \phi)\, dV \tag{9.3}$$

where the distance from Mars to the mass elements may be obtained by projecting the location of the mass elements onto the Mars-Phobos vector ignoring parallax,

$$|\mathbf{r}_p - \mathbf{r}| = r_p - \mathbf{r}_p \cdot \mathbf{r}$$

and the required inverse cube may be approximated by the first two terms of the Taylor series

$$\frac{1}{|\mathbf{r}_p - \mathbf{r}|^3} = \frac{1}{r_p^3}\left[\frac{r_p^2 + 3\mathbf{r}_p \cdot \mathbf{r}}{r_p^2}\right]$$

Replacing the vectors by components we obtain

$$\mathbf{M} = \frac{GM}{r_p^5} \iiint_V \begin{bmatrix} yz_p - zy_p \\ zx_p - xz_p \\ xy_p - yx_p \end{bmatrix} [r_p^2 + 3(xx_p + yy_p + zz_p)]\, \rho(r, \lambda, \phi) dV \tag{9.4}$$

Since the origin of the coordinate system is the center of mass, the first order terms in x, y, and z integrate to zero and the second order terms integrate to moments and products of inertia.

$$\mathbf{M} = \frac{3\, GM_m}{r_p^5} \begin{bmatrix} y_p z_p (I_{zz} - I_{yy}) + (y_p^2 - z_p^2) I_{yz} - x_p z_p I_{xy} + x_p y_p I_{xz} \\ x_p z_p (I_{xx} - I_{zz}) + (z_p^2 - x_p^2) I_{xz} - x_p y_p I_{yz} + y_p z_p I_{xy} \\ x_p y_p (I_{yy} - I_{xx}) + (x_p^2 - y_p^2) I_{xy} - y_p z_p I_{xz} + x_p z_p I_{yz} \end{bmatrix} \tag{9.5}$$

As Phobos rotates about Mars, the x coordinate axis very nearly points towards Mars and the moment about Phobos may be approximated by

$$\mathbf{M} \approx \frac{3\, GM_m}{r_p^5} \begin{bmatrix} 0 \\ 0 \\ x_p y_p (I_{yy} - I_{xx}) \end{bmatrix} \tag{9.6}$$

For small angular deviations of the x coordinate axis in longitude we have

$$\mathbf{M} \approx \frac{3\, GM_m}{r_p^3} \begin{bmatrix} 0 \\ 0 \\ \Delta\theta (I_{yy} - I_{xx}) \end{bmatrix} \tag{9.7}$$

The rotational equations of motion may also be simplified for the special case of rotation and moments only about the z axis. Thus Euler's equations of motion may be approximated by

$$\mathbf{M} \approx \begin{bmatrix} 0 \\ 0 \\ I_{zz}\, \ddot{\theta} \end{bmatrix} \tag{9.8}$$

We thus obtain the following second order differential equation for the rotation of Phobos.

$$\ddot{\theta} = \frac{3\, GM_m}{r_p^3} \frac{I_{yy} - I_{xx}}{I_{zz}} \Delta\theta$$

The forcing function is simply the difference between Phobos's x coordinate axis and the vector from Mars to Phobos which is defined by

$$\Delta\theta = \eta - \theta$$

where η is the true anomaly of Phobos orbital motion about Mars.

The orbital motion of Phobos about Mars may be described by Kepler's equation and we have

$$M = E - e \sin E$$

where M is now the mean anomaly, E is the eccentric anomaly, and e is the orbit eccentricity. We also have the relationships

$$M = \sqrt{\frac{GM_m}{a_p^3}}(t - t_p) = n(t - t_p)$$

where a_p is the semi major axis of Phobos orbit, t_p is the time of periapsis passage, and n is the mean motion. An approximate formula for η may be obtained by assuming the point on the circle that defines E is coincident with the point on the orbit that defined η. Thus we have

$$E - M = e \sin E \tag{9.9}$$

and from the geometry shown in Fig. 3.1

$$\sin(\eta - E) \approx \frac{c \sin E}{r}$$

If the point on the circle shown in Fig. 3.1 is coincident with the point on the ellipse, then $r \approx a \approx b, \quad \sin(\eta - E) \approx \eta - E, \quad M \approx E$ and since $c = ae$

$$\eta - E = e \sin M \tag{9.10}$$

Adding equations Eq. (9.9) and Eq. (9.10) an approximate equation for η as a function of time is obtained

$$\eta = nt + 2e \sin nt$$

The differential equation for the rotation of Phobos about the z axis thus becomes

$$\ddot{\theta} = 3\gamma n^2(nt + 2e \sin nt - \theta)$$

where

$$\gamma = \frac{I_{yy} - I_{xx}}{I_{zz}}$$

Taking the Laplace transformation we obtain

$$s^2\theta - s\theta_0 - \dot{\theta}_0 = \frac{3\gamma n^3}{s^2} + \frac{6\gamma e n^3}{(s^2+n^2)} - 3\gamma n^2\theta$$

Solving for θ as a function of s we obtain

$$\theta = \frac{s}{s^2+3\gamma n^2}\theta_0 + \frac{1}{s^2+3\gamma n^2}\dot{\theta}_0 + \frac{3\gamma n^3}{s^2(s^2+3\gamma n^2)} + \frac{6\gamma n^3 e}{(s^2+n^2)(s^2+3\gamma n^2)}$$

Transforming from the frequency domain back to the time domain we obtain

$$\theta = nt + \frac{6\gamma e}{3\gamma - 1}\sin nt + \theta_0 \cos(\sqrt{3\gamma}\, nt) + \frac{1}{\sqrt{3\gamma}}\sin(\sqrt{3\gamma}\, nt) \left[\frac{\dot{\theta}_0}{n} - 1 - \frac{6\gamma e}{3\gamma - 1}\right] \tag{9.11}$$

The above equation describes the rotation of Phobos as a function of the initial attitude(θ_0), initial attitude rate($\dot{\theta}_0$), gravity torque forcing function, and inertial properties of Phobos. Over many Phobos orbits, energy dissipation will result in the amplitude of the attitude oscillations attaining a minimum. This minimum energy condition imposes the following boundary condition on the initial attitude and attitude rate at periapsis.

$$\theta_0 = 0.$$

$$\dot{\theta}_0 = \left(1 + \frac{6\gamma e}{3\gamma - 1}\right) n$$

Substituting the boundary conditions into the equation of motion, we obtain for the attitude of Phobos

$$\theta = nt + \frac{6\gamma e}{3\gamma - 1}\sin nt + A_f \sin(nt\sqrt{3\gamma} + \theta_f) \tag{9.12}$$

Observe that this equation contains an additional term for the free libration of Phobos of amplitude A_f and phase θ_f. The minimum energy boundary condition results in this term vanishing except for a small residual that may be attributed to other external forcing functions and the initial attitude and rate that existed when Phobos became locked in rotation with Mars. The forced libration may be separated from the free libration through their respective frequency signatures. The amplitude of the forced libration from Eq. (9.12) is 0.978° where $\gamma = 0.12037$ and e = 0.0151. This compares very well with the result (0.9937°) obtained by numerical integration of the rotational equations of motion.

9.8 Summary

Navigation analyses are performed to verify that the navigation system will deliver a spacecraft to its destination and satisfy mission constraints associated with the spacecraft design and acquisition of science data. These objectives are satisfied by designing the trajectory and computing the probabilities of successful execution of the mission. A secondary purpose of analysis is to anticipate possible failure modes and design procedures for recovery. Analyses that have been performed in support of various missions have been described in some detail. These examples have been selected from a long list of navigation design and operations that have been experienced over many years. The criterion for selection of examples are those that provide insight into the navigation system and operations. Some of the more dramatic examples have not been included. Little is learned when a spacecraft blows up or falls in the ocean.

The first example of navigation analysis is associated with planetary quarantine. Since this was a major concern of the Viking mission to Mars, the analysis is described in detail. The next example is the Galileo probe delivery. This problem was of interest because of the narrow entry corridor that was targeted 150 days before arrival at Jupiter. The Pioneer (PVO) mission provided an opportunity to analyze atmospheric entry. The Magellan mission used aerodynamic braking to control the orbit around Venus. The NEAR mission was analyzed in detail since this mission presented many new challenges and the introduction of new data types. The MESSENGER post launch perturbation from outgassing provided an opportunity to analyze small nongravitational accelerations. The New Horizons spacecraft was the first to navigate to Pluto. The long round trip light time of 8 h presented many challenges in orbit determination and data acquisition. Finally, the numerous missions to Phobos, a satellite of Mars, provided data that was used to predict the libration of Phobos. Since small perturbations to the attitude of a body can have a large effect on the gravity field determination, future analysis of missions to binary comets or asteroids will benefit.

Exercises

9.1 During approach to Jupiter, the original navigation design involved separating the probe from the orbiter at encounter minus 50 days at a cost of 50 m/s. The 50 m/s was needed to deflect the orbiter from an entry trajectory to the orbit insertion aim point. In order to save ΔV, the separation was moved back to encounter minus 150 days. Determine the savings in ΔV.

9.2 Show that for a heading angle due East in the same direction as Jupiter's rotation, the relative entry angle is given by

$$\tan \gamma_e = \frac{\sin \gamma_I}{\cos \gamma_I - \frac{\omega_J\, r_e}{V_I}}$$

$$\cos \gamma_I = \frac{B}{\sqrt{r_e \left(\frac{2\,GM}{V_\infty^2} + r_e\right)}}$$

where ω_J is Jupiter's rotation rate, r_e is the entry radius, V_I is the inertial velocity at entry, V_∞ is the hyperbolic excess velocity, GM is Jupiter's gravitational constant, and γ_i is the inertial entry angle.

9.3 For the separation at minus 150 days in Exercise 9.1, the 99% impact parameter delivery error is ± 1300 km. Determine the 99% relative entry angle delivery error for $V_\infty = 5.86$ km/s, B = 724, 300 km, $r_e = 71,398$ km, $V_i = 59.9$ km/s, $\omega_j = 1.77 \times 10^{-4}$ rad/s and GM = 0.126×10^9 km^3/s^2.

9.4 For the estimation of atmospheric drag on a spacecraft in orbit about a planet with an atmosphere, the relationship between the velocity change at periapsis and the period of the orbit is needed. Show that

$$\frac{\partial P}{\partial v_p} = 6\pi \, \frac{v_p \, a^{\frac{5}{2}}}{GM^{\frac{3}{2}}}$$

9.5 A spacecraft is in a circular 30 km orbit about Eros. The plane of the orbit $(x-y)$ is perpendicular to the sun line. The sun is in the plus z direction. A maneuver is executed as the spacecraft crosses the x axis that places the spacecraft in a circular orbit that flies over the sub-solar point which is on the z axis. The spacecraft makes a 90° turn and flies over the subsolar point on the surface of Eros. Determine the maneuver components in the inertial x, y, z coordinate system. The gravitational parameter of Eros is 4.463×10^{-4} km^3/s^2.

9.6 In turning the spacecraft to the maneuver attitude in Exercise 9.4, an error results in the rocket motor being pointed in the direction of the spacecraft velocity vector. The thrust is in the opposite direction that the rocket motor is pointed. The magnitude of the burn or ΔV remains the same as for subsolar over-flight. Determine the periapsis radius of the resulting trajectory.

9.7 A model of Phobos is suspended on the surface of the Earth with an axel along the z axis and the x axis pointing down. If Phobos is perturbed by a small torque about the z axis, determine the period of oscillation. The moments of inertia of Phobos are $I_{xx} = 4.72$, $I_{yy} = 5.50$, $I_{zz} = 6.48$, $GM_e = 398,600$ km^3/s^2, $\mathrm{r_e} = 6378$ km. From the observed period, the gravity gradient may be determined and thus we have a crude gravity gradiometer.

Bibliography

Bursa, M., Z. Martinec, K. Pec 1990. Principal Moments of Inertia, Secular Love Number and Origin of Phobos. Adv. Space Res. Vol 10, No 3–4, pp.(3)67–(3)70.

Capen, E. B. and A. E. Joseph, "Software Requirements Document Viking Project - MOIOP", JPL Report, 1973.

Dunham, D. W., McAdams, 1. V., Mosher, L. E. and Helfrich, C. E., "Maneuver Strategy for NEAR's Rendezvous with 433 Eros", Paper IAF-97-A.4.01 presented at the 48th International Astronautical Federation Congress, Turin, Italy, Oct. 6–10,1997.

Dunham, D. W., et al, 1999. Recovery of NEARŠs ăMission to Eros. International Astronautical Congress Paper IAF-99-Q.5.05, Amsterdam, The Netherlands.

Duxbury, T. C. 1989. The Figure of Phobos.*Icarus* **78**, 169–180.

Farquhar, R. W., ed., 1995. Special Issue on theăNear Earth Asteroid Rendezvous Mission. J. Astronaut. Sci. 43.

Farquhar, R. W., Dunham, D. W. and McAdams,1. V., "NEAR Mission Overview and Trajectory Design," J Astron. Sciences, Vol. 43, No.4, Oct.-Dec. 1995, pp. 353–371.

Guo, Y., R. W. Farquhar, "New Horizons Mission Design for the Pluto-Kuiper Belt Mission", AIAA paper 2002–4722, AIAA/AAS Astrodynamics Specialist Conference, Monterey, California, August 5–8, 2002.

Guo, Y., R. W. Farquhar, "New Horizons Pluto-Kuiper Belt Mission: Design and Simulation of the Pluto-Charon Encounter", IAC paper 02-Q.2.07, 53rd International Astronautical Congress The World Space Congress-2002, Houston, Texas, October 5–8, 2002.

Guo, Y., R. W. Farquhar, "New Horizons Mission Design for the Pluto-Kuiper Belt Mission", AIAA paper 2002–4722, AIAA/AAS Astrodynamics Specialist Conference, Monterey, California, August 5–8, 2002.

Guo, Y., R. W. Farquhar, "New Horizons Pluto-Kuiper Belt Mission: Design and Simulation of the Pluto-Charon Encounter", IAC paper 02-Q.2.07, 53rd International Astronautical Congress The World Space Congress-2002, Houston, Texas, October 5–8, 2002.

Konopliv, A.S., W.B. Banerdt, and W.L. Sjogren 1999. Venus Gravity: 180th Degree and Order Model. Icarus 139, 3–18.

Keating, G. M., J. Y. NicholsonIII, and L. R. Lake, "Venus Upper Atmosphere Structure", *Journal of Geophysical Research*, vol. 85, no. A13, pp 7941–7956, December 30, 1980.

King-Hele, D., *Theory of Satellite Orbits in an Atmosphere*, Butterworths, London, 1964.

Kolyuka, Y. F., S. M. Kudryavtsev, V. P. Tarasov, V. F. Tikhonov, N. M. Ivanov, V. S. Polyakov, V. N. Potchukaev, O. V. Papkov, E. L. Akim, R. R. Nasirov 1990. Report of the International Project "Phobos" Experiment "Celestial Mechanics", in press.

McAdams, 1. V., Dunham, D. W., Helfrich, C. E., Mosher, L. E. and Ray, 1. c., "Maneuver History of the NEAR Mission: Launch through Earth Swingby Phase," Paper rSTS 98-c-23, 21st International Symposium on Space Technology and Science, Sonic City, Omiya Japan, May 24–31, 1998.

McAdams, J. V., D. W. Dunham, R. W. Farquar, T. H. Taylor, B. G. Williams, "Trajectory Design and Maneuver Strategy for the MESSENGER Mission to Mercury", Paper 05–173, 15th AAS/AIAA Space Flight Mechanics Conference, Copper Mountain,CO, January 23–27, 2005.

Miller, J.K. and F. T. Nicholson, "Galileo Jupiter Approach Orbit Determination", *The Journal of the Astronautical Sciences*, Vol 32, No 1, January-March, 1984, pp 63–79.

Miller, J. K., Weeks, C. J. and Wood, L. J., "Orbit Determination Strategy and Accuracy for a Comet Rendezvous Mission," J Guidance, Control, and Dynamics, Vol. 13, No.5, Sep.-Oct. 1990., pp. 775–784.

Miller, J. K., Williams, B. G., Bollman, W. E., Davis, R. P., Helfrich, C. E., Scheeres, D. 1., Synnott, S. P., Wang, T. C. and Yeomans, D. K., "Navigation Analysis for Eros Rendezvous and Orbital Phases," J Astron. Sciences, Vol. 43, No.4, Oct.-Dec. 1995, pp. 453–476.

Miller, J. K., et. al., “Navigation Analysis for Eros Rendezvous and Orbital Phases”, *The Journal of the Astronautical Sciences*, Vol 43, No 4, October-December 1995, pp 453–476

Miller, J. K., P. G. Antreasian, R. W. Gaskell, J. Giorgini, C. E. Helfrich, W. M. Owen, B. G. Williams and D. K. Yeomans, Determimation of Eros Physical ăParameters for NEAR Earth Asteroid RendezvousăOrbit Phase Navigation. AAS paper 99–463, AAS/AIAA Astrodynamics Specialist Conference, Girdwood, Alaska., 1999

Miller, J. K., P. G. Antreasian, J. J. Bordi, S. Chesley, C. E. Helfrich, A. Konopliv, W. M. Owen, T. C. Wang, B. G. Williams and D. K. Yeomans, Determination of ErosŠ Physical Parameters from Near Earth Asteroid Rendezvous Orbit Phase Navigation Data. AIAA paper 2000–4422, AAS/AIAA Astrodynamics Specialist Conference, Denver, Colorado, 2000.

Miller, J. K. *et alia*, ”Determination of the Shape, Gravity and Rotational State Of Asteroid 433 Eros“, *Icarus*, Vol. 155 Number 1, pp. 3–17, January 2002.

Miller, L. J., J. K. Miller and W. E. Kirhofer, “Navigation of the Galileo Mission,” Paper 83–1002, presented at the AIAA 21 st Aero- space Science Meeting, Reno, Nevada, January 1983.

Mottinger, N. A., “DSN Coverage Requirements for PVO Reentry”, IOM 314.7–165, 5 February 1992

O’Neil, W. J., and R. T. Mitchell, “Galileo Mission Overview,” Paper 83–0096, presented at the AIAA 21st Aerospace Science Meeting, Reno, Nevada, January, 1983.

Peale, S. J. 1977. Rotational histories of the natural satellites. In *Planetary Satellites* (J. Burns, Ed), pp. 87–112. Univ. of Arizona Press, Tucson.

Rourke, K. H., “Navigation of the 1982 Jupiter Orbiter-Probe Mission, ” paper presented at the AAS/AIAA Astrodynamics Specialists Conference, Jackson Hole, Wyoming, September, 1977.

Ryne, M. S., N. A. Mottinger, P. R. Menon, J. K. Miller, “Navigation of Pioneer 12 During Atmospheric Reentry at Venus”, AAS 93–712, AAS/AIAA Astrodynamics Specialist Conference, Victoria, B.C., Canada, August 16, 1993.

Russell, R. K., “The Effect of Atmospheric Drag on Position and Velocity Errors of Orbiting Spacecraft”, JPL

Russell, R. K., “Gravity Focussing of Hyperbolic Trajectories,” Technical Memorandum 391–424, Jet Propulsion Laboratory internal document, Pasadena, California, 30 March 1973.

Scheeres, D. J., “Analysis of Orbital Motion Around 433 Eros,” 1. Astron. Sciences, Vol. 43, No.4, Oct.-Dec. 1995, pp. 427–452.

Thomas, P. C. *et alia*, “Eros: Shape, Topography, and Slope Processes”, *Icarus*, Vol. 155 No 1, pp 18–37, January 2002.

Vojvodich, N. S., et al., Galileo Atmospheric Entry Probe Mission Description,ů Paper 83–0100, presented at the AIAA 21st Aerospace Science Meeting, Reno, Nevada, January, 1983.

Answers to Selected Exercises

Chapter 1

1.1 For an ideal basketball and golf ball, the golf ball would rebound to a height of 10.9 m. For a real basketball and golf ball dropped on a driveway, the height was only about 4 m which resulted in the golf ball being lost on the garage roof.

1.2 A 3×3 matrix containing the outer product of $\mathbf{r}$

$$\frac{\partial \mathbf{a}}{\partial \mathbf{r}} = \frac{-\mu}{r^3}\left[I - 3\,\frac{\mathbf{r}\otimes\mathbf{r}}{r^2}\right]$$

1.3

$$M_y(\alpha = 0) = \frac{3\,GM}{r^3}(I_{zz} - I_{xx})\frac{r_x r_z}{r^2} = \frac{3\,GM}{r^3}(I_{zz} - I_{xx})\sin\epsilon\cos\epsilon$$

1.4

$$r_l = k_z\cos\epsilon - k_x\sin\epsilon + k_z\cos\epsilon - k_x\sin\epsilon$$

$$\Delta r = k_z\sin\epsilon - k_x\cos\epsilon - k_z\sin\epsilon + k_x\cos\epsilon$$

1.5

$$\dot{\alpha} = \frac{3}{2}\left(\frac{GM}{r^3}\right)\left[\frac{I_{zz} - I_{xx}}{I_{zz}}\frac{\cos\epsilon}{\omega_e}\right] = 2.450\times 10^{-12}\ \text{rad/s}$$

J. Miller, *Planetary Spacecraft Navigation*, Space Technology Library 37,
https://doi.org/10.1007/978-3-319-78916-3

1.6 This problem makes use of the following:

$$\left[\frac{\partial(I\Omega)}{\partial I_e}\right]^T = \Omega^T \frac{\partial I^T}{\partial I_e}$$

$$\frac{\partial I^T}{\partial I_e} = \begin{bmatrix} 1\,0\,0\,1\,1\,0 \\ 0\,1\,0\,1\,0\,1 \\ 0\,0\,1\,0\,1\,1 \end{bmatrix}$$

1.7 $\bar{v} = 725\,\text{m/s}$

1.8 Assume that the volume swept out by all the molecules between collisions is equal to the volume of the container and the frequency of collisions is the reciprocal of the mean time between collisions. The mean free path is approximately 5.029×10^{-7} m and the number of collisions per second for one molecule is 9.639×10^8.

1.9 Tyrannosaurus rex's watch will have gained 0.61 s and will read Jan 1 2017 12:00:01 AD if we round up. Photon's watch will read Jan 1, 65,000,000 12:00:00 BC. Photon will have no memory of the trip and t rex was probably wiped out by an asteroid, but his watch survived.

1.10 For 30° integration step size and evaluating function on right side of interval,

$$\int_0^{90} \sin(x)dx \approx \frac{\pi}{6}[\sin(30) + \sin(60) + \sin(90)] = 1.23$$

For evaluation in middle of interval,

$$\int_0^{90} \sin(x)dx \approx \frac{\pi}{6}[\sin(15) + \sin(45) + \sin(75)] = 1.01$$

For 10° integration step size and evaluation on right side of interval the integral was 1.084 and for evaluation in middle of interval the integral was 1.0013

Chapter 2

2.1 R = 15,255 ft, H = 1800 ft

2.2 R = 5.80 miles

2.3 The thrust is 1164 pounds and the drag force is 19.8 pounds.

2.4 The equations of motion are

$$y = v_0 \sin(\theta)t - \frac{t^2}{2g_0}$$

$$x = v_0 \cos(\theta)\, t$$

and the trajectory is

$$y = \tan(\theta)\, x - \frac{g_0}{2v_0^2 \cos^2\theta}\, x^2$$

2.9

$$
\begin{aligned}
C_{20} &= \frac{1}{Ma^2} \iiint_V (-\frac{1}{2}x^2 - \frac{1}{2}y^2 + z^2)\, \rho(r, \lambda, \phi)\, dV \\
C_{21} &= \frac{1}{Ma^2} \iiint_V xz\, \rho(r, \lambda, \phi)\, dV \\
S_{21} &= \frac{1}{Ma^2} \iiint_V yz\, \rho(r, \lambda, \phi)\, dV \\
C_{22} &= \frac{1}{4Ma^2} \iiint_V (x^2 - y^2)\, \rho(r, \lambda, \phi)\, dV \\
S_{22} &= \frac{1}{2Ma^2} \iiint_V xy\, \rho(r, \lambda, \phi)\, dV \qquad (1.1)
\end{aligned}
$$

Chapter 3

3.2 For F negative, $\sinh(F) + \cosh(F) = e^F$ which for large negative F is very small. Since e^F is obtained by differencing two very large numbers $(\sinh(F),\ \cosh(F))$ their is a loss of significance.

3.3 $\sin\gamma = \left(\frac{GM}{vh}\right) e \sin(\eta)$

3.4 The first spacecraft had an orbit insertion maneuver of 1021 m/s and the second spacecraft had an orbit insertion maneuver of 975 m/s followed by a maneuver at apoapsis of 30 m/s for at total of 1005 m/s. The second strategy is more fuel efficient.

3.7 1.638 years

Chapter 4

4.1 The radius of the can is $h = \left(\frac{V}{2\pi}\right)$ and the height is twice the radius.

4.2 1/3

4.3 The relevant term that determines the sign of the Hessian is given by $16a^2U_2 - 16b^2U_1$. Since a is greater than b, the Hessian is positive and the solution is a minimum.

4.4 The critical plane is defined in the velocity space. A maneuver performed in this plane will acquire the target and minimize ΔV.

4.6 In computing the partial derivatives of v with respect to γ, the partial of r_a with resect to γ is zero. The terms that multiply $\partial v/\partial \gamma$ are factored out and divided to form a fraction. The denominator may be discarded and the numerator is zero only if γ is zero.

Chapter 5

5.1 $p = 4\binom{52}{2} = 1.539 \times 10^{-6}$

5.2 $p = \binom{500}{5} = 3.265 \times 10^{-14}$

5.3 Caesar's box is in a narrow annulus of width 3 yards where the PDF is constant.

$$\sigma = \frac{50}{1.17741} \qquad p = \left[e^{\frac{-98.5^2}{2\sigma^2}} - e^{\frac{-101.5^2}{2\sigma^2}} \right] \frac{3}{200\pi} = 4.96 \times 10^{-5}$$

5.4 The probability of hitting Caesar's box, if that is the target, is approximately

$$p = 1 - e^{\frac{-3^2}{2\sigma^2}} = 2.49 \times 10^{-3}$$

5.5 The binomial coefficients for $m = 2$ are obtained from

$$(1+x)^m = 1 + 2x + x^2$$

Since each coefficient for the next row of Pascal's triangle is the sum of the two coefficients in the row above, $B(m+1, k+1) = B(m,k) + B(m, k+1)$ and the solution is

$$B(m,k) = \frac{m!}{(m-k)!\,k!}$$

where

$$\binom{m}{k}\frac{m+1}{k+1} = \binom{m}{k} + \binom{m}{k}\frac{m-k}{k+1}$$

after factoring out B(m,k). The demonstration is complete if $B(2,k) = 1,\ 2,\ 1$ which it does.

Chapter 6

6.1 The rows of the matrix are $A(i, j)$. The measurement covariance (P_m), the inverse of the *a prior* matrix (P_0) and the estimated parameter *a prior* are set equal to zero. The number of measurements, which are assumed to be exact, is equal to the dimension of the matrix. Both the Kalman gain and the weighted least square gain give $A(i, j)^{-1}$.

6.2 Venus is the best guess since the maximum acceleration from Mars would be too small. In the real world, it was Mars. An early version of an orbit determination program left Mars out of the Equations of motion because the acceleration was believed to be too small to be detected. The actual ramp in the Doppler data was smaller than postulated for this problem and Venus was included in the equations of motion.

6.3 $E(X_2X_1^T) = K\ E(Z_{1,2}X_1^T) + (I - KA)\ E(X_1X_1^T)$

Since the data taken after t_1 is uncorrelated with X_1, $E(Z_{1,2}X_1^T) = 0$ and $P_{1,2} = P_2$

6.4 Draw a sample from $P_1 - P_2 = KAP_1$ and add it to X_1.

Chapter 9

9.1 33.3 m/s

9.4 $\Delta V_x = -3.858$ m/s, $\Delta V_y = 0$, $\Delta V_z = 3.858$ m/s

9.5 $r_p = 3.04$ km. The spacecraft crashes into Eros.

Index

J. Miller, *Planetary Spacecraft Navigation*, Space Technology Library 37,
https://doi.org/10.1007/978-3-319-78916-3

Printed in the United States
By Bookmasters